Muon-Catalyzed Fusion and Fusion with Polarized Nuclei

ETTORE MAJORANA
INTERNATIONAL SCIENCE SERIES
Series Editor:
Antonino Zichichi
European Physical Society
Geneva, Switzerland

(PHYSICAL SCIENCES)

Recent volumes in the series:

Muon-Catalyzed Fusion and Fusion with Polarized Nuclei

Edited by

B. Brunelli

National Commission for Nuclear and Alternative Energies—ENEA
CRE Frascati, Italy

and

G. G. Leotta

Commission for the European Communities
Directorate General XII—Fusion Programme
Brussels, Belgium

Springer Science+Business Media, LLC

Library of Congress Cataloging in Publication Data

Muon-catalyzed fusion and fusion with polarized nuclei.

(Ettore Majorana international science series. Physical sciences; v. 33)
"Proceedings of the eighth course of the International School of Fusion Reactor
Technology, held April 3–9, 1987, in Erice, Italy"—T.p. verso.
Includes bibliographies and index.
1. Fusion reactors—Congresses. I. Brunelli, B. (Bruno) II. Leotta, G. G. III. International
School of Fusion Reactor Technology. IV. Series.
TK9204.M86 1988   621.48′4   87-29171
ISBN 978-1-4757-5932-7   ISBN 978-1-4757-5930-3 (eBook)
DOI 10.1007/978-1-4757-5930-3

ECSC—EEC—EAEC(Euratom), Brussels and Luxembourg, 1987

EUR 11137 EN

Neither the Commission of the European Communities (CEC) nor any person acting on
behalf of the Commission is responsible for any use which might be made of the
following information.

Proceedings of the Eighth Course of the International School of Fusion
Reactor Technology, held April 3–9, 1987, in Erice, Italy

© 1987 Springer Science+Business Media New York
Originally published by Plenum Press, New York in 1987

FOREWORD

The *International School of Fusion Reactor Technology* started its courses 15 years ago and since then has mantained a biennial pace. Generally, each course has developed the subject which was announced in advance at the closing of the previous course.

The subject to which the present proceedings refer was chosen in violation of that rule so as to satisfy the recent and diffuse interest in cold fusion among the main European laboratories involved in controlled thermonuclear research (CTR).

In the second half of 1986 we started to prepare a workshop aimed at assessing the state of the art and possibly of the perspectives of muon- catalyzed fusion. Research in this field has recently produced exciting experimental results open to important practical applications. We thought it worthwhile to consider also the beneficial effects and problems of the polarization of the nuclei in both cold and thermonuclear fusion.

In preparing the 8th Course on Fusion Reactor Technology, it was necessary to abandon the traditional course format because the influence of the workshop procedure was inevitable: the participants were roughly equally divided into experts in cold fusion and experts in thermonuclear fusion. The course had largely an interdisciplinary character as many disciplines were involved: atomic and molecular physics, nuclear physics, accelerator technology, system analysis, etc. Plasma physics was excluded, with a sigh of relief from the experts in thermonuclear fusion.

The average participant could not pretend to be an expert in more than one or two disciplines; consequently most of the attendees felt that they had a right to be allowed to put naive questions, and the usual uninhibited atmosphere of the Erice courses was even more marked.

In the thermonuclear approach to controlled fusion it is common to define three milestones along the road toward the fusion reactor, i.e., the demonstration of three feasibilities: *scientific, technological,* and *economic.*

The consequent three areas of investigation must have a part in common if a fusion reactor is to be eventually viable. At present, world research on controlled thermonuclear fusion is "digging" vigorously in the scientific and technical areas and is not far from achieving compatible solutions for the related problems.

G. Grieger of the Max-Planck Institut für Plasmaphysik, Garching, at the unanimous request of the participants, described the status and strategy of CTR in a special lecture. As can be seen from Grieger's lecture (the last contribution before the round table) achieving demonstration of the scientific CTR feasibility means reaching a certain value of the product: temperature x density x confinement time of the burning fuel. The largest tokamaks operating at present in the world have reached values of the figure of merit not far (less than a factor 10) from the one required.

A similar situation seems to exist in muon catalyzed fusion, after the recent discovery of the resonant formation of the D,T muonic molecule, plus the unexpected limited loss of the muon by *sticking* to the alpha particle produced in the DT nuclear fusion reaction.

The remark that both hot and cold fusion are not far from the scientific breakeven point suggests an attempt, also far cold fusion, to define the typical problems to be solved far demonstrating the successive feasibilities toward the cold fusion reactor.

The attempt inevitably contains a great deal of arbitrariness; but, in spite of this, the workshop was organized keeping distinct the problems of physics, of technology, and reactor system engineering.

The scientific breakeven point in cold fusion can be considered achieved when the energy $E_{Fus} = X_\mu 17,6$ MeV produced by the nuclear fusion of X_μ meso-molecules D,T (successively formed by a single muon before dying or being captured by impurities e.g., alpha particle) equals the energy E_N of the light nuclei of the muon-generating beam. With the accelerator technology that can be envisaged today, $E_N \simeq 5000\text{-}10000$ MeV, can be achieved; hence

$$X_\mu = 300\text{-}600$$

The experimental results that were known before the workshop and that justified its organization, show that $X_\mu \simeq 150$ (in liquid DT at T = 100 K), which is not very far from the value required for the scientific breakeven point just defined.

There are indications that, optimizing the experimental conditions, X_μ can be substantially increased. During the workshop a lively debate developed among the experts of the three laboratories LASL (USA), SIN (Switzerland), KEK (Japan) on the most convenient recipes for the optimization. The relevant phenomenon in this respect is the loss of the muon as the catalyzer during its life (average life 2.2 μs), mainly due to its possible sticking to the alpha particle produced by the fusion reaction. In any case the best we can suppose is to have the sticking phenomenon and all the other possible losses reduced to zero; a value for X_μ of about 1500 corresponds to this ideal scenario, with a cycling rate of the order of nanoseconds (this average time elapsing between successive catalytic fusions induced by the muon is hardly reducible).

The efforts to achieve the *technological feasibility* of cold fusion must be concentrated mainly on developing accelerators and targets particularly efficient in muon production. The technological feasibility demands that the power from the grid needed for the generation of the beam of light nuclei of energy E_N equals the power which can be restituted to the grid from the fusion power produced and from the residual ($\sim\frac{1}{2}$) beam of light nuclei which have not produced muons. One finds:

$$\frac{X_\mu \cdot 17.6\,(\text{MeV})}{E_N(\text{MeV})} = 10$$

having made the following reasonable assumptions: 0.3 efficiency in transforming the thermal power produced into electric power; 0.6 efficiency in transforming the electric power into beam power.

In order to evaluate the minimum goal to be reached in accelerator technology, we can insert in the last formula the value $X_\mu = 1500$ which represents the maximum achievable. Thus,

$$E_n \leq 2.500\,\text{MeV}$$

The eventual *economic feasibility* requires for much more severe conditions:

- a power gain beyond the balance previously described;
- a very efficient heat extraction system for maintaning the DT fuel "cold";
- a blanket for the tritium production and a rapid separation system for the unburned tritium.

In order to evaluate the difficulties to be overcome and the steps to be taken toward the goal of an economic muon catalyzed fusion reactor, conceptual designs are needed, as demonstrated in the case of the thermonuclear approach.

During the course no conceptual designs of a pure cold fusion reactor were considered, but a sketch of a hybrid fusion-fission muon-catalyzed reactor was presented by S. Eliezer: the presence in the reactor blanket of fissile actinides amplifies the power production and consequently the requirements of the muon catalysis are relaxed to such an extent that the values of X_μ already achieved are sufficient for a positive power balance.

About one-third of the workshop was devoted to spin polarization of fusion fuel nuclei. Polarization technologies and the depolarization mechanisms were discussed: they pose problems whose solutions are a prerequisite for the three main possible applications of polarized fuel in the thermonuclear approach:

1)   to increase the fusion output power for a given *"beta"* of the plasma;
2)   to reduce to a negligible amount the DD reactions occurring in a D-^{3}He mixture;
3)   to take advantage of the preferential directions in which the fusion reaction products are emitted.

The concluding round table discussion was chaired by G. Grieger, and the summaries reported here are provided by to S. Eliezer and J.D. Jukes for muon catalyzed fusion and by R.M. Kulsrud for fusion with polarized nuclei. Stimulating questions and problems were posed by *old* hot-fusionists. It is too early to draw final conclusions on the potential of muon-catalyzed fusion for the generation of power.

Since it is the very accidental values of one energy level of the (DTμ-D) molecule which make its resonant formation and thus the high cycling rates possible, muon-catalyzed fusion is only open to the DT-process and not to advanced fuels.

The success of the workshop is based on the high quality of the invited lecturers.

Organization of the workshop proceeded smoothly thanks to the already traditional efficiency of the secretariat of the Ettore Majorana Centre and to the dedication of the secretaries, Gianna Albanese and Mina Misano, kindly provided by ENEA. I also thank Lucilla Crescentini for having prepared the manuscript of this proceedings in time.

B. Brunelli

CONTENTS

OPENING ADDRESSES

I. μ- CATALYZED FUSION

II. FUSION WITH POLARIZED NUCLEI

III. MAGNETIC CONTROLLED THERMONUCLEAR FUSION

ROUND TABLE

OPENING ADDRESSES

WAYS TOWARD FUSION

D. Palumbo

Accademia Nazionale di Scienze Lettere e Arti
90100 **Palermo**, Italy

Probably some of you will be surprised by my presence here. This is because I am partly responsible for the choice of the subjects of this *8th Erice International Course on Fusion Reactor Technology*. One or two years ago, when I was responsible for the Euratom Fusion Programme, which, as you know, is mainly devoted to magnetic fusion, some members of our Consultative Committee suggested that it would be wise to pay some attention to other routes toward fusion and thus in particular to muon-catalyzed fusion, and we decided to explore the possibility of a course or workshop. Prof. B. Brunelli, Director of this Fusion School in Erice, was the easiest target for me, and so I proposed that the course be devoted to muon-catalyzed fusion and fusion with polarized nuclei. As you know, Prof. Brunelli has done a lot of work to organize the present meeting and to get excellent speakers — I did nothing. So, he is justified in asking me, as a sort of compensation or punishment, to open this meeting even if my knowledge of these subjects is extremely modest. What seems less justified is to take up part of your time; therefore I will try to be very brief.

It is unnecessary to recall the importance of the energy problem. Fusion could offer an excellent solution. Three ways seem possible: magnetic, inertial and μ-catalyzed; at least for the first two, the utilization of polarized nuclei could offer clear advantages. The three ways are broadly comparable in age and nobility. Concerning magnetic fusion, at the end of the war, in 1945, Fermi had already started to discuss it with several collaborators like Alvarez, Teller [1], Tuck, etc. Several problems and modalities were already identified, such as the drift in the toroidal confinement and the minimum-B principle.

Inertial fusion comes directly from the H-bomb. As recalled by Jones [2], catalyzed fusion was proposed by Frank and Sakharov in 1947 and 1948.

We have to remember that the final aim of the ongoing R&D on fusion is the construction of reactors for the industrial production of energy in a way which should be economically competitive, ecologically acceptable, safe and reliable. In the case of magnetic fusion, which is in Europe and even in the world the most popular way, we have, rather arbitrarily, identified three main steps: the scientific, the technological, and the economic feasibility. The scientific feasibility can be represented by the products of three factors, the plasma density n (m^{-3}), the temperature T (keV), and the energy confinement time τ(s).The nτT product should be larger than some $3 \cdot 10^{21}$. At present the best value for nτt, $2 \cdot 10^{20}$, has been achieved on JET, and there is reasonable hope to approach the target in the next few years on JET (EUR) and on the other large tokamaks such as TFTR (USA) and JT-60 (Japan). I think that more or less similar considerations are possible for the two other ways. In the case of μ-catalyzed fusion, probably the target is represented by the number X_μ of DT reactions which have to be catalyzed by a muon for energy balance. X_μ is determined by the energy E required for the creation of one muon. It appears that after last year's results both in magnetic and in catalyzed fusion, what is still missing in the different parameters are modest numerical factors and not orders of magnitude. It is worthwhile noticing that while in magnetic fusion the present experimental results are not as good as the theoretical predictions, the reverse seems to be true for μ-catalyzed fusion.

But the scientific feasibility is just one step. The physical conditions have to be achieved making use of the available technology and with acceptable construction and operating costs. Unfortunately we know that achievement of the ecological and economic feasibility will require a rather long period, i.e., several decades, and we have never tried to hide these problems and delays.

On the one hand, specialists in long-term energy planning (e.g., Marchetti [3] at a recent meeting of the Italian Physical Society) consider that any radical new technological development, independently of its scientific difficulties, also takes several decades before entering the market. These facts should not discourage either us or the governments and agencies which are now supporting fusion research, because the energy problem is already something huge and its importance is certainly increasing. The amounts of money involved in the investment and operation of energy sources are enormous, much larger than those

2

devoted to R&D. On the other hand, this means that we have time. During this time the present leadership of one route over the others could be reversed. The same goes for the presently apparent advantages or handicaps concerning technology and economy. So it is wise to explore several possibilities in order to find the best solutions: relatively small differences in cost or ecological aspects could imply huge advantages and money saving in the end.

Let me now turn to the subject of our meeting. The aim is to have an assessment of the possibility offered by µ-catalyzed fusion and polarized nuclei. This coupling is not completely evident, except for the fact that, from the scientific point of view, the two have in common a large utilization of quantum physics. Of course, the utilization of polarized nuclei does not represent a new way toward fusion, but it can make the other ways easier. The potential advantages are important, both concerning the improvement in the reaction rate and the better distribution of the products of the reaction, in particular the neutrons. It is a relatively young action, but I should like to remember that, to my knowledge, it was first proposed in 1963 during the first meeting of the Euratom Consultative Committee for Fusion by E. Medi [4], the then Vice-President of Euratom.

At that moment a large production of polarized nuclei and even more the possibility of maintaining the polarization for a long time in a hot plasma seemed so remote that no consideration was given to this proposal. It seems to me that this difficulty has not yet been resolved, at least from the experimental point of view, even if the theory has confirmed possible advantages, and has foreseen its feasibility.

We have a different situation for µ-catalyzed fusion. The idea is as simple as it is brilliant: in any atomic system the natural units of length and energy are respectively the Bohr radius $\hbar^2/mc^2$ and the Rydberg constant $me^4/2c\hbar^2$. Usually m is the (reduced) mass of the electron. Of course, if the electron is replaced by a muon whose mass is two hundred times larger, the distances decrease, and the energies increase accordingly. Due to the enormous increase in the ratio between m and the nuclear masses, the isotopic effects will be enormously larger than in the ordinary atoms. These allow the fast formation of T atoms and the fast constitution of the DT biatomic ion. The DT nuclei will be at a very short distance in conditions similar to those existing in a white dwarf star, and the D + T = ^{4}He + n reaction takes place in a very short time, of the order of a picosecond. The problem is, considering the energy E required for the production of the muon, to maximize the number X_μ of fusion processes catalyzed by a single µ. X_μ, in turn, is determined by three quantities: the lifetime of the muon, the frequency of DT

3

fusion events catalyzed by a free muon, and the effective probability for the capture of a muon by the α-particles. Thirty years ago, this last event was considered as an insurmountable obstacle. For instance, in 1958, A.S. Bishop [5] wrote: *Unfortunately, as has been mentioned, the μ-meson has a very limited lifetime (several millionths of a second, or even less if, as is likely to happen, it is captured by and interacts with one of the helium nuclei produced in the fusion reaction). Thus even under optimum conditions, a μ-meson has time to catalyze only a few fusion reactions during its lifetime, and the total energy released is only a small fraction of that required to produce the meson in the first place. As a result, μ-meson fusion, while of great scientific interest, does not appear to be a practical means of releasing nuclear energy.*

Fortunately the experience has been much more encouraging than the predictions, especially concerning the trapping by the α-particles, and the present results on $X\mu$ seem in principle compatible with the requirements of a hybrid reactor, a result which has to be considered positively; but, at least in Europe, where pure fusion is preferred, it is still insufficient. However it seems that the last word has not yet been spoken and something can still be gained by optimizing the pressure and temperature of the fuel, even if the three quantities determining $X\mu$ are much less under control than the three quantities playing similar roles in magnetic fusion. Perhaps the best way could be in the attempt to reduce E and consequently increase $X\mu$, and hopefully this could be achieved by future technology. In the end the only fraction of the energy which really disappears is that taken away by the neutrinos produced in the $\pi \rightarrow \mu$-decay and in the desintegration of the muon.

I would like to remind you that the aim of the course is, as stated by Prof. Brunelli, to assess the state of the art as regards both the physics and the technology involved: specifically, the formation of meso-molecules, their properties, the vicissitudes of the muons, the technological aspects of accelerators, and the overall configuration of a possible energetic application. Finally, the beneficial effects and problems of the polarization of the nuclei in both cold and thermonuclear fusion will be discussed. In addition, we will enjoy the beauty of the physics involved.

Let me conclude with two short remarks: in magnetic fusion it has been possible to proceed with a succession of machines of increasing size and cost. In the case of tokamaks, we initially built and operated some small-sized machines, then some medium-sized machines, and, after that, JET, and now we are studying the Next Step (NET), which could become the object of worldwide cooperation. It should be interesting to discuss whether the situation is similar for the activities which are the subjects of the present course and the possibility of sharing the problems in space and time. In my opinion it makes political support and also international collaboration easier.

In the field of magnetic fusion, in addition to the informal international cooperation which normally takes place between scientists, we have concluded a series of agreements for general purposes or for specific aims involving two or three of the large world fusion programs. The most ambitious project concerns the joint construction, as a co-operative venture between the four large programs, of the Next Step, which in Brussels we call NET, and which in Vienna and Geneva has been successively called INTOR, ETR, or a few weeks ago, ITER. In the case of cold fusion, which we are considering now, I see that the informal collaboration is already excellent. I wonder if it is wise and possible to do more.

In the present course, in perfect agreement with the spirit of the Ettore Majorana Centre, which is not only to promote science and technology but also worldwide collaboration, islands and continents are represented. I regret that our Soviet colleagues are missing. Both Prof. Brunelli and I have done our best, at the highest possible levels, to have them participate, and theoretically they should be here, but it seems that in this case, as in magnetic fusion, experimental results are not as good as theoretical predictions.

REFERENCES

[1]   Teller, E. Proc. 2nd UN Internat. Conference *Peaceful Uses of Atomic Energy*, Geneva, September 1959, 31 , 27

[2]   Jones, S.E., *Nature* 321 (1986) 127, and Jones, S.E. et al., 4th International Conference ICENES, Madrid, June 1986, p. 173.

[3]   Marchetti, C., Proceedings del Conv. Nazionale Energia Sviluppo e Ambiente, Roma, gennaio 1987, SIF, in press.

[4]   *Physics Today,* August 1982, p. 17.

[5]   Bishop, A.S.,*Project Sherwood,* Addison Wesley Publ., 1958, p. 178.

I. μ-CATALYZED FUSION

MUON-CATALYZED FUSION: A SHORT INTRODUCTION AND A FEW COMMENTS

G. Fiorentini

I N F N Sezione di Pisa
56100 Pisa, Italy
and
Physics Department,
Università di Cagliari, Italy

μCF IN BRIEF

There are many reviews[1-8] on Muon Catalyzed Fusion (μCF) and most of this book is itself an extended presentation of the field, including the most recent achievements. The aim of this contribution is just to provide a short introduction and a few comments to the subject.

Schematically, the fate of a negative muon in a mixture of deuterium and tritium at about liquid Hydrogen density is depicted in Fig.1 and is summarized in Table 1. Remind that a muon is in any respect a "heavy electron" (m_μ =207m_e) and decays to evv with a lifetime τ_μ=2.2μs.

After the formation of dμ and tμ atoms (phase "a" in Fig.1) a mesomolecular ion dtμ is produced (c), where the two nuclei oscillate with a vibrational energy of about 100 eV around an equilibrium distance R≈10^{-11}cm. In such conditions nuclear synthesis reactions (d) occur easily,

$$(1) \qquad d+t \rightarrow n + \alpha$$

and the muon most likely is left as free (e) and is able to start again the series of reactions depicted in fig. 1, which represents the cycle of muon catalyzed fusion.

Note that all the rates quoted in table 1 are much larger than the muon decay rate, so that at least in principle a muon can catalyze several hundreds of nuclear fusion reactions in its lifetime. This is specific of d-t tritium mixtures, whereas for any other system (d-d, p-d, t-t,...) the rates are at most comparable to the muon decay rate.

Note also that the formation of the molecular ion dtμ is an essential ingredient of muon catalyzed fusion. A back-of-the-envelope calculation shows that the fusion in flight,

Table 1. The main steps of muon catalysed fusion in d-t mixtures.

Phase (Fig.1)	Process	Characteristic Quantity	Order of Magnitude	Notes
a	Slowing down and atomic capture	Slowing time	$\approx 10^{-6} \mu s$	Formation of μ atomic systems (size $\approx 10^{-11}$ cm binding $\approx$ KeV)
b	Transfer reaction	Transfer rate	$\approx 300 C_T \mu s^{-1}$	Depends on t-concentration.
c	Formation of muonic molecular ion	Formation rate	$\approx 500 \mu s^{-1}$	VERY HIGH for d-t mixtures. Depends on temperature and density.
d	Nuclear Fusion	Fusion rate	$\approx 10^{6} \mu s^{-1}$	14 MeV neutron produced and detectable
e OR				Muon free for new catalysis
f	Sticking	Sticking Probability ω_{in}	$\approx 0.9\%$	Muon sticks to α particle
g	Reactivation	Reactivation probability R	$\approx 1/4$	Muon can be shaken off. Effective sticking is $\omega_{eff} = \omega_{in}(1-R)$

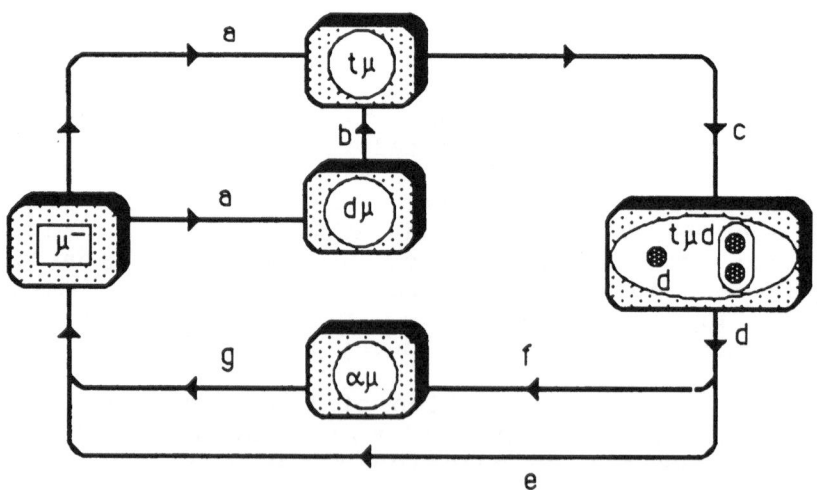

Fig.1 The main steps of μCF in d-t mixtures. See Table 1 for definitions.

(2) $t\mu + d \rightarrow \alpha + n + \mu$,

is unlikely to occur during the muon lifetime. The molecular ion, where the internuclear distance is about 200 times larger than in a ordinary liquid, acts as an amplifier by a factor of about $(200)^3$.

Beyond muon decay, another process can break the chain of muon catalyzed fusion: the muon may stick to the α-particle produced in reaction (1) (channel f of Fig.1). The $\alpha\mu$ system is positively charged and is repelled by other nuclei, so that the muon is then lost for the chain of muon catalyzed fusion. Theoretically, one estimates that a muon remains stuck to the α-particle about once out of 160 fusions; experimentally, the situation is controversial, with some experiment pointing towards smaller sticking losses.

As it is clear from table I, the cycling rate for μCF is such that a muon can catalyze several hundreds of nuclear reactions before it decays, unless it is trapped in the $\alpha\mu$ system. Furthermore it is possible to play with several parameters and make the cycling rate even larger. Thus sticking is the actual bottleneck of muon catalysed fusion in d-t mixtures. The author believes that a complete understanding, both theoretically and experimentally, of the sticking process is actually most important in the study of muon catalyzed fusion.

To many readers, the most important question is whether or not muon catalyzed fusion can be of interest for energy production.

For the problem of a positive energy output, no clear cut answer can be given at the moment. In principle, the minimum energy cost for muon production is equivalent to the muon mass, 106 MeV. Since in a d-t fusion reaction 17.6 MeV are released, a yield Y=6 fusions/muon is sufficient in principle for energy balance. Experimentally, yields in excess of 150 d-t fusions per muon have been reported so far. This looks very good, however:
i)at present, the energy cost per muon production is several order of magnitudes higher than the above estimate.
ii)in "gedanken experiments", the minimum energy cost per muon is estimated to be a few GeV.
ii)most of the energy output is carried away by 14 MeV neutrons (although schemes for use of neutrons as fertilizers of ^{238}U blankets, with consequent energy amplification, have been considered).
Schemes with a positive energy output (and with the design of a plant capable of 1 GW) have been presented in the literature in connection with a yield $Y \approx 200$ fusions per muon. As a personal opinion, the authors considers $Y = 1000$ as a reasonable value for practical applications in this sense.

Another application which might be considered is the use of muon catalyzed fusion as a mean for plasma heating toward ignition. There is however a major difficulty in this approach. The formation of the molecular ion $dt\mu$ is fast when the tμ collides with D_2 molecules and the binding energy is absorbed in vibrational excitations of the electronic molecule. If the molecules are dissociated, this mechanism is not efficient and the formation of dtμ is largely suppressed.

Before and beyond energetics, there are several reasons of interest for muon catalyzed fusion, and they are the main motivation for most people working on this subject. μCF is actually an arena where several fields of physics meet. The study of muonic atoms and muonic molecular ions is a challenging place for developing and testing theoretical techniques in few body problems governed by the Coulomb interaction. A sofisticated level of accuracy can be attained for example in comparing theoretical and experimental values of the binding energies of some levels of the muonic molecular ions. Nuclear reactions in the muonic molecular ions offer an almost unique possibility of investigating nuclear forces at very low energies. The nuclear capture of muons in hydrogen and deuterium - the muonic analogue of inverse β-decay - is a classical problem in the study of weak interactions. These and other topics in basic physics have provided the reason of a relatively constant interest for the field of muonic atoms and muonic molecules during several tens of years.

MAJOR STEPS IN THE STUDY OF μCF

There is some magic connection between μCF and the number 7:

1937:   the discovery of negative muons[9].

1947:   the first theoretical consideration[10] of μCF.

1957:   the first observation[11] of μCF.

1967:   the resonant mechanism[12] of formation of muonic molecular ions.

1977:   the prediction[13] of a large yield of muon catalysed fusions in d-t mixtures.

A more detailed hystorical approach - which does not spoil however the magic 7 - can be found elsewhere[5,8]. The importance of all the events marked above is evident, but for the 1967 event, which is worth introducing in some detail.

Formation of a muonic molecular ion - take ddμ as an example - requires transfer of the binding energy E_b and of the kinetic energy E_k of the dμ atom. This can be achieved in two ways, see fig.2:

a) an electron carries away the energy.

b) energy is given to the vibrational and rotational degrees of freedom of the complex molecular system.

In the latter case, since molecular levels are quantized,

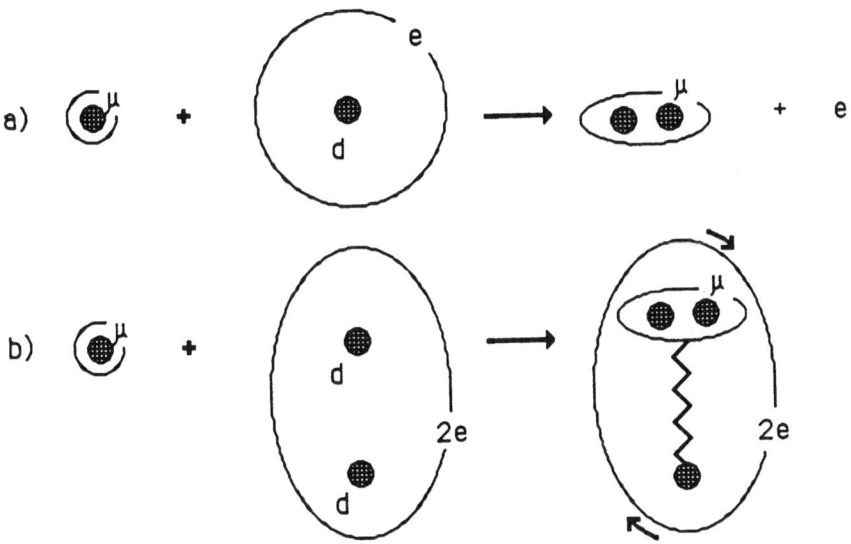

Fig. 2   Different mechanisms for the formation of muonic molecular ions: a) the non-resonant (Auger) process; b) the resonant process.

the transition can take place only if a resonance condition is satisfied:

$$(3) \qquad E_b + E_k = \Delta E_{vib,rot} \qquad ,$$

where $\Delta E_{vib,rot}$ is the excitation energy of some level of the molecular complex.

For example, if the $d\mu$ atoms have kinetic energies corresponding to the thermal distribution $\rho(E_k,T)$, only a fraction around $E_k = -E_b + \Delta E_{vib,rot}$ will induce muon catalyzed fusion through process b).

Three points are to be remarked about process b) - the so called resonant formation - and they will be extensively discussed through this book:

i) it offers the possibility of large formation rates of muonic molecular ions, as typical of a resonant process. This is the road to large fusion yields.

ii) it provides the basis for a method of accurate determinations of the binding energy of muonic molecular ions: by looking at the fusion yield while the target temperature is varied (and may be in the future other parameters can be varied) one can deduce the resonance kinetic energy and hence the value of E_b . In this way it has been possible in some case to determine experimentally the binding energy with accuracy of the order of meV on a typical scale of KeV.

iii)it requires the existence of weakly bound states of the muonic molecular ion, with a binding energy smaller than the dissociation energy of the hydrogen molecule, a few eV. Vesman's paper[12] in 1967 was at the origin of a big effort in the calculation of the energy levels, first to assess the existence of these weakly bound states - which were indicated by the experiment on ddμ[14], and then to push the accuracy of the calculation as it became clear that through ii) a useful method for the experimental determination of the energy levels had become available.

The process of resonant formation is at the basis of most research on μCF in the last twenty years.

The magic seven should bring some exciting news during 1987. It is hard to say what will be the new and by whom, however a quick look at the people presently involved in the study of μCF can of interest.

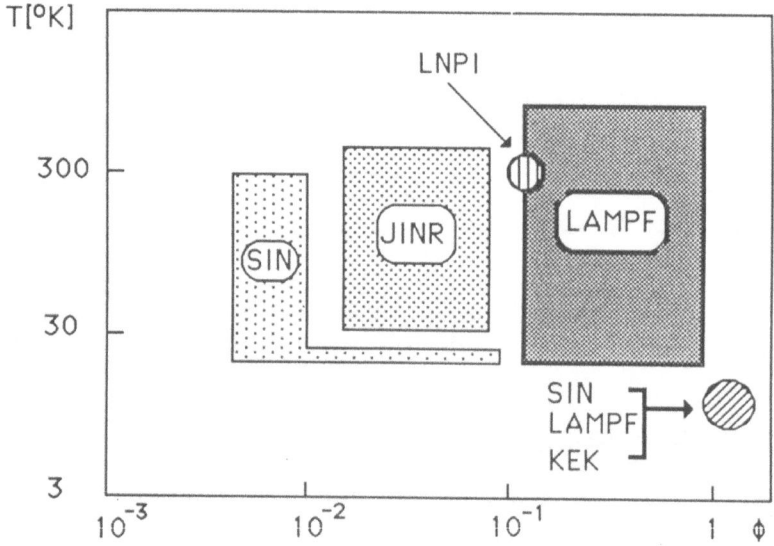

Fig.3 Sketch of the regions explored by different experiments on μCF in d-t mixtures in the density-temperature plane. ϕ is the density in units of liquid hydrogen density. JINR experiment worked with a fixed (low) tritium concentration. KEK experiment is working at a fixed 30 % tritium concentration. Other experiments can vary tritium concentration. LAMPF experiment is planned to work up to 1200 °K

On the experimental side, there are several groups investigating muon catalyzed fusion in d-t mixtures throughout the world: in the USA at LAMPF (Jones et al.), in Switzerland at SIN (Breunlich et al.), in Japan at KEK (Nagamine et al.), in the Soviet Union at LNPI-Gatchina (Vorobjev et al.) and at

JINR-Dubna (Zinov et al.). Fig. 3 shows schematically the regions of temperature and density which are being investigated. Other experiments are being considered/planned in Canada at TRIUMF and in the United Kingdom at the Rutherford Laboratory.

On the theoretical side, research is scattered throughout the world, with a major concentration in the Soviet Union around the group of Ponomarev et al., continuing a long tradition in that country.

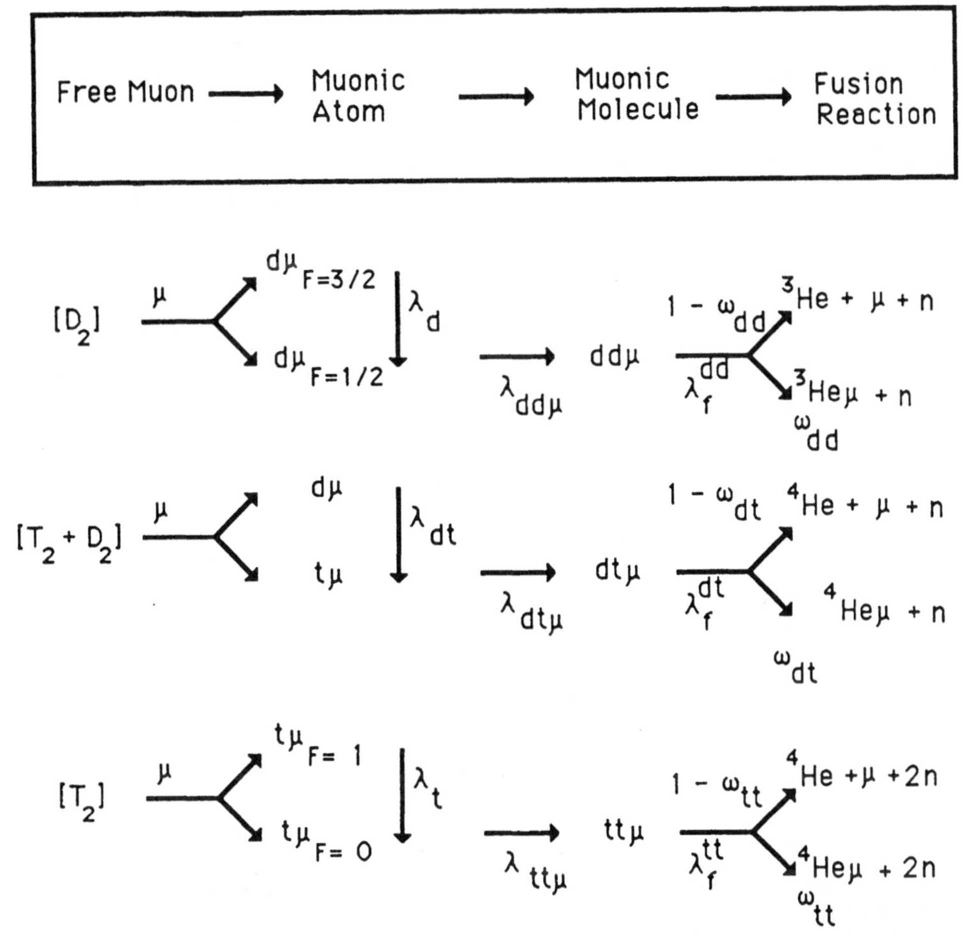

Fig. 4   Definition of quantities relevant to the kinetics of muon catalyzed fusion in mixtures of different hydrogen isotopes.

A rather large comunity is actively involved in this research. Tables 2,together with the accompanying fig. 4, summarizes the progress in the field in the last few years.

In the first two columns of Table 2 we show theoretical and experimental data on the processes occurring when a negative muon enters a mixture of D and T as they were

Table 2. Main characteristics of μCF in d-t mixtures

Quantity	1979		1987	
	Theory	Experiment	Theory	Experiment
λ_d, $10^7 s^{-1}$	0.7÷4.7	–	4.9 at 34°K	3.74±.15 at 34°K
λ_t, $10^9 s^{-1}$	0.9	–	0.91	–
λ_{dt}, $10^8 s^{-1}$	1.9	2.9±0.4	1.9	2.8±0.4
$\lambda_{dd\mu}$, $10^6 s^{-1}$ (at 300°K)	≈1	0.76±0.11	2.8	2.76±.08
$\lambda_{dt\mu}$, $10^8 s^{-1}$	≈1	>1	≈4 (at 100°K)	6.56±.58 at<130°K 3.5±0.5 at 30°K
$\lambda_{tt\mu}$, $10^6 s^{-1}$	3	–	3	$\lambda_{tt\mu}$=1.8±0.6 $\omega_t\lambda_{tt\mu}$=0.4±0.1
ω_{dt}, 10^{-2}	≈1	–	0.53±0.09	0.45±0.05*
ω_{dd}	0.13	<0.13	0.12	0.122±0.003
ω_{tt}	–	–	0.05÷0.18	0.14±0.03
λ_f^{dt}, $10^{12} s^{-1}$	≈1	–	1.1	–
λ_f^{dd}, $10^9 s^{-1}$	≈100	–	0.43	–
λ_f^{tt}, $10^6 s^{-1}$	–	–	–	15±2
ε_{dd}**	1.9±0.1		1.9749±0.0002	
ε_{dt}***	0.6±0.2		0.663±0.002	

*)Value at density φ=1.2. Density dependence is presently controversial.
**)Value of the Coulomb binding energy of the weakly bound ddμ level.
***)Value of the Coulomb binding energy of the weakly bound dtμ level.

available in 1979, i.e. after the first experiment in deuterium-trixitum mixtures[15]. It was absolutely evident that it was necessary to work out a complex program of μ-catalysis studies. The general outline of the program was presented in ref.16. From the theoretical point of view, the key points at that time were the calculation of ddμ and dtμ resonant formation rates and the calculation of the various sticking probabilities. Besides, it was necessary to calculate the fusion rates and the cross sections of many mesic atomic processes, namely elastic scattering and isotope exchange cross sections. The influence of the molecular electron shell, of the molecular vibrational and rotational structure should also be worked out. The muon fate from the time of muonic molecule formation to the fusion time should be followed. The loss of muon due to transfer to ^{3}He (which is continuously formed in Tritium decay) should also be known. During the last eight years the main part of this -roughly sketched- program has been accomplished.

REFERENCES

(See the contribution by Eliezer in this volume for an extensive bibliography on μCF).

1. Ya. B. Zel'dovich and S.S. Gerstein, Usp. Fiz. Nauk 71:581 (1960) [Sov. Phys. Uspekhi 3:593 (1961)].
2. S.S. Gerstein and L.I. Ponomarev, in Muon Physics (Eds. V.Hughes and C.S.Wu, Academic Press, New York, 1975, vol.III).
3. Proc. Muon Catalyzed Fusion Workshop, Jackson Hole 7-8 June 1984, Idaho National Lab, Idaho Falls 1984.
4. L.I. Ponomarev, Proc. of X European Conference on Controlled Fusion and Plasma Physics, Moscow 1981. L.I.Ponomarev, Atomkerenergie-Kernetechnik 43:175 (1983).
5. L. Bracci and G. Fiorentini, Phys. Reports 86:175(1982).
6. S.E. Jones, Nature 321:127 (1986).
7. L. Alvarez, Adventures in Physics, 4α:1(1972).
8. L.I. Ponomarev and G. Fiorentini, in Proceedings of μCF'87, to be published.
9. C.S. Anderson and S.H. Neddermeyer, Phys. Rev. 50:263 (1936).
10. F.C. Frank, Nature 160:525 (1947).
11. L.W.Alvarez et al., Phys. Rev. 105:1127 (1957).
12. E.A. Vesman, Zh. Eksp. Teor. Fiz.Pisma 5:113 (1967) [Sov. Phys. JETP Letters 5:91 (1967)].
13. S.S. Gerstein and L.I. Ponomarev, Phys. Lett. 72B:80 (1977).
14. V.P. Dzhelepov et al., Zh. Eksp. Teor. Fiz. 50:1235 (1966) [Sov. Phys. JETP 23:820 (1966)].
15. V.M.Bystritsky et al., Phys. Lett. 94B:476 (1980).
16. S.S. Gerstein et al. Zh. Eksp. Teor. Fiz.78:2099 (1980).

MUON-CATALYZED NUCLEAR FUSION FOR ENERGY PRODUCTION

S. Eliezer

Plasma Physics Department
Soreq Nuclear Research Center
Yavne, 70600, Israel

ABSTRACT

The physics of muon-catalyzed fusion is summarized and discussed in the

perspective of energy production.

1. INTRODUCTION

The muon (μ) was discovered by Anderson and Neddermeyer (1936) [1] and by Street and Stevenson (1937) [2]). In 1947 after the discovery of the pion (π) and the π decay into μ, it was realized that the muon is actually a *heavy electron* with a mass (m_μ)

$$m_\mu = 206.77 \, m_e = 105.66 \, \frac{MeV}{c^2} \tag{1.1}$$

where m_e is the electron mass. μ and e also differ by their lepton quantum number. Moreover, the muon is an unstable particle with a lifetime (τ_μ) of

$$\tau_\mu = 2.2 \ 10^{-6} \ \text{sec} \tag{1.2}$$

The size of a μ-atom (e.g. the hydrogen atom where the electron is replaced by a muon) is of the order of the muon Bohr radius a_μ given by

$$a_\mu = \frac{m_e}{m_\mu} \, a_e = 2.6 \ 10^{-11} \text{cm} \tag{1.3}$$

where a_e is the Bohr radius of an electron atom.

When an energetic negative muon (μ^-) enters a compressed hydrogen gas (e.g. hydrogen liquid density defined by $n_0 = 4.25 \ 10^{22}$ atoms/cm^3), the following chain of reactions occurs:

a) Slowing down of μ^-.

b) Capture of μ^- into atomic levels and the cascade to the atomic ground state (1s).

c) μ-*molecular* formation and the de-excitation to the molecular ground state.

d) Nuclear fusion of the nuclei (usually hydrogen isotopes) which are kept close together by the negative muon. The μ^- is either released or captured by the nuclear fusion products (μ^- sticking).

If the lifetime of the muon is long compared to the time scale of the other processes (a to d), then many fusion reactions can occur during the lifetime of a muon. This chain of reactions, resulting in the nuclear fusion process, is usually called **muon catalyzed fusion.**

The idea that negative muons might be able to catalyze proton-deuteron (p-d) fusion was first considered by Frank (1947) [3]. Deuterium-deuterium and deuterium-tritium fusions catalyzed by muons were suggested by Sakharov and Lebedev (1948) [4] and further rediscovered by Zeldovich (1954) [5] and analyzed by Jackson (1957) [6]. The first experimental observation by Alvarez et al., (1957) [7] of (pdμ) fusion

$$(pd\mu) \rightarrow {}^3He + \mu^- (5.4 \ MeV) \tag{1.4}$$

was a rediscovery and for a short period of time the Berkeley group thought that they had *solved all the fuel problems of mankind for the rest of the time* (Alvarez, 1968) [8]. After this discovery the muon catalyzed fusion was studied in England (Ashmore et al., 1958 [9] through the reaction

$$(pd\mu) \rightarrow {}^3He\mu + \gamma(5.4 \ MeV) \tag{1.5}$$

These experiments were followed by a series of theoretical calculations, mainly by Soviet scientists (Gershtein and Zeldovich, 1961 [10]. The main research resulted in the prediction that one could not expect more than few catalyzed fusions per muon during its lifetime. It thus appeared that muon-catalyzed fusion reactions were useless as an energy source. This conclusion was questioned after Dzhelepov et al., (1966) [11] experiments which showed that the rate of ddμ molecule formation depends strongly on temperature. Following this experiment it was suggested by Vesman (1967) [12] that the muon-molecules can be formed resonantly if there are weakly bound states of these molecules (i.e. the

20

rate of ddμ molecule formation depends strongly on temperature. Following this experiment it was suggested by Vesman (1967) [12] that the muon-molecules can be formed resonantly if there are weakly bound states of these molecules (i.e. the binding energy is smaller than the dissociation energy of the hydrogen molecules ∼ few eV).

This idea had already been mentioned by Zeldovich (1954) [5]. However, Dzhelepov et al.,(1966) [11] experiment, Vesman's paper (1967) [12] and the following calculations (Gershtein and Ponomarev, 1975 [13]); Bracci and Fiorentini, 1982, and Jones (1986) [14]) and experimental measurements (Breunlich, 1981 [15]; Breunlich et al., 1984 [16], of the energy levels of the muon molecules were crucial steps in renewing the interest in muon catalyzed fusion. Using the same resonance theory formation for the dtμ molecule, as for the ddμ molecule, there was predicted (Gershtein and Ponomarev, 1977 [17]) a very large dtμ formation rate which was confirmed experimentally (Bystritsky et al. 1980 [18]; Jones et al. 1983 [19]). However, not everything seems to be understood on muon catalyzed fusion. For example, the crucial problem of muon sticking (Jones et al., 1986 [20]), as observed at Los Alamos, at high density is not explained theoretically in a satisfactory way. Before 1985 theorists (Jackson 1957) [6]; Gershtein et al., 1981) [21], Bracci and Fiorentini, 1981 [22] predicted muon alpha sticking loss fraction to be about 0.9% which implies that the possible maximum numbers (i.e., upper limit) of d-t fusions per muon is 110. Los Alamos experiment (Jones et al., 1986) [20] measured 150 d-t fusions per muon.

A comprehensive understanding of muon catalyzed fusion comprises the knowledge of atomic physics, molecular chemistry, nuclear and particle physics, accelerator science and possibly reactor technologies. At this stage one can conclude only that better theories and more experimental evidence are necessary to complete our understanding.

2. ENERGY LEVELS OF μ-ATOMS AND μ-MOLECULES

The binding energy of the muonic atoms in the simple Bohr model are given by

$$E_B = - \bar{m}_\mu c^2 \frac{(Za)^2}{2n^2} \qquad (2.1)$$

where $\bar{m}_\mu$ is the muon- nucleus reduced mass; Z is the nucleus electric charge, n is the principal quantum number and a is the fine-structure constant $a = 1/137$. The corresponding Bohr radius is

$$r_n = \frac{\hbar^2 n^2}{\overline{m}_\mu c^2 Z}$$

(2.2)

The binding energies of the μ-*atom* are two orders of magnitude larger than the corresponding binding energies of the electronic atoms, while the radii of the Bohr orbits are two orders of magnitude smaller. For example, the radius of the 1s level of *lead-muon* (Pbμ) is 4 fm (1 fm = 10^{-13} cm) which is smaller than the 7 fm nuclear radius of Pb. From this example it is evident that one has to take into account relativistic effects. In this case the Dirac equation yields, to first order in αZ, for the muonic energy levels the following expression

$$E_{nj} = -\overline{m}_\mu c^2 \frac{(\alpha Z)^2}{2n^2}\left\{1 + \frac{(\alpha Z)^2}{n^2}\left(\frac{n}{j + \frac{1}{2}} - \frac{3}{4}\right)\right\}$$

(2.3)

The binding energy of the 1s level of Pbμ is 21 MeV, while the fine structure splitting of its 2p level is 550 keV!

For the muon-hydrogen atoms the energy levels of the ground states are (Gershtein and Ponomarev, 1975 [13])

$$\varepsilon(p\mu) = -2528.52 \text{ eV}; \quad \varepsilon(d\mu) = -2663.23 \text{ eV}; \quad \varepsilon(t\mu) = -2711.27 \text{ eV}$$

(2.4)

while the hyperfine splitting is given by

$$\Delta\varepsilon(p\mu) = 0.183 \text{ eV}; \quad \Delta\varepsilon(d\mu) = 0.049 \text{ eV}; \quad \Delta\varepsilon(t\mu) = 0.241 \text{ eV}$$

(2.5)

In the study of μ-*molecules* one encounters a difficulty which is not met in the study of ordinary molecules, namely the relatively large value of the ratio m_μ/M, where M is the mass of the hydrogen (isotope) constituting the nuclei of the μ-molecules. This fact implies that the Born-Oppenheimer approximation is not a good method of calculations for the μ-molecules. For these molecules one needs to know the binding energy of a muon up to 0.1 eV out of about 3 keV [see Eq. (2.4)]. Therefore a relative precision of better than 10^{-4} is needed. Moreover, since the muon is relatively *heavy* one must consider also the nuclear recoil due to the muon motion. These facts make the numerical calculations extremely difficult for these molecules. A comprehensive review on the methods and results of the theoretical calculations of μ-molecule binding energies can be found in Gershtein and Ponomarev (1975) [13] and Bracci and Fiorentini (1982) [14]. In **Table I** we show the μ-molecules energy spectrum for the hydrogen isotopes (Vinitskii et al., 1978) [29]; Bogdanova et al., (1982) [23]) ε_{JV} (in electronvolt) is the difference between the molecular energy and the atomic energy of the heaviest hydrogen isotope. J and V are the rotational and the vibrational

Table I

Binding energy ε_{JV} (eV) of the μ-molecules of the hydrogen isotopes with respect to the appropriate mesic atom (i.e. the difference between the molecular energy and the atomic energy of the heaviest isotope).

$-\varepsilon_{JV}$ (eV)

Rotational (J) and Vibrational (V) quantum numbers	(ppμ)	(pdμ)	(ptμ)	(ddμ)	(dtμ)	(ttμ)
J = 1, V = 1				2.0	0.7	44.9
J = 3, V = 0						47.7
J = 0, V = 1				35.6	34.7	83.7
J = 2, V = 0				85.6	102.3	172.0
J = 1, V = 0	105.6	96.3	97.5	226.3	232.2	288.9
J = 0, V = 0	253.0	221.5	213.3	325.0	319.1	362.9

quantum numbers respectively. It is important to notice (from **Table I**) that ddμ and dtμ have weakly bound states with binding energies of -2.0 eV and -0.7 eV respectively.

In these cases a resonance molecular formation is possible (Vesman, 1967 [12]) i.e., the energy released from the μ-molecules formation is absorbed by the electronic molecule. Another interesting feature of the μ-molecules is the fact that very few levels can satisfy $\varepsilon_{JV} < 0$. This situation is significantly different to the case of ordinary molecules where a rich vibrational and rotational spectrum exists.

In calculating the spectrum of the μ-molecules the spin effects must also be taken into account. The hyperfine structure of the energy levels has been calculated to lowest order in α (Bakalov and Vinitskii 1980 [24]). In Fig. 1 we show a schematic hyperfine structure of the levels of the tμ atom and the dtμ molecule (Bogdanova et al., 1982 [23]). Denoting by F the spin quantum number of the tμ system (in the dtμ molecule), by S the molecular total spin quantum number $\vec{S} = \vec{F} + \vec{S_d} = \vec{S_t} + \vec{S_\mu} + \vec{S_d})$ and by j the total angular momentum $(\vec{j} = \vec{S} + \vec{J})$, one can see from Fig. 1 that $\Delta\varepsilon_F \sim 10^{-1}$ eV, $\Delta\varepsilon_S \sim 10^{-2}$ eV and $\Delta\varepsilon_j \sim 10^{-3}$ eV. $(\Delta\varepsilon_F = \varepsilon(J, V; F = 1) - \varepsilon(J, V; F = 0)$, etc.).

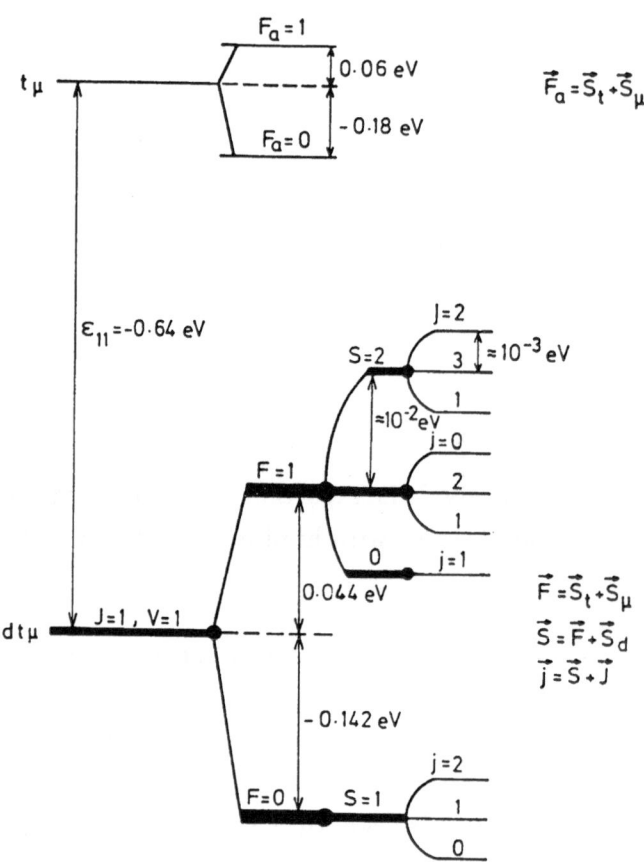

Fig. 1: *Schematic hyperfine structure of the levels of the atom tμ and the molecule dtμ*

In order to appreciate the difficulty of the energy level calculations we end this section by pointing out the **effects of vacuum polarization,** the strong interactions between the nuclei of the μ-molecule and the electron screening effects. For example, the vacuum polarization shifts the energies upwards (i.e. increase the binding energy) for the ls states in pμ , dμ and tμ by 1.89 eV, 2.13 eV and 2.21 eV respectively (Melezhik and Ponomarev, 1978 [25]). The energy level shifts of the dtμ molecule induced by the nuclear interaction (Bracci and Fiorentini, 1982 [14]) are about 10^{-3} eV for the (J,V) = (0,0) and (0,1) states and are as small as 3.10^{-8} eV for the (J,V) = (1,1) state. The electron screening effects may contribute a shift in the energy levels of about 10^{-2} eV for the dtμ and ddμ molecules. Since there are errors of the order of 0.1 eV in the calculations of the Coulomb energies, most of the non Coulomb contributions just discussed are of academic interest only.

3. MUON-CATALYZED p-d FUSION

The slowing down and absorption of μ− in hydrogen were calculated by Fermi and Teller (1947) [26] assuming a mechanism of adiabatic ionization where the electron leaves the atom and the negative muon is captured into a highly excited orbit of the pμ atom, in less than 10^{-9} sec. The deexcitation time of an isolated pμ atom is very large

$$\mu^- + H \rightarrow p\mu + e^- \tag{3.1}$$

where H denotes an hydrogen atom. In this model (Wightman, 1950 [27],) a μ− slows down from relativistic velocities (v ~ c) to v ~ αc (α = 1/137) in a liquid hydrogen density (n $\ell_{h.d.}$ = 4.25 × 10^{22} /cm³) and is captured into a pμ atom in less than 10^{-9} sec.. The deexcitation time of an isolated pμ atom is very large. However, in liquid hydrogen density the mu-hydrogen atom decays to its n = 4 level state during a time of about 10^{-12} sec. (Leon and Bethe, 1962 [28]) due to Auger effect on the electrons of another atom. The μ− decays from n = 4 to the ground state by radiative transitions during 10^{-9} sec,.

The charge neutrality and the small dimensions of the pμ atom enable it to penetrate freely through the electron shells of atoms and molecules and to approach the nuclei within a distance of about the muon Bohr radius a_μ ≃ 2.0 × 10^{-10} cm (Eq. (1.3)). The typical elastic cross sections of these atoms are about $\sigma \sim 4\pi a^2_{\mu'} \simeq 10^{-20}$ cm².

The muon from the pμ atom can be transferred, irreversibly to a deuteron

$$p\mu + d \rightarrow d\mu + p \qquad (3.2)$$

because the binding energy for the dμ ground state is 135 eV lower than that of the pμ atom. This transfer rate λ_{pd} is about 10^{10} sec^{-1}. Even for a very low deuteron concetration, $C_d = 4.10^{-5}$, the yield of Eq. (3.2) is 4.10^5 sec^{-1}, which is not too different than the μ$^-$ decay probability rate $\sim \tau^{-1}_\mu \simeq 4.5.10^5$ sec^{-1}.

Mu-molecules (sometimes called mesonic molecules, however, since the muon is a lepton and not a meson, we avoid the nomenclature *mesoatoms* and *mesomolecules* for the muon atoms and the muon molecules) are created during collisions of dμ with the hydrogen and deuterium nucleus

$$\mu d + H \rightarrow (pd\mu)^+ + e^- \qquad (3.3)$$

In the pdμ molecule, the proton and the deuteron are separated by a distance of about 10^{-10} cm (few times a_μ). Therefore, there is a large probability per unit time ($\lambda_f \simeq 3.10^5$ sec^{-1}) for the proton and deuteron to overcome the Coulomb barrier and to fuse into ^{3}He and μ- (Eq. (1.4)) or into ^{3}He μ + γ (Eq. (1.5)). For a liquid hydrogen density with 1% deuterium the sticking probability (i.e. for reaction (1.5)) is 85%! In this case only 15% of the muons which did not decay are available to repeat the muon catalysis cycle. The sequence of reactions leading to muon catalyzed pd fusion, in liquid hydrogen density containing 1% deuterium can be repeated less than one fusion per muon.

4. MUON CATALYZED d-d FUSION

When a negative muon enters a condensed deuterium medium it quickly forms a dμ atom ($\lambda_a \sim 4.10^{12}$ sec^{-1} for liquid hydrogen density (ℓ.h.d.) of deuterium). Subsequently a muonic molecular ion (ddμ)$^+$ is formed with a rate $\lambda_{dd\mu} \sim 3 \times 10^6$ sec^{-1} (for ℓ.h.d.). The deuterons are tightly bound (radius 5.10^{-11} cm) and therefore the following muon induced fusions can occur

$$(dd\mu) \rightarrow \begin{cases} ^3\text{He} + n + \mu \, (+\, 3.3 \,\text{MeV}) \\[4pt] ^3\text{He}\mu + n \\[10pt] t + p + \mu \, (+\, 4\,\text{MeV}) \\[4pt] t\mu + p \end{cases} \qquad (4.1)$$

About 88% of the muons repeat the cycle while 12% are lost to 3He by forming the ^{3}Heμ atom. This last effect is called the *sticking* loss.

The ddμ molecule has a rotational-vibrational level (J = 1, V = 1) with a very small binding energy ($\approx$ 2 eV). This fact causes a resonance formation of this molecule (Vesman, 1967 [12]), namely, during the E1 transition to the J = 1, V = 1 level, the binding energy (of ddμ) is transferred to the entire D_2 molecule. This phenomenon can happen as long as the binding energy of ddμ is larger than the dissociation energy of the D_2 molecule ($\sim$ 4.7 eV).

The energy released in the formation of the ddμ molecule is transferred to the excited levels of the [(ddμ) d2e] molecule (an H_2 type of molecule), which is produced during the collision of a dμ atom with a D_2 molecule,

$$d\mu + D_2 \rightarrow [(dd\mu)\, d2e]^*$$ (4.2)

The rate of the resonant formation is given by

$$\lambda = N_0 \sigma v \ \text{sec}^{-1}$$ (4.3)

where N_0 is the D_2 density (of the order of the ℓ.h.d.), v is the relative velocity of the colliding particles [in (4.2)] and σ is the appropriate cross section. Following Vinitskii et al., (1978) [29], the rate for resonant production of ddμ (or dtμ) is given by

$$\lambda\left(sec^{-1} \right) = \beta \frac{8\pi^2 m_e e^4}{3\hbar^3} = \left(N_0 a_0^3 \right)\left(\frac{m_e}{m_\mu} \right)^5 \left(\frac{m_\mu}{m_a} \right)^2 I_v^2 \left| d_{fi} \right|^2 \gamma\, (\varepsilon_o, \varepsilon_T)$$ (4.4)

where β is a statistical factor which equals 1/3 for ddμ formation (and 1 for dtμ), a_0 is the Bohr radius ($a_0 = \hbar^2/m_e\, e^2$); $m_a^{-1} = m_\mu^{-1} + M_a^{-1}$ where M_a is the deuteron (or triton) mass; I_v is the dipole interaction matrix element describing the transition from the ground state D_2 (J = 0, V = 0) molecule to the excited [(ddμ) d2e] molecule (J = 1, V = 8); d_{fi} is the dipole matrix element calculated between the functions of the initial state i of the dμ$^+$ d system and the final state f(J = 1, V = 1) of the ddμ molecule; γ is the Maxwellian distribution factor

$$\gamma(\varepsilon_o, \varepsilon_T) = \left(\frac{27\,\varepsilon_o}{2\pi} \right)^{1/2} \varepsilon_T^{-3/2} exp(-3\varepsilon_o/2\,\varepsilon_T),$$ (4.5)

$\varepsilon_T = 3/2\ kT$ is the average kinetic energy of the dμ atoms at temperature T and ε_T satisfies the resonant condition

$$\varepsilon_0 = \varepsilon_{11} + E_v - E_0 \tag{4.6}$$

where ε_{11} is the binding energy of the molecule ddμ (or dtμ) for the J = 1, V = 1 state, E_v is the vibrational energy of the [(ddμ) d2e] (or [(dtμ) d2e]) molecule. Equation (4.5) indicates a temperature dependence of λ(ddμ) as given in (4.4). The maximum value of λ(ddμ) is achieved at $\varepsilon_0 = \varepsilon_T = 0.053$ eV (~ 600 °K) and equals 0.8×10^6 sec^{-1} in good agreement with the measured experimental value. (The rates are calculated at ℓ.h.d., $N_0 = 4.25 \times 10^{22}$ cm^{-3}). Moreover, further experiments have confirmed the temperature dependence of the formation rate λ(ddμ); e.g., when the temperature of the deuterium changes from -160°C to +100°C ($\Delta\varepsilon \simeq 0.04$ eV corresponds to 260°C) the rate λ(ddμ) increases by a factor of four (Vinitskii et al., 1978, [29], 1980 [22]). The nonresonant formation of the ddμ molecule occurs according to the reaction

$$d\mu + D_2 \rightarrow [(dd\mu)\,de]^+ + e^- \tag{4.7}$$

and this process has a reaction rate of about 5×10^4 sec^{-1}, which differs by a factor of 16 from the measured resonance formation rate (Dzhelepov et al. 1966 [11]).

The muon catalysis in pure deuterium can be summarized by the following chain of reactions:

a) μ^- is slowed down and captured by the deuterium (dμ) during a about 10^{-9} sec.

b) The dμ atom collides with the ℓ.h.d. of D_2 molecules and a [(ddμ) d2e] molecule is formed resonantly. The time scale for this process at $\sim$ 600°K is about 10^{-6}sec.

c) The dd induced fusion occurs in $\sim 10^{-9}$sec where ^{3}He + n (60%) and t + p (40%) are created. About 12% of the muons are lost by the sticking process to ^{3}Heμ, while 88% of the μ^- repeat the cycle.

5. MUON CATALYZED d-t FUSION

When a negative muon enters a dense deuterium (D) - tritium (T) target it starts a chain of reactions: (we denote by capital letters D and T the deuterium and tritium atoms while the small letter denotes the nuclei).

a) stopping and capture of μ^- by d or t and the cascade to the ground state of the hydrogen like atom,

$$\mu^- + D \quad \overset{\lambda_a}{\text{----------}} > (d\mu) + e- \tag{5.1}$$

$$\mu^- + T \quad \overset{\lambda_a}{\text{----------}} > (t\mu) + e- \tag{5.2}$$

where e- denotes an electron and λ_a is the rate of the processes (5.1) and (5.2). In general, the rate λ (sec^{-1}) is given by

$$\lambda = n\sigma v \, [\text{sec}^{-1}] \tag{5.3}$$

where n[sec^{-3}] is the density of the target, σ[cm^2] is the cross section describing the process under consideration and v[cm/sec] is the velocity of the projectile. The rate of the muonic atomic formation in Eqs (5.1) and (5.2) for liquid hydrogen density, n $= 4.25 \times 10^{22}$ cm^{-3}, is 4×10^{12} sec^{-1} (Markushin, 1981 [30])

b) μ^- transfer from the deuterium to the tritium

$$(d\mu) + t \quad \overset{\lambda_{dt}}{\text{-----------}} > (t\mu) + d \tag{5.4}$$

The rate for this process at liquid density is estimated (Jones et al., 1983 [19]) to be 2×10^8 sec^{-1}. The 48 eV difference in the binding energies of (dμ) and (tμ) causes the transfer of μ in an irreversible process described by Eq. (5.4).

c) At this stage, a (dtμ) molecular ion may be formed at the center of an H_2 type molecule. The formation of the relevant hydrogen type mumolecules are described by

$$(d\mu) + D_2 \xrightarrow{\lambda_{dd\mu}} [(dd\mu)\; d2e] \tag{5.5}$$

$$(t\mu) + D_2 \xrightarrow{\lambda_{dt\mu\text{-}d}} [(dt\mu)\; d2e] \tag{5.6}$$

$$(t\mu) + DT \xrightarrow{\lambda_{dt\mu\text{-}t}} [(dt\mu)\; t2e] \tag{5.7}$$

$$(t\mu) + T_2 \xrightarrow{\lambda_{tt\mu\text{-}t}} [(tt\mu)\; t2e] \tag{5.8}$$

where in general the mumolecule is in an excited state. The values of the rates describing Eqs (5.5) - (5.8) where measured (Jones et al., 1983 [19]) to be $\lambda_{dd\mu} \simeq 3 \times 10^6\,\mathrm{sec}^{-1}, \lambda_{tt\mu} \simeq 3 \times 10^6\,\mathrm{sec}^{-1}$ and $\lambda_{dt\mu} \gtrsim 10^8\,\mathrm{sec}^{-1}$, where $\lambda_{dt\mu}$ is defined by

$$\lambda_{dt\mu} = \lambda_{dt\mu\text{-}d}c_d + \lambda_{tt\mu\text{-}t}c_t \tag{5.9}$$

c_d and c_t are the concentrations of the deuterium and tritium nuclei, so that if the presence of He and other impurities are neglected one has,

$$c_d + c_t = 1. \tag{5.10}$$

The value of the dtμ rate is almost two orders of magnitude larger than the ddμ and ttμ rates. This phenomena can be explained by the resonant formation of these molecules (Vinitskii et al., 1978 [29]). A degeneracy in the excited state of the dtμ ion and the excited state of the electron molecular complex is causing a strong resonance effect. The rate, $\lambda_{dt\mu}$ is found to be dependent on the temperature since the kinetic energy of the tμ atom, which forms the dtμ ion, is temperature dependent. This resonance condition Fig. 2 can be described by the energy conservation

$$\varepsilon_e = \varepsilon_T + \varepsilon_\mu \tag{5.11}$$

where $\varepsilon_\mu \simeq 0.7$ eV is the energy of dtμ ion in the quantum state $J = 1, V = 1$ (J is the rotational quantum number and V is the vibrational quantum number of the dt nuclei forming the ion $(dt\mu)^+$), ε_e is the electron energy in the H_2-type molecule and $\varepsilon_T = 3/2\, kT$ is the thermal energy of the tμ atom at a temperature T (for T = 500° one has $\varepsilon_T \simeq 0.04$ eV). The energy released $(\varepsilon_T + \varepsilon_\mu)$ during the formation of the $J = 1$, $V = 1$ dtμ ion is absorbed resonantly by the electronic excited states of the mumolecule [(dtμ) d2e]. Since the thermal and the electronic molecular excitation have broad energy distributions, the resonance condition can be satisfied over a large range of temperatures. This effect has been observed experimentally (Jones et al., 1983 [19], 1986 [20]).

RESONANCE: $\epsilon_r = \epsilon_T + \epsilon_\mu$ (J=1, V=1)

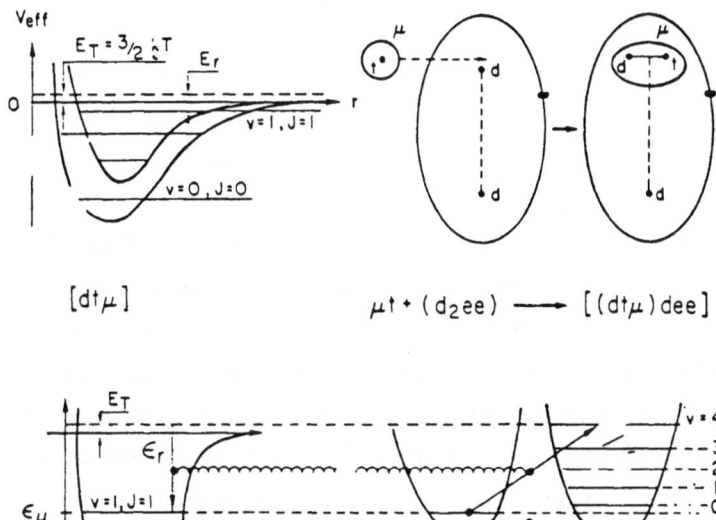

Fig. 2 *Resonance condition*

d) Next in the chain of reactions one has the deexcitation of dtμ, ddμ and ttμ ions to their ground states and the occurrence of the nuclear fusion reactions

$$d + t \xrightarrow{\lambda_f} {}^4He + n + 17.6 \, MeV \tag{5.12}$$

$$d + d \xrightarrow{\frac{1}{2}\lambda_{fd}} {}^3He + n + 3.3 \, MeV \tag{5.13}$$

$$d + d \xrightarrow{\frac{1}{2}\lambda_{fd}} t + p + 4 \, MeV \tag{5.14}$$

$$t + t \xrightarrow{\lambda_{ft}} {}^4He + 2n + 10 \, MeV \tag{5.15}$$

The fusion rates are estimated to be $\lambda_f \simeq 10^{12}$ sec^{-1} for the dt nuclear reaction while $\lambda_{fd} \sim \lambda_{ft} \simeq 10^{11}$ sec^{-1} (Bogdanova, Markushin and Melezhik, 1981 [31]).

e)   During the fusion reaction there is a possibility that the muon sticks to a charged product. In particular in reaction (5.12) the muon can stick to the ^4He by forming a *muonic helium ion* ($^4He\mu$). In this case, if the muon stays bound to the He particle it is lost for the chain of reactions described above [a) to d)]. The sticking probabilities for the reactions given by Eqs (5.12)-(5.15) were estimated to be (Gershtein, et al., 1981, [21]), $\omega_s \simeq 0.9 \times 10^{-2}$, $\omega_d \simeq 0.13$ [for Eq. (5.13)], $\omega_d' \simeq 0.003$ [for Eq. (5.14)] and $\omega_t \simeq 0.05$, implying that the sticking fraction to the He particle in the dt fusion is about 0.9% of the events, in the dd fusion about 15% and in the tt fusion about 5%. For the most important dt case, the value of ω_s implies that the muon sticks to the He particle after catalyzing $1/\omega_s \simeq$ 110 fusions, no matter how fast are the other processes leading to the mesomolecule formation and fusion. The sticking process may be the bottle neck of muon catalyzed fusion idea. However, new Los Alamos experiments (Jones et al., 1986 [20] had measured 150 fusions per muon indicating that the sticking probability is about 0.6%. Moreover, a density dependent analysis of the alpha-muon sticking might suggest $\omega_s = (0.2 \pm 0.1)\%$ (Jones et al., 1986 [20]) for $\phi = 1.2$ (i.e. 1.2 ℓ.h.d.). At present, neither the smallness of ω_s nor its density dependence is understood. Jackson, (1957) [6], was the first to estimate $\omega_s \simeq 0.9\%$ by using a Born-Oppenheimer approximation. A Monte Carlo calculation by Ceperley and Alder, (1985) [32], yielded $\omega_s \simeq 0.6\%$. Mueller and Rafelski, (1985) [33], suggested that the nuclear d-t interaction affects the wavefunctions describing the dtμ system in such a way that ω is reduced to 0.1%. It has also been suggested by Cohen, (1985); Cohen and Leon (1985) and Cohen (1987)

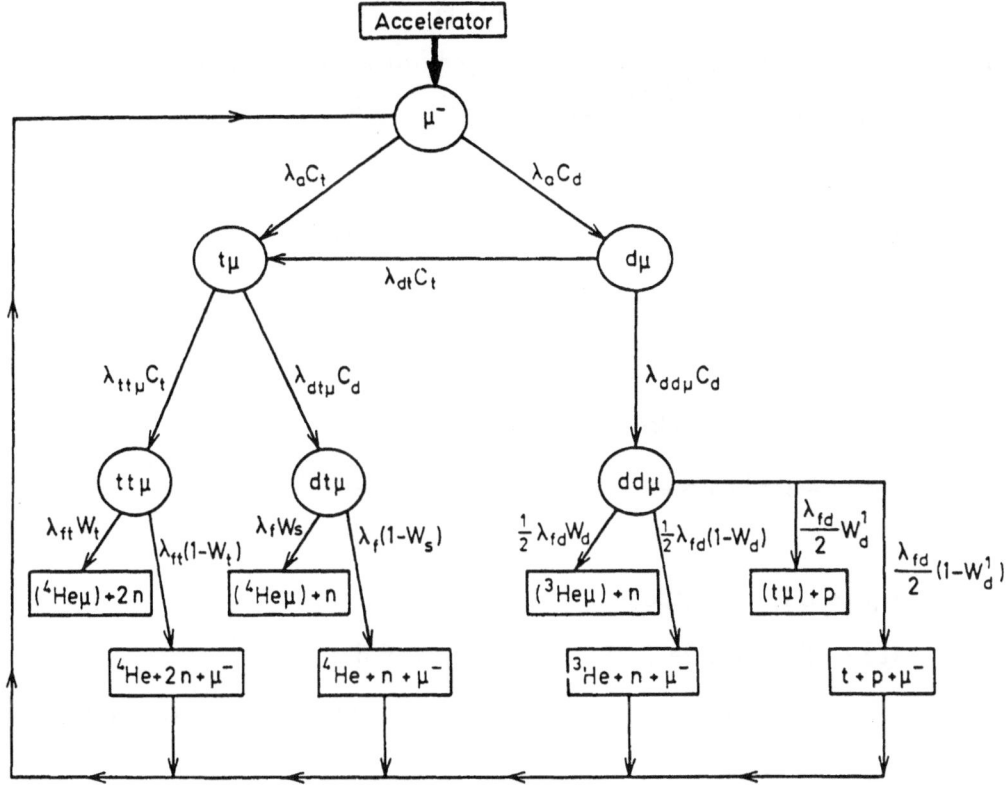

Fig. 3 *The kinetics of muon catalyzed fusion in a mixture of deuterium and tritium. The values for λ and ω are given in Table II*

[34], that a fraction of the muons may be delayed significantly during the deexcitation of the dtμ to the ground state, thus increasing ω_s (eff) above a very small initial value (Tajima and Eliezer 1987, [35]).

A set of rate equations can be written to describe the kinetics (Vinitskii et al., 1978, [29]) of the chain reactions described above from a) to e). This chain of reactions is described schematically in Fig. 3 . The rates for these reactions and the sticking probabilities are summarized in **Table II** for liquid density of the hydrogen isotopes ($n = 4.25 \times 10^{22}$ cm^{-3}). The solution of the rate equations results in an expression for the fusion neutron yield-X_μ, namely the average number of dt fusion catalyzed by one muon. For high density mixtures of d and t and neglecting the small effects, one obtains that the value of X_μ^{-1} is given by the sum of probabilities of muon decay during a catalysis cycle and the muon capture by ^{3}He or ^{4}He (sticking)

$$X_\mu^{-1} = \frac{\lambda_\mu}{\lambda_c} + W \qquad (5.16)$$

where $\lambda_\mu = 1/\tau_\mu = 0.45 \times 10^6$ sec^{-1}, λ_c, is the muon cycling rate, estimated to be

33

Table II

Estimates of the rate coefficients at hydrogen liquid density ($n = 4.25 \times 10^{22} cm^{-3}$) *normalized to the rate of muon decay* ($\tau_\mu = 2.2 \times 10^{-6} sec$). *The sticking probabilities are given by their inverse* $1/\omega$.

Process	$\lambda \tau_\mu$	$1/\omega$
$\mu^- + D \rightarrow (d\mu) + e^-$	$\lambda_a \tau_\mu = 8.8 \times 10^6$	----------
$\mu^- + T \rightarrow (t\mu) + e^-$		
$d\mu + t \rightarrow t\mu + d$	$\lambda_{dt} \tau_\mu = 4.4 \times 10^2$	----------
$(d\mu) + D_2 \rightarrow (dd\mu)d2e$	$\lambda_{dd\mu} \tau_\mu = 6.6$	---------
$(t\mu) + D_2 \rightarrow (dt\mu)d2e$	$\lambda_{dt\mu} \tau_\mu \geq 2.2 \times 10^2$	----------
$+$		
$(t\mu) + DT \rightarrow (dt\mu)t2e$		
$(t\mu) + T_2 \rightarrow (tt\mu)t2e$	$\lambda_{tt\mu} \tau_\mu = 6.6$	---------
$d + t \rightarrow {}^4He + n$	$\lambda_f \tau_\mu = 2.2 \times 10^6$	$1/\omega_s = 300$
$d + d$ $\begin{cases} (50\%) \rightarrow {}^3He + n \\ \\ (50\%) \rightarrow t + p \end{cases}$	$\lambda_{fd} \tau_\mu = 2.2 \times 10^5$	$1/\omega_d = 7.7$
$t + t \rightarrow {}^4He + 2n$	$\lambda_{ft} \tau_\mu = 2.2 \times 10^5$	$1/\omega_t = 20$

$$\frac{1}{\lambda_c} = \frac{c_d}{\lambda_{dt} c_t} + \frac{1}{\lambda_{dt\mu} c_d} \tag{5.17}$$

and W is the probability of muon capture by He, and can be approximated for the chain of reactions given in Fig. 2 by

$$W = \omega_s + \frac{0.5 \lambda_{dd\mu} \omega_d c_d}{\lambda_{dd\mu} c_d + \lambda_{dt} c_t} + \frac{\lambda_{tt\mu} \omega_t c_t}{\lambda_{dt\mu} c_d}. \tag{5.18}$$

The rate equations described by Fig. 2 are more complex if one takes into account the existence of ${}^3He$ and ${}^4He$ concentrations in the deuterium-tritium target. In this case Eq. (5.10) is replaced by

$$c_d + c_t + c_{He} = 1 \qquad\qquad (5.19)$$

Where $c_{He} = c_{3He} + c_{4He}$. The role of the He-sink and its complexity is described schematically in Fig. 4.

Although the muon catalyzis of d-t is not yet completely understood, the Los Alamos recent experiments (Jones, et al., 1986 [20]) suggest that it may be possible to achieve 300 fusion per muon in deuterium-tritium mixtures.

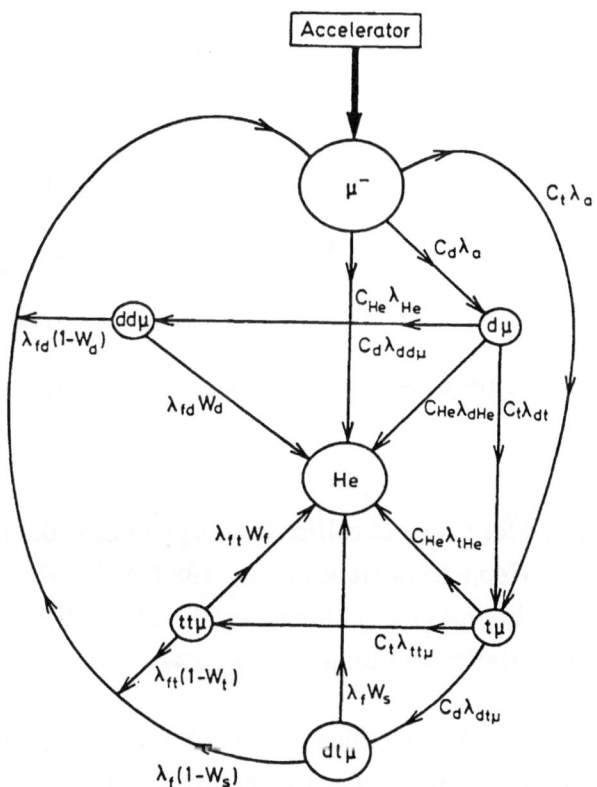

Fig. 4 *The He sink. He denotes* ^3He, ^4He, $^3He\mu$ *and* $^4He\mu$*. The concentrations* c_d, c_t *and* c_{He} *satisfy the relation* $c_d + c_t + c_{He} = 1$

6. MUON-CATALYZED FUSION FOR NON-HYDROGEN ELEMENTS

It is very interesting to analyze the muon-catalysis of nuclei with charge $Z > 1$ in general, and those leading to nuclear reactions without neutrons in the final state in particular. For example, the following reactions are of major interest in achieving *very clean fusion*:

$$d + {}^3He \quad \text{-----} > \quad {}^4He + p + 18.3 \text{ MeV} \qquad (6.1)$$

$$d + {}^6Li \quad \text{------} > 2\,{}^4He + \quad 22.4 \text{ MeV} \qquad (6.2)$$

$$p + {}^{11}B \quad \text{------} > 3\,{}^4He + \quad 8.7 \text{ MeV} \qquad (6.3)$$

However, in order to achieve muon catalysis, the muon molecules must be formed and the nuclear fusion must occur faster than the muon decay time. It turns out that although the rates for some molecular formations have been calculated to be faster than the muon decay, the rates of the nuclear fusion reactions are too small (Kumar, 1983, [36]; Squigna and Harms, 1983 [37]; Aristov et al., 1981 [38]; Kravtsov et al., 1981 [39]; 1982 [40]; 1984, [41]).

The formation rates $\lambda_p \equiv \lambda(Zp\mu)$, $\lambda_d \equiv \lambda(Zd\mu)$, and $\lambda_t \equiv \lambda(Zt\mu)$ for the appropriate molecular formation were calculated to be (Kravtsov et al.,1984 [41])

$$Z = {}^3He \qquad \lambda_p = 0.87 \times 10^8 \text{ sec}^{-1}; \quad \lambda_d = 1.48.10^8 \text{ sec}^{-1}; \quad \lambda_t = 5.62 \times 10^8 \text{ sec}^{-1}$$

$$Z = {}^4He \qquad \lambda_p = 0.44 \times 10^8 \text{ sec}^{-1}; \quad \lambda_d = 2.03.10^8 \text{ sec}^{-1}; \quad \lambda_t = 1.98 \times 10^8 \text{ sec}^{-1}$$

$$Z = {}^6Li \qquad \lambda_p = 22.1 \times 10^6 \text{ sec}^{-1}; \quad \lambda_d = 3.45.10^6 \text{ sec}^{-1}; \quad \lambda_t = 2.08 \times 10^6 \text{ sec}^{-1}$$

$$Z = {}^7Li \qquad \lambda_p = 10.8 \times 10^6 \text{ sec}^{-1}; \quad \lambda_d = 1.85.10^6 \text{ sec}^{-1}; \quad \lambda_t = 0.81 \times 10^6 \text{ sec}^{-1}$$

$$(6.4)$$

The rates are calculated for thermal collision energy of $\varepsilon_0 \simeq 0.04$ eV ($\sim 450°K$). Experimental data for ${}^4He p\mu$ (Bystritskii et al., 1983 [42]), ${}^4He d\mu$ (Balin D.V. et al., 1984, [43]) and ${}^3He d\mu$ and ${}^3He t\mu$ (Jones, et al., 1983 [19]) are in agreement with the theoretical calculations summarized in (6.4).

The low rates of the fusion reactions of these molecules in comparison with the appropriate *hydrogen* muon molecules are primarily caused by the large nuclear separation. While $R \sim 2a_\mu$ for muonic hydrogen molecules, $R \sim 4a_\mu$ for the $He p\mu$ molecule and $R \sim 6a_\mu$ for the $Li p\mu$ molecule, where a_μ is the Bohr radius of muonic hydrogen. Due to the increase in the Coulomb barrier one gets for the Bpμ molecules a value of R ∼ 15a_μ. It was estimated (Kravtsov et al., 1984 [41])that the fusion reaction rate for ^{3}Hedμ is about 100 sec^{-1}, while for ^{6}Lidμ it drops down to 0.01 sec^{-1}. Therefore due to the large dimensions of these molecules the fusion reactions rate are small in comparison with the muon decay rate. Moreover, the Zpμ molecules are formed in an excited state which dissociate very fast (∼ 10^{-12} sec) by Auger or radiative transitions

$$p_\mu + Z \rightarrow Zp\mu \rightarrow Z\mu + p. \tag{6.5}$$

In this respect the Zpμ and Ztμ are distinguished from the Zpμ molecules.

In conclusion, muon catalyzed fusion for energy application might be used only for *clean* deuterium-tritium mixtures.

7. COLD FUSION REACTORS

7.1 Introduction

Although it appears at this moment not feasible to achieve energy gain by pure fusion, it is possible to gain energy by combining the catalyzed fusion with fission blankets. The *energy cost* for the creation of muons is analyzed. Furthermore, two nuclear hybrid reactor schemes are described:

a)   Petrov's pioneering ideas,

b)   a reactor where the target, the converter and the synthesizer were unified into one vessel. The d-t condensed gas fuel is the target and the produced pions and muons are trapped in a magnetic mirror configuration. By developing compact accelerators and by recollecting, reaccelerating and recirculating the part of the beam which does not interact strongly with the target. This scheme might have the way to a commercial hybrid reactor.

Muon absorption in matter and the induced fusion of deuterium and tritium is a remarkable phenomenon since the negative muon is capable during its lifetime (τ_μ = 2.2 × 10^{-6} sec) of inducing about 150 nuclear dt fusion reactions in a liquid density medium [20]. This end result of nuclear fusion occurs after a chain of atomic and molecular processes. The most crucial step in the physics of the muon catalysis cycle is the resonant formation [13], [14] of the dtμ molecule which increases the probability of the end result by at least two orders of magnitude (relative to nonresonant dtμ formation). However, this high production rate of the dtμ molecule is disturbed by muons lost in the catalysis cycle. There seems to be a probability of about 0.3% that the helium ion created during the nuclear fusion interaction will capture the negative muon [20].

The *energy cost* for the creation of a muon is one of the most important practical parameters in analyzing the relevance of muon catalyzed fusion for energy production. The muons are produced during the decay of pions which can be created in nucleon-nucleon collisions. The energy threshold for pion generation is about 500 MeV of projectile kinetic energy and the process of negative muon generation seems to be most effective for nucleonic projectiles with a kinetic energy of 1 GeV per nucleon. An optimistic estimate [44] requires 5 GeV of energy to produce one negative muon. Therefore if one muon catalyzes about two hundred dt fusions the energy output is $\sim$ 3.5 GeV (one dt fusion gives 17.6 MeV) per muon, so that no energy gain seems to be possible from *pure fusion* nuclear reactions. Due to the crucial role of the negative pions and their energy cost, section 7.2 analyzes the nuclear physics and the energy balance for the relevant pion production.

For a pure fusion reactor one can define the scientific gain G, as the ratio between the output energy $X_\mu q$ and the input energy E_μ, $G = qX_\mu/E_\mu$, where X_μ is the number of fusion reactions per muon, $q = 17.6$ MeV, and E_μ is the energy invested in creating one muon. Taking into account the thermal to electricity efficiency η_{th}, the accelerator efficiency η_A, and the fraction η_r of the total electric power required to recirculate (in order to operate the accelerator and the auxiliary facilities) one gets the basic equation for the energy balance

$$X_\mu = \left(\frac{E_\mu}{17.6~MeV} \right) \left(\eta_A \eta_r \eta_{th} \right)^{-1} \tag{7.1}$$

For example, taking available (optimistic) efficiencies, $\eta_A = 60\%$, $\eta_{th} = 35\%$ and assuming that $\eta_r = 15\%$ is an economic possibility, one needs $X_\mu = 9000$ fusions per muon in a muon catalyzed reactor. This number seems to be unrealistic for the present status of knowledge. However, by improving technologies, for example, to $\eta_{th} = 80\%$, $\eta_A = 70\%$, and $\eta_r = 20\%$ together with $E_\mu = 2$ GeV, one reduces X_μ to 1000.

Since pure fusion devices using muon catalyzed fusion seem to lose rather than to gain energy, Petrov [45] suggested using muon catalysis in a hybrid fusion-fission reactor [46]. This reactor scheme includes *an accelerator* (of tritium or deuterium), *a target* (of tritium or beryllium) where the pions are created, *a converter,* where pions are confined (in vacuum) by strong magnetic fields, *a synthesizer* with the d-t fuel and *a blanket* where the fission and fissile materials are produced. In section 7.3 this reactor concept is summarized.

Following Petrov's concept of a muon catalyzed fusion-fission reactor, Eliezer-Tajima and Rosenbluth [47], suggested a reactor concept based on two new main ideas:

a) The high energy beam of tritium or deuterium (~ 1 GeV/nucleon) is injected into a target of tritium (or beryllium) with dimensions smaller than the mean free path for strong interactions. After passage through the target, the bulk of the beam is collected for re-use and only the small portion of the beam which suffered strong interactions is directed into an electronuclear blanket.

b) The pions created in the target are surrounded by the fuel of deuterium-tritium and are magnetically confined until they slow down and decay into muons which catalyze the fusion in situ. In this scheme *the converter and synthesizer are combined into one vessel.* The d-t fuel is the target and the produced pions are trapped in the fuel which slows down the pions, so that the necessary conditions for trapping the pions are sufficient conditions for stopping the muons. These muons cause the catalyzed chain reactions leading to nuclear dt fusion. In this way an efficient trapping of muons in a relatively small physical volume is achieved. This scheme is described in section 7.4. Section 7.5 concludes this paper with a discussion on energetics and a comparison with other fusion-fission schemes.

7.2 μ^- Production

The μ^- particles are produced by the decay of the negative pion, $\pi^- \rightarrow \mu^- \, v^-_\mu$. The π^- mesons are generated by using protons, deuterium or tritium accelerators. For kinetic energies per nucleon T_0 smaller than 0.7 GeV/nucleon, the probability of inelastic nucleon-nucleon collisions, $V_{ab}(T_0)$, (a,b = p (proton) or n (neutron)) is very small so that a π^- cannot effectively be produced. V_{ab} is usually defined by the ratio of inelastic cross-section, σ_{ab}^{in}, to the total cross-section, σ_{ab}^{tot}, for strong interactions $\sigma_{ab}^{tot} = \sigma_{ab}^{in} + \sigma_{ab}^{el}$ where σ^{el} is the strong elastic cross-section). We do not consider initial energies larger than 2 GeV per nucleon since in this case the number of undesirable particles (e.g., neutral pions) that are created increases. Therefore, in the energy domain under consideration, $0.7 \leq T_0 \leq 2$ GeV, the negative pions are produced as shown in Fig. 5, with cross-sections $\sigma_{pp}^{\pi^-}$, $\sigma_{pn}^{\pi^-}$ and $\sigma_{nn}^{\pi^-}$. The probabilities W_{ab} of producing a π^- once the inelastic collision occurs are given by $W_{ab}(T_0)$. The multiplicities $Y_{AB}\pi$ were calculated (see Fig. 5 (C^{AB} give the fraction of pp, pn and nn collisions between a projectile A and a target B (A,B = p,d) $(y_{AB}\pi(x))^{-1}$ is the number of nucleonic projectiles necessary to create one pion while the projectile A passes through x cm of target B, and $(y_{AB}\pi^{-1} T_0$ is the energy yielded for this type of π^{-1} creation. $y_{AB}\pi(x)$ is given by

$$y_{AB}\pi(x) = n_0\phi x\{Z_A Z_B \sigma_{pp}\pi^- + [Z_A(A_B - Z_B) + Z_B(A_A - Z_A)]\sigma_{pn}\pi^-$$
$$+ (A_A - Z_A)(A_B - Z_B)\sigma_{nn}\pi^- \tag{7.2}$$

where Z and A are the number of protons and nucleons respectively for the appropriate collisions (A-target; B-beam). ϕ is the density in liquid hydrogen units ($\phi = n/n_0$ where $n_0 = 4.25\ 10^{22}$ atoms/cm^3). $\sigma_{ab}\pi^-$ (a,b, = p or n) is calculated from Fig. 5 by using $\sigma_{pn}^{tot} \simeq 0.8\ \sigma_{pp}^{tot}$ (which is satisfied for our range of energies), $\sigma_{pp}^{tot} = \sigma_{nn}^{tot}$ and σ_{pp}^{tot} (1 GeV) $\simeq$ 45 mb (1mb = 10^{-27} cm^2). The mean free path $\ell_{S,AB}$ is calculated from

$$\ell_{S,AB}^{-1} = n\sigma_{AB}^{tot} \equiv \phi\, n_0\sigma_{AB}^{tot}, \tag{7.3}$$

where σ_{AB}^{tot} is given by

$$\sigma_{AB}^{tot} = Z_A Z_B \sigma_{pp}^{tot} + (A_A - Z_A)(A_B - Z_B)\sigma_{nn}^{tot}$$
$$+ [Z_A(A_B - Z_B) + Z_B(A_A - Z_A)]\sigma_{pn}^{tot} \tag{7.4}$$

One obtains, for example, from (7.2) - (7.4) that in tritium-tritium collision along 64 cm (x = $\ell_{s,tt}$ = 64 cm) one π- is created by four nucleons having a total kinetic energy of 4 GeV. The number of nucleonic projectiles necessary to create one π- are given in Fig 5 for a single collision. This situation is achieved for target dimensions much smaller than the mean free path for strong interactions. For large target dimensions, the nucleon projectile has multiple collisions in the target, about 50% of the incident nucleons have an elastic scattering in the first collision. These nucleons can have in their second or third scattering an inelastic collision if the transverse and longitudinal dimensions are large enough. However, since the probability of a strong inelastic interaction (where a π- can be created) is very small for energies less than 0.6 GeV, all the nucleons with energies below this value are ineffective and their energy is actually lost as far as the creation of negative pions is concerned. Therefore the multiplication factor is calculated by adding the probabilities of the first collision being

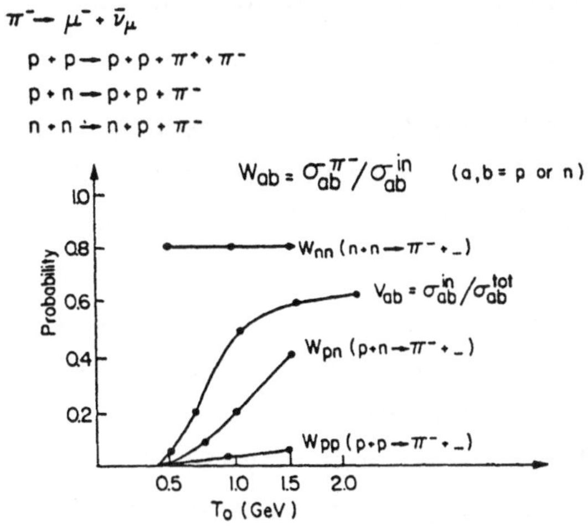

$\pi^- \rightarrow \mu^- + \bar{\nu}_\mu$

$p + p \rightarrow p + p + \pi^+ + \pi^-$

$p + n \rightarrow p + p + \pi^-$

$n + n \rightarrow n + p + \pi^-$

$\left(Y_{AB}^\pi\right)^{-1}$ = Number of nucleonic projectiles (passing a ∞ target) necessary to create one π^-. (in $A + B \rightarrow \pi^- + \ldots$)

$Y_{AB}^\pi = C_{pp}^{AB} V_{pp} W_{pp} + C_{pn}^{AB} V_{pn} W_{pn} + C_{nn}^{AB} V_{nn} W_{nn}$

Target	(B) Beam		
(A)	p	d	t
p	∞	20	14.3
d	20	6	4.8
t	14.3	4.8	4

Y_π^{-1} VALUES AT $T_0 = 1$ GeV

Fig. 5 *The probabilities for* μ^- *production*

inelastic and the appropriate probabilities that the possible second and third collisions are inelastic. Taking into account multiple scattering we obtain the following energies E_π required to create one π^- (for projectile kinetic energy of 1 GeV/nucleon).

$E_\pi(\text{d-d}) \simeq 4.5 \text{ GeV}, \quad E_\pi(\text{d-t}) \simeq 3.7 \text{ GeV},$

$$(7.5)$$

$E_\pi(\text{t-t}) \simeq 2.0 \text{ GeV}, \quad E_\pi(\text{t-50\%d} + 50\%\text{t})) \simeq 2.85 \text{ GeV},$

In the ETR scheme (section 7.4), the π- mesons are created inside the d-t fuel and therefore the pions created by multiple collisions are not lost into an undesirable surrounding material.

We end this section by pointing out the possibility to create π- by triton-triton colliding beams. A laboratory kinetic energy of 1 GeV per nucleon corresponds to a center of mass energy of 225 MeV/nucleon. In this case one needs 8×225 MeV = 1.8 GeV of energy in order to produce a π- in triton triton colliding beams [48]. In this scheme the production of desirable numbers of π- , e.g. 10^{18} sec^{-1} will require very large aperture storage rings [48] , something similar to Tokamak rings. The minimum luminosity in this case is about 10^{44} cm^{-2} sec^{-1} (number of events = luminosity (L) × cross section (σ)). This luminosity is many orders of magnitude larger than that of any existing storage ring. However, it is interesting to point out that such a luminosity is anticipated in Tokamak devices.

7.3 Petrov's hybrid reactor

In 1980, Petrov [45] suggested a power reactor based on muon catalyzed fusion combined with nuclear fission processes. This reactor scheme includes an *accelerator* (of tritium or deuterium), a *target* where pions are created, a *converter* where the pions are confined in vacuum by strong magnetic fields, a *synthesizer* with the d-t fuel and a *blanket* where the fission and fissile materials are produced. The converter is a cylinder about 40 m long and 20 cm radius (5 m^3 volume) having a longitudinal magnetic field of 11 to 16 tesla and an applied d.c. electric field of 7.5×10^5 volt/m along the converter. Inside this vessel there is a cylinder 2 m long with a 2 cm radius target where the pions are created. The synthesizer is a second cylinder, about 20 m long with an average radius of 20 cm surrounded by a (longitudinal) magnetic field coil of 11 tesla. The density of the dt fuel is 0.5 liquid hydrogen density with 30% tritium, so that the synthesizer has 80 kg of tritium. The main result of the converter-synthesizer is the conversion of about 75% of the created pions into muons that participate in the catalyzed dt fusion. This means that the energy cost of a stopped negative muon in the dt mixture is about 6 GeV (using 4.5 GeV to create one negative pion) for tritium projectiles and 8 GeV for a deuterium projectile beam. These results seem to be optimistic in this model reactor, mainly because pion and muon drifts to the wall due to collisions were neglected. Also, the influence of the magnetic field on the scattered proton beam was not considered. In fact, the proton Larmor radius is of the same order of magnitude as that of pions, therefore a significant portion of the proton beam is lost in the vessel in this concept reactor. The fissions and the fissile material are produced in the blanket not only by the neutrons derived from d-t fusion but also from the fast nucleons of the incident beam which have about 80% of their initial energy after the creation of the pions. Taking into account the losses of the projectiles in the converter-synthesizer vessels due to the magnetic field will reduce the energy and the fissile material obtained from direct beam-blanket collisions. Moreover, by taking into account the diffusion due to collisions, the necessary quantity of tritium might increase significantly.

Petrov's reactor is actually a *neutron factory* which breeds a thermal fissile isotope 233U or 239Pu in order to use these materials in satellite fission reactors. In this scheme [45], [46] it was estimated that a commercial reactor would require about 100 fusion per muon. On the other hand a pure fusion reactor seems to acquire 1000 fusion per muon in order to be economically viable [49].

7.4 Eliezer - Tajima - Rosenbluth (ETR) hybrid reactor [47]

ETR reactor concept is based on three ideas:

a)   The high energy beam of tritium or deuterium is injected into the deuterium-tritium (d-t) fuel and, after passage through the fuel, part of the beam is collected for re-use while the portion of the beam which suffered a strong interaction is directed into an electronuclear blanket (see Fig. 6).

b)   The pions created in the target are surrounded by the fuel of deuterium and tritium and are magnetically confined until they slow down and decay into muons which catalyze the fusion in situ.

c)   The fusion created neutrons are absorbed by blankets to breed fissile matter for energy production.

In comparison to Petrov's concept the highlights of our concept are a) and b). Instead of a separate target, converter and synthesizer we combine these three functions into one. The D-T fuel is the target and the produced pions are trapped in the fuel which slows them down before they decay into muons. The muons are created in the fuel and trapped there, catalyzing the D-T fusion through the atomic and molecular processes until they decay. In this way one of the most difficult problems of muon catalyzed fusion is solved, i.e., efficient trapping of muons in relatively small physical volume. The present concept solves this problem by creating mesons in the fuel and by confining them magnetically.

Figure 6 sketches a version of the reactor concept. The driver is a tritium (or deuterium) beam which is retrieved in part after passing through the target. The significantly scattered portion of the beam feeds into the electronuclear blanket. The fusion created neutrons are captured in the fissile blanket surrounding the fusion fuel. Figure 6 shows a magnetic mirror configuration. The mirror is filled with a pressurized gas mixture of deuterium and tritium gas. The gas is circulated through the mirror, with a cooling section between traversals. The mirror is enclosed by ^{233}U or ^{232}Th blankets with admixed lithium for breeding of Pu, ^{233}U and tritium. Magnets provide a magnetic field configuration with a mirror ratio R_m. The field at the mirror throats is typically 10 tesla.

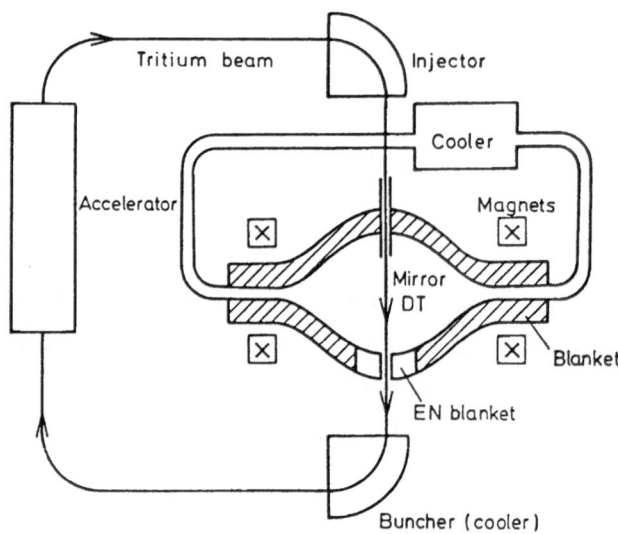

Fig. 6 *The mirror reactor concept*

An accelerated tritium beam of approximately one GeV per nucleon and a current of about 0.1 ampere is injected through a small tube perpendicular to the axial magnetic field into the fuel. When the high energy beam (1 GeV/nucleon) strongly interacts with the target, an effect of proton-proton, proton-neutron, neutron-neutron scatterings take place.

The fraction (f_0) of pions created in the loss cone (for perpendicular injection) was calculated [47],

$$f_0 \le \begin{cases} 15\% \text{ for } R_m = 1.5 \\ 10\% \text{ for } R_m = 2 \\ 5\% \text{ for } R_m \ge 3 \end{cases} \tag{7.6}$$

where $R_m = B_{max}/B_{min}$ is the ratio of reflecting magnetic field at the mirror throat to the field in the injection (target) region. The muon loss cone is included in the pion loss cone since the nonrelativistic μ^- are absorbed in a short range while the relativistic μ^- are created mainly in the direction of the μ^-.

Another factor in considering this scheme is the fact that the pions which come to rest in the high density hydrogen fuel are absorbed by the protons and therefore lost by strong interactions before converting into a muon. The fraction of pions lost by nucleon capture is given by

$$f = \int_0^\infty \frac{dN}{dt} e^{-t/t_n} dt \int_0^\infty / \frac{dN}{dt} dt \tag{7.7}$$

Since the initial distribution of pions is given as a function of their kinetic energy T, we express dN/dt by

$$\tag{7.8}$$

$$\frac{dN}{dt} = \left(\frac{dN}{dT} \right) \left(\frac{dT}{dx} \right) \beta c \tag{7.8}$$

where dT/dx is given by the Bethe-Bloch formula,

$$-\frac{dT}{dx} = \frac{4\pi Z_A^2 e^4 n Z_B}{mc^2 \beta^2} \left\{ \ell n \left[\frac{2mc^2\beta^2\gamma^2}{I} \right] - \beta^2 \right\} \tag{7.9}$$

($\beta = v/c$, $\gamma = 1 + T/Mc^2$, M and Z_A are the mass and charge of the projectile with velocity v, e and m are the electron charge and mass respectively, n is the target density in atoms/cm^3 and I is a phenomenological constant which characterizes the binding energy of the electrons of the medium) and dN/dT is the experimental data for the initial pion distribution. The appropriate time, t, in Eq. (7.7) is derived from

$$\frac{t}{\tau_n} = \int_o^t \frac{dt}{\tau_n(\beta)} = \int_o^T \frac{dT}{\beta c \left(\dfrac{dT}{dx} \right) \tau_n(\beta)}$$ (7.10)

where $\tau_n(\beta) = [\tau_n/(1 - \beta^2)^{1/2}]$, $\tau_n = 2.6 \times 10^{-8}$ sec, and dT/dx is given by (7.9).

Performing the above calculations the fraction of lost pions were calculated [47], to be

$$f = \begin{cases} 0.32 \text{ for } \phi = 0.3 \\ 0.48 \quad \phi = 0.5 \\ \\ 0.68 \quad \phi = 1.0 \\ 0.77 \quad \phi = 1.5 \end{cases}$$ (7.11)

where $\phi = n/n_o$ is the d-t fuel density in liquid hydrogen units. Taking into account other losses it was obtained [47] that the efficiency of converting the created pions into muons seems to be 50% for half liquid hydrogen density.

As was stated earlier, the most important consideration is how not to lose pions and muons that are created by the high energy tritium (or deuterium) beam. Here we analyze the confinement of pions and subsequently muons in the mirror magnetic fields. The Bethe and Bloch formula (7.9) or its derivative gives the differential energy loss of a charged particle per unit length in a condensed matter. The mean free path or the stopping length is obtained by

$$\ell_{mfp} = \left(\frac{d}{dx} \ell nT \right)^{-1},$$ (7.12)

which is a function of the energy. The mean free path is longer for higher energy particles. The collision frequency is defined as

$$v = \frac{\beta c}{\ell_{mfp}}$$ (7.13)

For typical energies of pions created by the injected tritium (or deuterium) beam colliding the target, the collision frequency is approximately $\sim 10^{+8}$ sec^{-1}. A more exact concept is the range, numerical values of which may be found in Ref. [50]. The range for pions under consideration typically varies from less than 1 m to 50 m for 0.5 liquid hydrogen density of the d-t fuel. It is clear that if we do not mirror particles (pions), we need a fairly long containment system. If we apply the longitudinal magnetic field, the transverse size (the radius) of the fuel container would be at least $2\rho_n^{max}$, where ρ_n^{max} is the maximum (within 95% population) of the pion Larmor radius

$$\rho_{\pi} = \frac{\gamma \beta_{\perp} m_{\pi} c^2}{eB}$$

where m_{π} is the pion rest mass, B is the applied longitudinal magnetic field, and $\beta_{\perp}$ is $v_{\perp}/c$ with $v_{\perp}$ being the velocity perpendicular to the magnetic field. With B of the order of 10 tesla, $\rho_{\pi}^{max} \simeq 10$ cm so that a ≥ 20 cm would suffice for $T_{\pi} \leq 500$ MeV. The mirror fields can reduce the system length by a substantial amount. Such a reduction of the system length will cut down the volume of the fuel container and thus the amount of the fuel tritium.

Unfortunately, some of the pions in the mirror fields are untrapped and will be lost through the loss cone in velocity space

$$\frac{v_{\perp}}{v_{\parallel}} \leq \frac{1}{(R_m - 1)^{1/2}} \tag{7.15}$$

The typical scattering angle $<\theta^2>^{1/2}$ is of the order of $2°$. Therefore, the electromagnetic scattering does not much deflect the beam. The pion diffusion due to strong interactions in the d-t fuel is negligible since the rate of strong interaction is $n\sigma v \sim 10^5 - 10^6 \, \text{sec}^{-1}$ ($\sigma \sim 2.10^{-27} \, \text{cm}^2$ and $n = 4.10^{22} \, \text{cm}^{-3}$) while the pion lifetime is only $\tau_{\pi} \simeq 2.6 \, 10^{-8}$ sec. Since we have shown the angular scattering to be small, one can see that collisional radial diffusion is much less than ρ_{π} and thus negligible. Although to a zero order approximation, the magnetic moment μ of pions in the mirror is conserved as an adiabatic invariant, there are nonadiabatic changes to the magnetic moment as a particle bounces back and forth in the mirror field. Since the pitch angle scattering of pions by the DT fuel is relatively minor, one should also consider the nonadiabatic change of μ as a primary concern. Let $\Delta\mu$ be the jump in the magnetic moment per one passage of the midplane (i.e., per half bounce of the trapped oscillations) in the mirror. A typical number of bounces in the mirror is given by

$$n = \frac{R}{2L\cos\theta'} \simeq \frac{R}{2L\sqrt{1 - \frac{1}{R_m}}} \tag{7.16}$$

where R is the range of the pion in the mirror; n is 43 for $\varepsilon_m = 1$, ($\varepsilon_m = a/L$ is the inverse aspect ratio) R $= 5$ m case, while n $= 430$ for $\varepsilon_m = 1$, R $= 50$ m case. For these long ranged pions, it may be more apropriate to use the stochastic approximation to evaluate the time of flight for the pion to diffuse in velocity space, which yields the condition

$$\gamma\tau_{\pi} < \left(\frac{\mu}{\Delta\mu}\right)^2 \frac{2L}{v_{\parallel}} \tag{7.17}$$

where $v_{\parallel}$ is the parallel velocity of the pion. The jump $\Delta\mu$ due to the nonadiabatic effect has been given as [51]

$$\frac{\Delta\mu}{\mu} = A \exp\left(\frac{3L}{2\rho_n}\right) \qquad (7.18)$$

with A being a numerical constant, approximately 5. Solving Eqs (7.17) and (7.18) for $v_{\parallel} = c \cos\theta'$, c being the velocity of light $\cos\theta' = [1 - (1/R_m)]^{1/2}$, we get the inequality

$$L \geq \frac{\rho_{\parallel}}{3} \ell n \left[A^2 \left(1 - \frac{1}{R_m}\right)^{1/2} \frac{^{Y\iota_n c}}{^{L}} \right] \qquad (7.19)$$

which yields for L the following estimate for a mirror ratio $R_m = 1.5$,

$$L \simeq 2a = 4\rho_n^{max} \qquad (7.20)$$

where $\rho_n^{max} \simeq 10$ cm is the maximum pion Larmor radius for a magnetic field of 10 tesla and pion kinetic energies of 500 MeV. At this stage the mirror volume and the tritium mass can be estimated. Since we have shown that an aspect ratio $L/a \simeq 2$ can be chosen with the accepted mirror ratio $R_m \simeq 1.5$, the mirror volume is

$$V \simeq \pi a^2 L \simeq 16\pi \left(\rho_n^{max}\right)^3 \simeq 5.10^4 \, cm^3 \qquad (7.21)$$

where the maximum Larmor radius is 10 cm for the most energetic pions. The density of the d-t fuel is 0.5 liquid hydrogen density with about 30% tritium (apparently the muon catalysis is most efficient for 30% tritium concentration) implying a tritium mass M_t in the vessel of $M_t \simeq 1.5$ kg. A total tritium inventory including the cooling and recirculating sections is perhaps twice as much.

The fusion generated alpha particles have a range of about 0.1 mm and, therefore, they are totally confined in and heat the fuel. The rate of heating is related to the amount of muons created in the fuel, $\varepsilon_\alpha \chi_\mu N_\mu$, where ε_α is the alpha particle energy of 3.5 MeV, χ_μ is the number of fusion per muon and N_μ is the number of muons supplied by the accelerator per unit time. Equating the temperature rise to the power input, one gets

$$nC_p \frac{dT}{dt} = \frac{\varepsilon_\alpha \chi_\mu N_\mu}{V}$$

where n is the average density of deuterium and tritium, C_p is the specific heat under constant pressure, T is the fuel temperature and V is the volume containing the d-t fuel. By recirculating the fuel through the gas flow, the heat generation can be carried away so that the fuel vessel stays at the optimal temperature, say 500 °K. This condition is expressed approximately by

$$\frac{n T v_g}{L} = \frac{\varepsilon_\alpha X_\mu N_\mu}{V}$$

where v_g is the gas flow velocity of the fuel in the vessel. For a typical estimate of the flow velocity we take, $n \simeq 4.10^{22}$ cm^{-3}, $T \simeq 0.1$ eV, $V \simeq 10^5$ cm^3, $L \simeq 100$ cm, $\varepsilon_\alpha = 3.6 \times 10^6$ eV, $X_\mu = 200$ and $N_\mu \simeq 10^{18}$ sec^{-1}, implying $v_g \simeq 2$ m/sec. The d-t fuel is cooled, and cleaned from He, outside the mirror vessel. In this case one has to cool the fuel very efficiently and very fast, so that the tritium inventory is not significantly increased.

The ETR reactor scheme does not discuss the materials problems related to the design of the mirror vessel in particular, and the whole reactor in general. Particular consideration should be given to the material strength and endurance of the mirror vessel, taking into account a vessel with about 1500 atmosphere pressure ($p \sim nkT$ for $T = 500°$ and $n \sim 2.10^{22}$ cm^{-3}, i.e., $\phi \sim 0.5$), and a typical internal curvature of about 20 to 30 cm. In particular, the injection of the beam, the stresses on the vessel, the heat removal and the cooling technologies need serious engineering considerations and seem to be at the edge of the technological possibilities.

7.5 Discussion

In a muon-catalyzed fusion reaction the main energy expended is in the production of muons. It is therefore of paramount interest to minimize the necessary energy for muon creation and, once created, to minimize the loss of muons, in order to introduce a reactor concept utilizing muon fusion catalysis. One must devise an effective configuration to utilize created muons as efficiently as possible with minimal loss. Following Petrov's concept [45] of a muon catalyzed fusion-fission reactor, one may break down the above required tasks for a reactor concept into (i) an accelerator, (ii) a target, (iii) a converter, and (iv) a synthesizer for the catalyzed fusion. In addition we need (v) a blanket for fission and breeding driven by fusion created neutrons. One notes that these five components involve different levels of physics: (i) involves accelerator physics, (ii) nuclear physics, (iii) dynamics of charged particles (accelerator physics and plasma physics), (iv) atomic and molecular physics, and (v) nuclear physics. It may be fair to say that any reactor concept must optimize some or all of these five areas. We note that there are not particularly difficult physical requirements for the confinement and ignition of the fuel as posed in conventional magnetic or inertial fusion reactor concepts. Because the fuel is not ignited, *there is no requirement for alpha particle confinement,* either.

For a muon catalysis with a 100 to 200 fusions/muon the reactors described in this paper are actually neutron factories which breed fissile isotopes in order to use these materials in satellite fission reactors.

Ther total gain factor is given by

$$K \equiv \frac{E_{blanket,\,out}}{E_{target,\,in}} = \frac{E_{blanket,\,out}(EN) + E_{blanket,\,out}(\mu c)}{y_\mu^{-1} y_n^{-1} T_o} \tag{7.22}$$

EN denotes the energy gain from *direct* beam-blanket interaction and μc denotes the energy gain from muon catalytic processes. The value of K is given in terms of the physical qualities X_μ, y_μ and y_n by the formula

$$K = \frac{Z_e T_{o.f.} \varepsilon_{fis}}{T_o} + \frac{y_\mu y_n X_\mu \delta_f \varepsilon_{fis}}{T_o} \equiv K_{EN} + K_{\mu c} \tag{7.23}$$

where y_μ is the π⁻ to μ⁻ conversion efficiency (~ 50%), y^{-1}_n is the number of nucleonic projectiles necessary to produce a π⁻, the EN contribution is $K_{EN} = Z_e T_{of} \varepsilon_{fis} / T_o$ and the muon catalyzed gain is given by the second term. $\varepsilon_{fis} \simeq 0.2$ GeV is the uranium fission energy [52] (for an 238U blanket), $Z_e \simeq 20$ fission/GeV in the *direct* beam-blanket interaction [45] (i.e., the EN process), $T_{o.f}$ is the beam kinetic energy before colliding with the blanket while T_o is their initial kinetic energy before hitting the target (for pion → muon creation). $\delta_f \simeq 1$ is the number of fissions in the blanket caused by one 14 MeV neutron from the fusion process [52] . X_μ was measured experimentally [20], to be about 150 for liquid hydrogen d-t targets and y_n and y_μ were estimated in sections (7.2) and (7.4). The above reactor produces a considerable power on its own. However, the fissile material produced in the blanket (e.g., 239Pu) can be used to run between 3 to 6 satellite fission plants [53] of equal output. X_R denotes the number of these satellite fission plants. The value of X_R ,of course, depends on the design and breeding ratio of those plants and is thus somewhat arbitrary.

The total electrical gain K_{tot} is given by

$$K_{tot} = \frac{\eta_{th} P_{out}}{P_A} = \eta_{th}(1 + X_R)K \qquad (7.24)$$

where P_A is the accelerator power, $\eta_{th} \simeq 0.35$ is the thermal to electric power conversion efficiency and K is given in Eq. (7.23). Defining a as the fraction of the total electric power required to recirculate in order to operate the accelerator and the auxiliary facilities, then the electric net output is

$$P_E = (1 - a)\,\eta_{th}(1 + X_R)KP_A \qquad (7.25)$$

One gets the following relation

$$P_A/\eta_A = a\eta_{th}P_{out} \qquad (7.26)$$

implying

$$a = \frac{1}{\eta_A \eta_{th} K(1 + X_R)} = \frac{1}{\eta_A K_{tot}} \qquad (7.27)$$

Substituting in Eqs (7.23) and (7.27) the following parameters for a fuel containing 50% tritium and 50% deuterium at 0.5 liquid hydrogen density, in the ETR reactor scheme it is conceivable to take:

$$Z_e = 20\ GeV^{-1},\ \varepsilon_{fis} \simeq 0.2\ GeV,\ \delta_f \simeq 1, T_0 = 1\ GeV,\ X_R = 5,\ X_\mu = 200$$

$$\eta_A = 0.6,\ \eta_{th} = 0.35,\ y_\mu = 0.5,\ y_n = 0.35,\ T_{o.f.} = 0.8\ GeV \qquad (7.28)$$

which yields:

$$K_{EN} = 3.2,\ K_{\mu c} = 7.0,\ K = 10.2,\ K_{tot} = 21.4,\ a = 7.8\% \qquad (7.29)$$

An important result is the value of a, which can serve as a figure of merit in a driven nuclear reactor. If we assume as usual an output electricity of $P_E = (1 + X_R)\ 10^9$ W we need an accelerator power of $P_A = 3.0\ 10^8$ W which for 1 GeV/nucleon projectile implies an average current of 0.3 ampere.

In order to compare our scheme with other driven fusion-fission reactors, we define as usual a gain Q for these reactors by (compare with K_{tot} in Eq. (7.27))

$$Q = \frac{1}{\eta_{th}\eta_D a(1 + X_R)} \qquad (7.30)$$

where ηD is the driver efficiency (taken optimistically) as 0.6 for magnetic confinement and for ion inertial fusion devices and 0.05 for laser inertial fusion. Equation (7.30) implies that the μCFF (muon catalyzed fusion-fission) reactors are equivalent to MFF (magnetic fusion-fission) or ion ICFF (inertial confinement fusion - fission) which have a gain of $Q = 10$ or a laser ICFF with a gain of $Q = 120$.

ACKNOWLEDGMENT

I would like to thank my colleagues Profs T. Tajima and M.N. Rosenbluth for the collaboration on the ideas summarized in this paper.

REFERENCES

/1/ Anderson, C.D. and S.H. Neddermeyer, *Phys. Rev.* **50** (1936) 263.

/2/ Street, J.C. and E.C. Stevenson, *Phys. Rev.* **52** (1937) 1003.

/3/ Frank, F.C., *Nature (London),* **160** (1947) 525.

/4/ Sakharov, A.D. and P.N. Lebedev (1948), Report of the Physics Institute Academy of Sciences, USSR.

/5/ Zeldovich, Ya. B., *Dokl Akad Nauk SSSR* **95** (1954) 493.

/6/ Jackson, J.D., *Phys. Rev.* **106** (1957) 330.

/7/ Alvarez, L.W. *et al.*, *Phys. Rev.* **105** (1957) 1127.

/8/ Alvarez, L.W. (1968), Nobel prize acceptance lecture.

/9/ Ashmore, A. *et al.*, *Proc Phys Soc (London)* **71** (1958) 161.

/10/ Gershtein, S.S. and Ya. B. Zeldovich, *Sov Phys Usp* **3** (1961) 593.

/11/ Dzhelepov, V.P. *et al.*, *Zh Eksp Teor Fiz,* **50** (1966) 1235, *Sov Phys JETP* **23** (1966) 820.

/12/ Vesman, E.A., *Pis'ma Zh Eksp Teor Fiz* **5** (1967) 113, *JETP Lett* **5** (1967) 91.

/13/ Gershtein, S.S. and L.I. Ponomarev, *Muon Physics,* Eds. V.W. Hughes and C.S. Wu. Publ. Academic Press, N.Y. Vol.III (1975) p. 141.

/14/ Bracci, L. and G. Fiorentini, *Phys. Rep* **86** (1982) 169 *and* Jones S.E., *Nature (London),* **321** (1985) 127.

/15/ Breunlich, W.H., *Nucl Phys* **A353** (1981) 201.

/16/ Breunlich, W.H. *et al.*, *Phys Rev Lett* **53** (1984) 1137.

/17/ Gershtein, S.S. and L.I. Ponomarev, *Phys Lett* **72B** (1977) 80.

/18/ Bystritsky, V.M. *et al.*, *Phys Lett* **94B** (1980) 476.

/19/ Jones, S.E. *et al.*, *Phys Rev Lett* **51** (1983) 1757.

/20/ Jones, S.E. *et al.*, *Phys Rev Lett* **56** (1986) 588.

/21/ Gershtein, S.S. *et al.*, *Sov Phys JETP* **53** (1981) 782.

/22/ Vinitskii, S.I. *et al.*, *Sov Phys JETP* **52** (1980) 353.

/22/ Bracci, L. and G. Fiorentini, *Nucl Phys* **364A** (1981) 383.

/23/ Bogdanova, L.N. *et al.*, *Zh Eksp Teor Fiz* **83** (1982) 1615 *Sov Phys JETP* **56** (1982) 931.

/24/ Bakalov, D.D. and S.I. Vinitskii. *Nucl Phys* **32** (1980) 372.

/25/ Melezhik, V.S. and L.I. Ponomarev, *Phys Lett* **77B** (1978) 217.

/26/ Fermi, E. and E. Teller, *Phys Rev* **72** (1947) 399.

/27/ Wightman, A, *Phys Rev* **80/4** (1950) 766.

/28/ Leon, M. and H.A. Bethe, *Phys Rev* **127** (1962) 636.

/29/ Vinitskii, S.I. *et al.*, *Zh Eskp Teor Fiz* **74** (1978) 849, *Sov Phys JEPT*, **47** (1978) 444.

/30/ Markushin, V., *Zh Eksp Teor Fiz* **80** (1981) 35, *Sov Phys JETP* **53** (1981) 16.

/31/ Bogdanova, L.N. V.E. Markushin and V.S. Melezhik, *Zh Eksp Teor Fiz,* **81** (1981) 829, *Sov Phys JETP* **54** (1981) 442

/32/ Ceperley, D. and B.J. Adler, *Phys Rev* **A31** (1985) 1999.

/33/ Mueller, B. and J. Rafelski, *Phys Lett* **164B** (1985) 223.

/34/ Leon, M. and J.S. Cohen, *Phys Rev A,* **31/4** (1985) 2680, Cohen, J.S. and M. Leon, *Phys Rev Lett* **55/1** (1985) 52, and Cohen, J.S., *Phys Rev Lett* **58** (1987) 1407.

/35/ Tajima, T. and S. Eliezer (1987), *Laser and Particle Beams.* To be published.

/36/ Kumar, A., *Atomkernenerg/Kerntechn,* **43** (1983) 203.

/37/ Squigna, A.P. and A.A. Harms, *Atomkerneneg/Kerntechn,* **43** (1983) 207.

/38/ Aristov Yu A. *et al.*, *Yad Fiz,* **33** (1981) 1066, *Sov J Nucl Phys* **33** (1981) 564.

/39/ Kravtsov, A.V. et al., *Phys Lett* **83A** (1981) 379.

/40/ Kravtsov, A.V. et al., *Yad Fiz* **35** (1982) 1496, *Sov J Nucl Phys* **35** (1982) 876.

/41/ Kravtsov, A.V., N.P. Popov and G.E. Solyakin, *Pis'ma Zh Eksp Teor Fiz,* **40** (1984) 124, *JEPT Lett* **40** (1984) 875.

/42/ Bystritskii, V.M. et al., *Zh Eksp Teor Fiz,* **84** (1983) 1257, *Sov Phys JETP* **57** (1983) 728.

/43/ Balin, D.V. et al., *Phys Lett* **141B** (1984) 173.

/44/ Petrov, Yu. V. and Yu. M. Shabelskii, *Sov J Nucl Phys* **30** (1979) 66 (*Yad Fiz* **30** (1979) 129

/45/ Petrov, Yu. V., *Nature (London)* **285** (1980) 466.

/46/ Petrov, Yu. V. and E.G. Sakhnovsky, Proc. *Muon-Catalized Fusion Workshop*, Jackson Hole, Wyoming, Ed. S.E. Jones (E.G. & G. Idaho, Idaho Falls, 1984) p. 302.

/47/ Eliezer, S., T. Tajima and M.N. Rosenbluth, *Muon-Catalized Fusion-Fission Reactor Driven by a Recirculating Beam*, Preprint IFSR-223 (1986).

/48/ Chapline, G. Jr. and R. Moir, (1986), UCRL - Preprint 93611 (to be published in *Journal of Fusion Energy*).

/49/ Jackson, J.D., Preprint LBL-18266 (1984).

/50/ Particle Data Group, *Rev Mod Phys* **56/2** (1984) Part II.

/51/ Cohen, R.M., G. Rowlands and J.H. Foote, *Phys Fluids,* **21** (1978) 627.

/52/ Moir, R.W., *Fusion*, Ed. E. Teller, Publ. Academic Press, N.Y., Vol.1 (1981) Part B, p. 411.

/53/ Miller, R.L. and R.A. Krakowski, Proc. *Muon-Catalyzed Fusion Workshop*, Jackson Hole, Wyoming, Ed. S.E. Jones (EG & G. Idaho, Idaho Falls, 1984), p. 177.

THEORY OF CROSS-SECTION FOR FORMATION OF MESO-MOLECULES

A.M. Lane

U K Atomic Energy Authority
T.P. 424.4, Harwell
Oxon OX11 ORA, United Kingdom

SURVEY OF ROUTES TO FUSION

Let us recapitulate some basic facts about the fusion reactions initiated by muons in a chamber of hydrogen isotopes. We consider the case of pure deuterium, in which the muon forms a $(d\mu)$ atom which leads to fusion:

$$d + (d\mu) \rightarrow \begin{cases} t + p + \mu \\ {}^3\text{He} + n + \mu, \end{cases}$$

Also we consider the case of a deuterium-tritium mixture in which the muon can also form a $(t\mu)$ atom, leading to fusion:

$$d + (t\mu) \rightarrow \alpha + n + \mu$$

In each case, there are three routes to fusion. It may occur directly, or it may occur via the formation of states of the meso-molecule (mm); further, mm formation may occur by an Auger process leading to strongly-bound states of mm:

$$(d\mu) + D_2 \rightarrow \left[(dd\mu)de\right] + e$$

or it may occur by resonant formation of lightly-bound states of mm in an overall molecule:

$$d\mu + D_2 \rightarrow \left[(dd\mu)dee\right]_{K\nu} \equiv \text{"}D_2\text{"}$$

K and ν are the rotational and vibration quantum numbers of "D_2", which is like D_2 but with $(dd\mu)_{11}$ in place of a deuteron. The rate for the first route is easily estimated from the usual formula $\lambda = \rho v \sigma$ = density x velocity x cross-section. The cross-sections for bare nuclei are, at E_{lab}= 100 keV:

$$\sigma(d+d) \sim 60\text{mb}; \quad \sigma(d+t) \sim 5\text{b}$$

Extrapolating to low energies for a Coulomb barrier cut-off beyond separation $\sim a_\mu = 2.5\text{x}10^{-11}\text{cms}$, we find:

$$\lambda(dd\mu,\text{dir.}) \sim 30 \text{ s}^{-1}; \quad \lambda(dt\mu,\text{dir.}) \sim 4\text{x}10^3 \text{ s}^{-1}$$

at liquid hydrogen density (LHD, p=4.25x10²² atoms/cc), independent of temperature.

The rate for the fusions occurring via the intermediate formation of mm is the geometric mean of the rate for formation of (ddμ) and the rate for its decay by fusion. The second rate is much larger than the first, so the first determines the fusion rate. Calculations by the Dubna group (Ponomarev and Faifman 1976, Vinitsky et al. 1978) give:

$$\lambda(dd\mu, Auger) \sim 1.2x10^5 s^{-1}; \quad \lambda(dt\mu, Auger) \sim 5.8x10^5 s^{-1}$$

at T = 300°K and LHD. Finally the third rate has been estimated:

$$\lambda(dd\mu, res) \sim 1.3x10^7 s^{-1}; \quad \lambda(dt\mu, res) \sim 2x10^8 s^{-1}$$

(by Faifman et al. 1986a and Vinitsky et al. 1978 respectively). Comparing the various rates, one sees that the resonance formation of mm dominates, by a factor of order 100. Therefore we consider only this route. From now on, everything will be concerned with the evaluation of λ(res). For a resonance r at energy E_r, and a thermalised muonic atom, the rate is related to the resonance formation width Γ_r (assumed $\ll E_r$) by:

$$\lambda_r = \rho v \frac{\pi^2 \hbar^2}{mE_r} f(E_r, T) g_r \Gamma_r \tag{1}$$

obtained by integration over energy with $f(E_r, T)$ as the Boltzman factor:

$$f(\varepsilon, T) = \frac{2}{kT} (\frac{\varepsilon}{\pi kT})^{\frac{1}{2}} e^{-\varepsilon/kT} \tag{2}$$

which integrates to unity. g_r is the spin-statistical factor, and m is the reduced mass. Resonances contribute additively, $\lambda = \sum_r \lambda_r$. Clearly, before evaluating Γ_r (which is our main problem), we need to specify the resonances.

Numerical Values of Resonance Energies

ddμ: Ignoring rotational and hyperfine effects for the moment, the best estimate of energy is E_r=34meV, made up as (-1964 + 2026 - 28). These are respectively the energy of the (11) state of ddμ relative to (dμ) threshold (Gocheva et al. 1985, Frolov and Efros 1985), the excitation energy of ν=7 in "D_2", and the difference in ground state energies of D_2 and "D2" (Faifman et al. 1986b). Hyperfine effects split this into a quartet with extra energies -32, -8, 16, 40 meV corresponding to spins F(dμ),S(ddμ) = (3/2, 1/2), (3/2, 3/2), (1/2, 1/2) and (1/2, 3/2). Rotational effects add $2.2K(K+1) - 3.7K_i(K_i+1)$ in meV.

For (dtμ), formed by (tμ) + D_2 the basic resonance energy for ν=2 is -49 meV (= - 628 + 613 - 34), while the hyperfine quartet (F(tμ),S(dtμ)) = (11), (1,T), (0,1), (0,T) have extra energies -202, -13, 38, 227, where T represents the near-degenerate triplet of states S(dtμ) = 0,1,2. The basic energy for ν=3 is 236meV. Rotational effects add $2.4K(K+1) - 3.7K_i(K_i+1)$, (Assuming process (tμ) + (DT) changes this to $1.8K(K+1) - 3.1K_i(K_i+1)$, while the basic resonance energy for ν=2 is -127 = -628 + 538 -37).

Note the differences between (ddμ) and (dtμ) energies. The hf splitting is small in the (ddμ) case and all four members of the quartet (at E_r=2,26,50,74 meV) are appreciably involved in the rate at 400°K. In the (dtμ) case, the hf splitting is as large as the vibrational splitting, and the only low energy resonances are from [ν=2,(FS)=(0,1)] at -11meV, and $[\nu=3,(FS)=(11)]$ at 34 meV. Further, while the four resonance widths of the quartet contain comparable spin factors $<F1S>^2$ in $(dd\mu)$, those in $(dt\mu)$ have $<F1S>^2=1$ for $(FS) = (1,T)$ and $(0,1)$, while $<F1S>^2<0.01$ for $(FS) = (1,1)$ and $(0.T)$. Thus the (11) resonance at 34 meV is weak; further, it is excited only by a small fraction of $(t\mu)$ viz. those in the triplet state, which have not been thermalised by decay to the singlet state. It follows that the rate for $(dt\mu)$ at $T\lesssim400°$ is dominated by the rotational sub-resonances on the basic state $(0,1)$ at -11meV, viz. $(K,K_i") = (2,0),(3,1),(4,2),(5,3),(3,0),(4,1),(5,2)(4,0)(5,1),(5,0)$ spread over the range 0-60meV with the first four below 18 meV (Leon 1984)

FORMALISMS FOR RESONANCE WIDTH CALCULATION

The first thing we need is a formal framework for calculation. Remarkably, in view of the sixty years of history since Gamow's theory of alpha decay, the appropriate framework is not trivially available. Many calculations of widths in recent times use variants of Feshbach's projection operator formula:

$$\Gamma(E) = 2\pi\rho \ <\psi_E|QHP|\phi_o>^2 \tag{3}$$

where $P\equiv \phi_o><\phi_o$, and ψ_E are the continuum states of QHQ with $Q\equiv1-P$. This formula is essentially perturbative. The initial state ϕ_0 is assumed to contain no open channel components, $<\psi_E|\phi_0>=0$, and QHP does not occur in the Hamiltonian (PHP+QHQ) of ϕ_0 and ψ_E. As an example, ϕ_0 could be a 2p-1h state of a shell-model system, where the open channel contains a continuum particle and a closed shell). ϕ_0 and ψ_E are both members of the set of shell-model states. In contrast to this, the resonance state and the continuum states may have basically different character with different (re-arranged) clustering. An example is that of decay of "D_2" into $D_2+(t\mu)$:

$$\left[(dt\mu)_{11}dee\right]_{k\nu} \rightarrow \left[ddee\right]_{K_i\nu=o}+(t\mu)$$

We can schematise this, retaining the essence of the process, by ignoring the electrons (and assuming the d-d Born-Oppenheimer interaction is "given") and regarding $(t\mu)$ as a unit. Let us label it by 3, and the deuteron it combines with by 1, and the spectator by 2, then the process is:

$$\left[(31)2\right] \rightarrow (12) + 3$$

which displays the rearrangement character, implying that Feshbach-type perturbation theory is not suitable.

Non-Perturbative Formalism for Width Calculation

A resonance formalism that makes no reference to perturbation is R-Matrix theory, including Kapur-Peierls theory; see, e.g. Lane & Thomas 1958). In the case of one open channel, the real resonance energy E_r and width Γ have the forms:

$$E_r = Re.E_\lambda + \left|\gamma_\lambda\right|^2(Re.b-S) \tag{4}$$

$$\Gamma = 2P\left|\gamma_\lambda\right|^2$$

where:

$$\gamma_\lambda \equiv (\frac{\hbar^2}{2ma^3})^{\frac{1}{2}} <\delta(r-a)Y_\ell(\Omega)\phi|X_\lambda> \qquad (5)$$

We suppress angular momentum coupling for the sake of brevity. ϕ denotes the product of the internal wave-functions of the two channel fragments, and $\underline{a}$ is an interaction radius chosen such that there are no significant polarising forces between the fragments for separations $r>a$. Ω,ℓ are the angles and angular momentum of the relative motion of the fragments. S,P are real quantities such that $a^{-1}(S+iP)$ is the logarithmic derivative of the radial outgoing wave-function at $r=a$. X_λ is the solution of the full Hamiltonian H with eigenvalue E_λ and V prescribed log.der. at $r=a$ equal to $a^{-1}b$. Even if b (and so X_λ) is complex, it can be shown that, for discrete resonances, γ_λ is real to corrections of order width-to-spacing. We wish to write Γ as the square of an interaction matrix element. If $F(r)Y_\ell(\Omega)\phi$ is the solution at energy E of $H_o+(T+v)\equiv H-\Delta H$ where H_o is the Hamiltonian of ϕ and (T+v) is that of the relative motion, then Green's Theorem gives, for $E=E_\lambda$:

$$<\bar{F}Y_\ell\phi|\Delta H|X_\lambda> = (\frac{\hbar^2 a}{2m})^{\frac{1}{2}}\gamma_\lambda \ (b\bar{F}-a\bar{F}')_{r=a} \qquad (6)$$

This gives γ_λ as an interaction m.e. but leaves the question of whether the factor P in Γ can also be incorporated. There are at least two choices that achieve this (i) If v is a finite potential that makes F have a node at $r=a$, $F(a)=0$, then, if F has unit amplitude as $r\to\infty$, it follows that $(aF'(a))^2=akP$ where $E=\hbar^2k^2/2m$ so:

$$\Gamma = \frac{4m}{\hbar^2a^2k} \ |<\phi Y_\ell\bar{F}|\Delta H|X_\lambda>|^2 \qquad (7)$$

Note that Γ does not depend directly on b, and is insensitive to the surface value of X_λ.
(ii) If b is equal to (S+iP) taken at $E=E_r$, (the Kapur-Peierls condition) then

$$|(b\bar{F}-a\bar{F}')_{r=a}|^2 = kaP \qquad (8)$$

and the same equation (7) for Γ follows. However, there is an important difference. The condition (i) implies a different v for each ℓ, whereas the present one does not, so we can write further:

$$\Gamma = \frac{4mk}{\hbar^2} (\frac{1}{4\pi})^2 \sum\int d\Omega_k |<\phi u_{\underline{k}}|\Delta H|X_\lambda>|^2 \qquad (9)$$

where the sum is over magnetic quantum numbers (that of X_λ being fixed), u_k is the solution of (T+v) corresponding to initial plane wave $\exp(i\underline{k}.\underline{r})$. Although the complex value of b implies complex X_λ and E_λ, in the isolated resonance situation, X_λ is almost real and $\mathrm{Im}.E_\lambda = -\frac{1}{2}\Gamma$. As a special case, we can choose v=0, so that u_k equals $\exp(i\underline{k}.\underline{r})$. This has an important advantage in our problem as we will see, since it facilitates separation of coordinates. Note several points: (a) if the energy is below barrier, then a third choice giving (7) or (8) is that b is chosen equal to S. (7) follows with any choice of v, so $\bar{F}$ does not necessarily have a surface node. Note that, for $(t\mu)+D_2$, the energy is only below A.M. barrier for $E_r \lesssim 4meV$, so this approach is not useful. However, it applies to $(t\mu)+d$ for $E_r \lesssim 10eV$. (b) the form of Γ is non-perturbative since X_λ is a solution of the full Hamiltonian. (c) the form for Γ gives $\Gamma(E)$ if $u_{\underline{k}}$ is taken at energy E instead of E_λ. The extra terms containing

$(E-E_\lambda)$ in the Green's Theorem result can be shown to be small (Lane 1987), (d) Γ, being a physical quantity, cannot depend on the choice of a. In (10), a occurs only in X_λ through the imposition of boundary condition at r=a, but for outgoing waves this is essentially independent of a. (e) the normalisation region of X_λ is out to the radius beyond which there is free motion, i.e. to the outer turning-point if the energy is below barrier, otherwise to r=a, where a should be chosen at its minimum value, (f) if the resonance corresponds to a bound state, it can nevertheless be observed, if lightly bound, since its width above threshold is non-zero. It is given by (9), with the bound state wave-function for X_λ (normalised over all space). We will apply this result to the width of the bound $(dt\mu)_{11}$ state as reflected in $(t\mu)$+d scattering above threshold.

APPLICATION TO DECAY OF "D_2"

For definiteness, we will consider the case "D_2" $\to (t\mu)+D_2$, so 3 represents particles t and μ. Choosing v=0, the matrix element in (9) is:

$$\text{m.e.} = \langle \eta(\underline{r}_{t\mu})\Psi_{ok_i}(\underline{r}_{21})e^{i\underline{k}_i \cdot (\underline{r}_3-\underline{R}_{21})} | V_{31}+V_{32} | \Phi\Psi_{vk}(\underline{r}_2-\underline{R}_{31}) \rangle \quad (10)$$

where:

$$V_{3i} \equiv V_{ti} + V_{\mu i} \quad (i=1,2)$$
$$\underline{r}_{ij} \equiv \underline{r}_i-\underline{r}_j; \ \underline{R}_{ij} \equiv (m_i\underline{r}_i+m_j\underline{r}_j)(m_i+m_j)^{-1} \quad (11)$$

(Note: $\underline{r}_3 \equiv \underline{R}_{t\mu}$). η is the state of the $(t\mu)$ atom. K, K_i are the rotational A.M. of the resonance and target. Ψ_{vK}, Ψ_{OK} are the resonance and target molecular states, v being vibrational quantum number. We have approximated X_λ in (9) by the product of Ψ_{vK} with the meso-molecule state $\Phi\equiv(dt\mu)_{11}$ which has unit A.M. (Explicit reference to A.M. coupling is suppressed for notational simplicity). An evaluation of the accuracy of the approximate form of X_λ (which is evidently not exact since it is bound) gives an error of $\lesssim 10^{-4}$ in intensity (Lane 1987). Thus we expect it to yield a good result for Γ when used in (9), although clearly not when used in (4), (3). With the approximate form of X_λ (but not with the exact form), Green's Theorem shows that the interaction form $(V_{31}+V_{32})$ in (10) may be replaced by the form $(\Delta V_{21}+V_{23})$ where ΔV_{21} is defined as $V_{21}(\underline{r}_{21})-V_{21}(\underline{r}_2-\underline{R}_{13})$. As we will see, this identity gives a useful numerical check on calculations.

The first form is the more "intuitive", because the V_{31} part can be identified as simple recoil excitation of D_2 by the deuteron recoiling against the emission of $(t\mu)$ from $(dt\mu)$. Indeed, one can develop (10) and derive a form (Lane, 1987) for Γ which contains the width of the underlying process $(dt\mu)\to d+(t\mu)$ as a factor. Referring to Lane (1987) for details:

$$\Gamma_J(KS,K_iF) = \Gamma_J(K,K_i) \langle S| F \rangle^2$$
$$\Gamma_J(K,K_i) = \sum_{L=0}^{\infty} \gamma_J^2 (K,K_iL) \quad (12)$$

Here J is the total orbital A.M. = $\underline{K}+\underline{l}$ = $\underline{K_i}+\underline{L}$, while $\underline{S}$, $\underline{F}$ are the total intrinsic spins of $(dt\mu)_{11}$ and $(t\mu)$ respectively, satisfying $\underline{S}=\underline{F}+\underline{1}$, and $\langle S| F \rangle$ is the spin overlap factor. Further:

$$\gamma_J(K,K_iL) = \sum_{\alpha=L\pm1}^{J+K_i} (-)^{J+K_i} U(KK_i1L,\mathcal{L}J)\left(\frac{2\mathcal{L}+1}{2J+1}\right)^{\frac{1}{2}} \gamma_{\mathcal{L}}(K,K_iL) \tag{13}$$

$$\gamma_{\mathcal{L}}(K,K_iL) = \left(\frac{\pi k_i m_i}{3\hbar^2}\right)^{\frac{1}{2}} 4I\theta \frac{<\mathcal{L}\|1\|L><K_i\|\mathcal{L}\|K>}{(2\mathcal{L}+1)} (hM_L' - i^{\mathcal{L}+1-L} f k_i M_{\mathcal{L}}) \tag{14}$$

where the various quantities are:

$$h \equiv m_3/(m_1+m_3); \quad g \equiv m_2/(m_1+m_3); \quad f=m_1 h/m_i$$

$$m_1 = m_2 = m_d; \quad m_3 = m_t + m_\mu; \quad m_i \equiv m_3(m_1+m_2)(m_1+m_2+m_3)^{-1}$$

$$<a\|b\|c> \equiv <Y_a\|Y_b\|Y_c> \tag{15}$$

$$\left.\begin{array}{l} M_L' \equiv <\bar{\phi}_v(R)|\, j_L(gk_iR)\left|\dfrac{d\psi_o}{dR}\right.> \\[3mm] M_{\mathcal{L}} \equiv <\bar{\phi}_v(R)|\, j_{\mathcal{L}}(gk_iR)|\, \psi_o(R)> \end{array}\right\} \quad \begin{array}{l}\text{"Debye-Waller" factors representing} \\ \text{vibration excitation of "}D_2\text{" due to} \\ \text{recoil of (dt}\mu\text{)}\end{array}$$

$$\Psi_{0K_i}(\underline{R}) \equiv \psi_o(R)Y_{K_i}(\Omega_R); \quad \bar{\Psi}_{vK}(\underline{\rho}) = \bar{\psi}_v(\rho)Y_K(\Omega_\rho)$$

$$I \equiv <z_{13}\eta(r_{t\mu})|\, V_{1t} + V_{1\mu}|\Phi_o> \tag{16}$$

where subscript zero on Φ_o denotes zero component of rotational spin unity along the z direction. θ is 1 for (dtμ) but $\sqrt{2}$ for (ddμ) if η in (16) refers to a specific deuteron. The only assumption in deriving the above is that the range of $(V_{1t} + V_{1\mu})\Phi_o$, about $8a_\mu$, is small compared to the other lengths (size of D_2, $(fk_i)^{-1}$). The latter is satisfied for collision energies <<20eV.

I determines the width of the sub-problem $(dt\mu)_{11} \to (t\mu)+d$. Although $(dt\mu)_{11}$ is bound, its width is non-zero above threshold and manifests itself in $(t\mu)+d$ scattering. For energy $E = \hbar^2k^2/2m$, it is, from (9) at small k, using definition (16):

$$\Gamma^{(o)}(E) = \frac{mk^3}{3\pi\hbar^2} (I\theta)^2 \tag{17}$$

Small k means $k^{-1} >>$ range of $(V_{1t}+V_{1\mu})$, viz. $8a_\mu$, (implying E<<7eV).

Evaluation of Width $\Gamma^{(o)}$

We now give values for $\Gamma^{(o)}$ and I deduced from the literature. There are two sources (which fortunately are consistent with each other). First, one can use the formula (4), which means $\Gamma^{(o)}(E) = 2(ka)^3\gamma^2$ for E << 10eV, where γ is obtained from the wave-function plot (Fig. 4) of Vinitsky et al. (1978). The wave-function Φ has the Born-Oppenheimer form, for dtμ:

$$\Phi(dt\mu) = 2^{-\frac{1}{2}} r_{1t}^{-1}[\phi_g(\bar{\chi}_a(r_{1t}) + \bar{\chi}_b(r_{1t})) + \phi_u(\bar{\chi}_a(r_{1t}) - \bar{\chi}_b(r_{1t}))]Y_1(\Omega_{1t})$$

$$\to r_{1t}^{-1} \bar{\chi}_a(r_{1t})\eta(r_{t\mu})Y(\Omega_{1t}) \text{ as } r_{1t} \to \infty \tag{18}$$

with

$$\int dr_{1t}(\bar{\chi}_a^2(r_{1t}) + \bar{\chi}_b^2(r_{1t})) = 1 \tag{19}$$

$\phi_{g,u}$ are the even, odd solutions of muon motion for fixed nuclei. From

60

(5), it follows that:

$$\gamma^2 = \frac{\hbar^2}{2ma^2}\, a\, \bar{\chi}_a^{-2}(a) \tag{20}$$

Note that Fig. 4 plots the quantity $\bar{\chi}_\mu\, a^{-\frac{1}{2}}$.

From (20), we obtain $\Gamma^{(o)}(eV) = 0.43\, E^{3/2}(eV)$, corresponding to $I \approx 3200\ eV\ a_\mu^{5/2}$. This agrees with the value of $\Gamma^{(o)}$ that can be obtained from the second route, the calculation of scattering phase-shift for $(t\mu)+d$ by Melezhik et al. (1983). For the $dd\mu$ case:

$$a_\mu^{\frac{1}{2}}\Phi(dd\mu) = r_{1d}^{-1}\, \phi_g\, \bar{\chi}_a(r_{1d})Y_1(\Omega_{1d}) \tag{21}$$

$$\xrightarrow[r_{1d}\to\infty]{} \sqrt{\tfrac{1}{2}}\, r_{1d}^{-1}\, \bar{\chi}_a(r_{1d})Y_1(\Omega_{1d})(\eta(r_{1\mu})+\eta(r_{d\mu}))$$

The plot of Fig. 4 of Vinitsky et al. gives $(2a_\mu)^{-\frac{1}{2}}\,\bar{\chi}_a$. Equation (20) applies to $(dd\mu)$ when the factor $\theta^2=2$ is taken into account. We get $\Gamma^{(o)}(eV) = 0.50\, E^{3/2}(eV)$, or $I \approx 3100\ eV\ a_\mu^{5/2}$. From the scattering calculations of Melezhik and Wozniak (1986) we obtain the very rough value $\Gamma^{(o)}(eV)\sim E^{3/2}(eV)$.

The Contribution from V_{32}

Now we consider the contribution of V_{32} in (10). We must use the Born-Oppenheimer potential, which takes account of electron screening, for the d-d interaction V_{21}, and similarly for $V_{32}\equiv(V_{2t} + V_{2\mu})$. The hope is that, when this is done, we can ignore the electrons (and thereby keep further complications out of the problem). Actually there will be an overlap effect because the electrons on the two sides of (10) are aligned along different axes, viz. $\underline{r}_{21}$ and $(\underline{r}_2-\underline{R}_{31})$ on the left and right. Since $\underline{r}_{31}$ is small, the overlap factor is ≈ 1 and we ignore it. (Note that the period of internal $(dt\mu)$ motion is $<<$ period of electron orbit $<<$ molecular period). Since $r_{t\mu}(\approx a_\mu)$ is much smaller than the separation of 2 from $(dt\mu)$ we can approximate:

$$V_{32} \approx \underline{r}_{t\mu}\cdot \underline{\nabla}\, V_{BO}(\underline{r}_2-\underline{R}_{31}) \tag{22}$$

where we have also used $r_{13} << r_{21}$. The $\underline{r}_\mu$ integration in the matrix element of (10) is then, for fixed $\underline{r}_{13}$:

$$\underline{\nabla}\, V_{BO}(\underline{r}_2-\underline{R}_{13})\cdot{<}\eta(\underline{r}_{t\mu})|\,\underline{r}_{t\mu}|\Phi{>} \tag{23}$$

This can be compared to the corresponding expression from ΔV_{12} when used in (10):

$$h\underline{\nabla}V_{BO}(\underline{r}_2-\underline{R}_{13})\cdot\underline{r}_{13}\,{<}\eta(\underline{r}_{t\mu})|\Phi{>} \tag{24}$$

In terms of χ_a in (18), we have, for $r_{1t} \gtrsim 3a_\mu$ and fixed $\underline{r}_{13}$

$${<}\eta(r_{t\mu})|\Phi{>} \approx r_{13}^{-1}\chi_a(r_{13})Y_1(\Omega_{13})\equiv\zeta(\underline{r}_{13}) \tag{25}$$

It follows that

$${<}\eta(\underline{r}_{t\mu})|\,\underline{r}_{t\mu}|\Phi{>} \equiv {<}r_{t\mu}^2{>}\,\underline{\nabla}\zeta(\underline{r}_{13}) + {<}\eta(\underline{r}_{t\mu})|\,\underline{r}_{t\mu}|\,\eta(\underline{r}_{1\mu}){>}\zeta(\underline{r}_{13}) \tag{26}$$

(25) times $\underline{r}_{13}$ is larger than the first term of (26) by $\sim 20(r_{13}/a_\mu)$. The second term of (26) is comparable to (25) times $\underline{r}_{13}$ for $r_{13} \sim a_\mu$, but falls off exponentially with increasing r_{13}. (Using an oscillator form

for η, the extra factor is $(1/2)\exp(-r_{13}^2/4b^2)$ for all r_{13}, where b is the oscillator size constant). For small $\underline{k_i}$, the matrix element (10) with the ΔV_{12} form is, using (24):

$$\text{m.e.} \sim h\langle\Psi_{0K_i}|\underline{\nabla}V_{B0}|\bar{\Psi}_{\nu K}\rangle\cdot\langle\underline{r}_{13}\eta(\underline{r}_{t\mu})|\Phi_0\rangle \tag{27}$$

In passing, we note that the corresponding form from V_{13} is, taking the largest term as $k_i\to 0$:

$$\text{m.e.} \sim h\langle\Psi_{0K_i}|\underline{\nabla}\,\bar{\Psi}_{\nu K}\rangle\cdot\langle\underline{r}_{13}\eta(r_{t\mu})|V_{1t}+V_{1\mu}|\Phi_0\rangle \tag{28}$$

The last factor in (27) is $\sim \langle\underline{r}_{13}\zeta(\underline{r}_{13})\rangle$ so we see that the extra exponential cut-off factor implied by V_{32} causes a large reduction in m.e. relative to the ΔV_{12} form. Thus, at least at small $\underline{k_i}$, we deduce that V_{32} may be ignored relative to ΔV_{21} or V_{13}.

Numerical Equivalence of V_{13} and ΔV_{12}

When (10) is developed with ΔV_{12} instead of V_{13}, we find, using (24), that the form of Γ is just as given before by (12)-(15) except that M_L' and I are replaced (and $M_{\mathcal{L}}$ dropped for small k_i). In M_L', $d\phi_0/dR$ is replaced by $(dV_{B0}/dR)\phi_0$. In I, $(V_{1t} + \bar{V}_{1\mu})$ is dropped, giving:

$$I(\Delta V_{12}) = \langle(\tfrac{4\pi}{3})^{\frac{1}{2}} Y_{10}(\Omega_{13})(3j_1(fk_i r_{13})/fk_i)\eta(\underline{r}_{t\mu})|\Phi_0\rangle \tag{29}$$

$$\approx \langle z_{13}\,\eta(r_{t\mu})|\Phi_0\rangle$$

The absence of $(V_{1t}+V_{1\mu})$ means that the approximation applies over a smaller range of energy than for $I(V_{13})$ of (1b), viz. $<<2eV$ instead of $<<10eV$.

From (29), with (18) or (21):

$$I(\Delta V_{12}) = \sqrt{\tfrac{4\pi}{3}}\, a_\mu^{-\frac{1}{2}}\, \theta^{-1}\int r_{13}^2\, \bar{\chi}_a(r_{13})dr_{13} \tag{30}$$

For (ddμ), using Fig. 4 of Vinitsky et al. (1978) which plots $(2a_\mu)^{-\frac{1}{2}}\bar{\chi}_a$, we find $I(\Delta V_{12}) \approx 3120\, a_\mu^{5/2}$. The ratio of matrix elements (8) for small k_i is:

$$\frac{\text{m.e.}(\Delta V_{12})}{\text{m.e.}(V_{13})} \approx \left(\frac{M_0'(\Delta V_{12})}{M_0'(V_{13})}\right)\left(\frac{I(\Delta V_{12})}{I(V_{13})}\right) \tag{31}$$

The previously quoted value of $I(V_{13})$, 3100 $eVa_\mu^{5/2}$ makes the last ratio $(1.0eV)^{-1}$ while Tables of M_L' (Leon 1986) gives the first ratio = $(1.0eV)$, so the matrix elements are equal, as expected.

Absolute Evaluation of Formation Rate

From (1), with $\bar{g} \equiv (1/3)(\vec{J}\hat{S}/\hat{k_i}\hat{F})^2$ where $\hat{x} \equiv (2x+1)^{\frac{1}{2}}$, we find, using (13) that the formation rate summed over the J-multiplet is:

$$\lambda(KS,K_i F) = (3\alpha^0)^{-1}(\hat{S}/\hat{k_i}\hat{F})^2 \sum_J \hat{J}^2\Gamma_J(KS,K_i F) \tag{32}$$

where:

$$(\alpha^0)^{-1} \equiv \pi^2 h^2\rho v_i(m_i E_r)^{-1}f(E_r,T) = \rho(2\pi h^2/mkT)^{3/2}\exp(-\epsilon/kT) \tag{33}$$

$$\sum_J \hat{J}^2\Gamma_J(KS,K_i F) = \langle S|F\rangle^2\sum_{L\mathcal{L}} (2\mathcal{L}+1)\gamma_{\mathcal{L}}^2(K,K_i L)$$

At low energies, the sum is dominated by L=0, $\mathcal{L}$=1. From (14), dropping $M_\mathcal{L}$:

$$\sum_J \hat{J}^2 \Gamma_J(KS,K_iF) \approx \frac{m_i k_i (\theta I)^2}{\pi h^2} (\frac{K+K_i+1}{2}) (hM'_0)^2 <S| F>^2 \qquad (34)$$

where $\theta=\sqrt{2}$ for ddμ, 1 for dtμ.

Application to (dμ)+D$_2$ at kT = 30°

We assume that there is a resonance at E_r = 4 meV due to the transition F=3/2, K_i=0 → S=1/2, K=1. The spin factor in the (ddμ) case is:

$$<S| F>^2 = U^2(11S\tfrac{1}{2},1F) \qquad (35)$$

(In the dtμ case, the factors are all 0 or 1; the difference arises because, due to statistics, the two deuterons in (ddμ)$_{11}$ must couple to unit spin). We obtain, with $I \approx 3100 eVa_\mu^{5/2}$, and Leon (1986) Tables for $M'_0(\nu=7)$:

$$\sum_J \hat{J}^2 \Gamma_J(KS,K_iF) = \Gamma_0(1\tfrac{1}{2},03/2) \approx 1.4 \times 10^{-7} eV \qquad (36)$$

$$\approx 2.1 \; 10^8 s^{-1} \hbar$$

This leads (with f=0.115(meV)$^{-1}$, ρ=4.22x10^{22}cm^{-3}, $\alpha^0 \approx 5.5$) to:

$$\lambda(KS,K_iF) \approx 0.65 \times 10^7 s^{-1} \qquad (37)$$

Unfortunately there is no experimental value for this quantity. Instead there is a measure of the fusion rate which is reduced below the formation rate because of back decay. In the low density limit where collisions are negligible, the correction factor depends on details of the A.M. coupling in the "D$_2$" system, total A.M. being conserved (Lane 1987). In practice, for densities used in experiment, collisions thermalise spins (Ostrovkii and Ustimov, 1980). Not only is total A.M. not conserved, but neither is rotational A.M. The correction factor is thus $\Gamma_f(\Gamma_f+\Gamma(S))^{-1}$ where Γ_f is the fusion decay rate of the mm and $\Gamma(S)$ is the back-decay rate.

$$\Gamma(S) = \sum_{F,K_i,K} \Gamma(SK,FK_i)\omega(K) \qquad (38)$$

where $\omega(K)$ is the Boltzman factor $\propto(2K+1)\exp(-E_K/kT)$ satisfying $\sum_K \omega(K)=1$.

(For ddμ, deuteron statistics introduce a further factor into $\omega(K)$, 2 for even K, 1 for odd K).

PUBLISHED WORK

In chronological order:

Vinitsky et al. 1978. This was the pioneering work, but is now seen to have flaws. The interaction form was $(\Delta V_{12}+V_{23})$, but this was taken in its unscreened form, $e^2 d.\rho \; \rho^{-3}$ where d is the charge dipole moment of (dtμ) and $\rho = (\underline{r}_2-\underline{R}_{13})$. Further the plane-wave state in (10) was unfortunately replaced by a scattering solution of the (tμ)+d problem. The rotational motion was limited to K_i=0, K=1 and hyperfine splitting was ignored. For the dμ+D$_2$ case with ε(ddμ)=-2,196eV, E_r=53meV and ν=8 the value of λ, taken at kT = 2/3 E_r is 0.63x10^{6}s^{-1}. For the tμ+D$_2$ case, with ε(dtμ) = -1.09eV, E_r=70meV and ν=4, λ is 94x10^{6}s^{-1}.

Leon 1984. This took account of rotational excitations and hyperfine effects, and used $e^2 d.\rho \, \rho^{-3}$ with a plane wave. For $t\mu + D_2$ with $\varepsilon(dt\mu)=0.64eV$, $E_r= -18meV$ for $K_i=0$, $K=1$ $F=0$, $S=1$ and $v=2$, λ was found to be $\sim 2 \times 10^8 s^{-1}$ for $T \gtrsim 50°$ and increases strongly if E_r is raised.

Cohen and Martin 1984. This commented that screening effects reduced Leon's estimates by a factor 6.

Cohen and Leon 1985. This mentioned a new mechanism whereby the recoil of the decay $(dt\mu) \rightarrow (t\mu)+d$ excites the D_2 molecule. In fact, as we have seen, this corresponds to the V_{13} form of interaction in (10), and this is actually equivalent to the previous mechanism, which corresponds to the ΔV_{12} form. This heuristic approach lead to Γ depending on incident wave-number as k^3 (like that for the underlying process) instead of the correct k following from the M'_o term in (14). Their Figure 1 shows different values for λ from the two mechanisms. As we have seen these should coincide when calculated properly.

Menshikov and Faifman 1986. They consider the question of a reliable calculational framework for Γ when rearrangement is involved. They do this in terms of model problems. However, as we have seen in paragraph 2, the Kapur-Peierls theory gives the correct form.

Menshikov 1985. This calculates the $(t\mu)+D_2$ case with the ΔV_{12} form with screening, but using oscillator forms for the vibrational states of the molecules, and summing over rotational states with analytic approximations. With $E_r= -12meV$ for $K=K_i=0$, he finds $\lambda(380°) \approx 15 \times 10^8 s^{-1}$.

Faifman et al. 1986. They calculate the $d\mu+D_2$ case with the ΔV_{12} form with screening, including all rotational and hyperfine effects. They give formulae for $\lambda(SK,FK_i)$ (their (6)) and $\Gamma(SK,FK_i)$ (their (15)). I agree with (6) but believe that (15) should be divided by 2. Figures 4 and 5 give plots of λ and Γ for various FS. There seems to be an inconsistency between the formulae and the Figures. The ratio Γ/λ at 300° from the formulae is about 4 x that from the Figures. It is found that the back decay reduction factor is $\approx 1/4$ i.e. $\Gamma(S) \approx 3\Gamma_f$, so that back decay is crucial in reducing the formation rate at 300° (of order $11 \times 10^6 s^{-1}$) to the observed fusion rate $(\lambda=2.76 \times 10^6 s^{-1})$. The formation rate at 46°, using $E_r=5.9meV$ for $K_i=0$, $K=1$, $F=3/2$, $S=1/2$ is $16 \times 10^6 s^{-1}$. In view of the above inconsistencies, there is some doubt about these values. Note that the fusion rate, for $\Gamma(S)>\Gamma_f$, is more sensitive to Γ_f than to $\Gamma(S)$, and is not a sensitive measure of formation rate. Fitting observed λ_f at 300° and the observed ratio of $\sum_s \lambda(S,F)$ for $F=3/2$, $1/2$ at 30°(=80) determines E(ddμ) and also $\Gamma_f = 4.2 \times 10^9 s^{-1}$.

QUADRATIC DENSITY DEPENDENCE OF FORMATION RATE

There have been recent experimental indications that the fusion rate in the (dtμ) case contains an increasing quadratic term in the density, as well as the basic linear term that we have been discussing (embodied by (32), (33)). This raises the question of the origin of such dependence. Does it require an entirely new, three-body mechanism, not at all easy or reliable to calculate because of its necessarily complex nature (Menshikov and Ponomarev 1986), or is there a straightforward effect that can produce the new density dependence (Petrov 1985)?

Menshikov and Ponomarev (1986) have calculated the three-body process $(t\mu)+D_2+D_2 \rightarrow "D_2"+D_2'$, where "$D_2$" is the $v=2$ state at $-11meV$. All rotational motion is ignored and molecules are treated as points so the calculation is schematic. They claim that the formation rate is $\sim 10^8 s^{-1}$

for T=150-800° (falling off at low T) and LHD. This quadratic rate is not much smaller than estimates of linear rate. The experimental value at LHD is about $6 \times 10^8 s^{-1}$ (see talks by Jones and Petitjean).

Petrov (1985) has suggested that an important effect arises from resonance broadening due to rotation-changing collisions. This affects not only the positive-energy resonances (K=2,3,4... at E_r=3,18,37 meV) but also the negative-energy states (K=0 at -11meV, K=1 at -6meV). Ostrovskii and Ustimov (1980) find that the K=1 relaxation rate at LHD and 300° is $\Gamma \hbar \sim 2.5 \times 10^{13} s^{-1}$, corresponding to width 16meV. Thus large fractions of the K=0,1 states are spread to the region above threshold (about 20%, 30% respectively.

Let us first note an effect for positive-energy resonances that has been ignored so far (i.e. in (32)). The correct formation rate involves the quantity:

$$\frac{\Gamma}{kT} \int_0^\infty dx \; \frac{x^{\frac{1}{2}} e^{-x}}{(x-x_r)^2 + (\Gamma/2kT)^2} \qquad (39)$$

where $x_r \equiv E_r/kT$ and we have assumed that the formation width (12) – (14) varies with E as $E^{\frac{1}{2}}$ (as it does for dominant L=0 to a good approximation). Underlying (32) is the assumption that $\Gamma \ll kT$ and x_r is not much larger than unity. This gives (39) the value $2\pi\sqrt{x_r}\exp(-x_r)$, which is of order 1 (for $x_r > 0.01$). If x_r or Γ/kT is $\gg 1$, then (39) becomes

$$\sqrt{\pi} \; \frac{(\Gamma/2kT)}{x_r^2 + (\Gamma/2kT)^2} \qquad (40)$$

which is $\ll 1$, especially if Γ/kT is $\ll 1$. This is the basis of the usual form (32). For a negative energy resonance, form (39) applies with x_r replaced by $-x_r$. For x_r or $\Gamma/kT \gg 1$, (40) holds, i.e. positive and negative-energy resonances of the same resonance parameters (except $E_r \to -E_r$), give the same rate. The numbers given above imply, for the K=1 state (which is excited by L=0 from K_i=0): x_r = 0.24, $(\Gamma/2kT)$ = 0.32. These imply that, in units of the 'normal' value $2\pi\sqrt{x_r}\exp(-x_r)$ = 2.4, (39) is 0.6, 0, 0.20 for positive, negative x_r. From (12) – (14), at E_r = 6meV, we find formation width $\Gamma_o(11,00)$ = 0.1meV. With α^o evaluated at 300° and E_r = 6meV: $\alpha^o \approx 60$, the "normal" formation rate is Γ_o/α^o = $2.4 \times 10^9 s^{-1}$. Thus the quadratic term in the rate due to collision-broadening of the negative-energy (K=1) state is:

$$\lambda(K_i=0, \; K=1) \approx 4.8 \times 10^8 s^{-1} \; (\frac{\rho}{\rho_o})^2 P_o \qquad (41)$$

where ρ_o is the atom density of liquid hydrogen, and P_o is the Boltzman probability for the target D_2 to have K_i=0, at 300°.

Now we take account of $K_i \neq 0$. At 300°, the populations of K_i=0,1,2... are 19,21,39,11,8,1%. The states of K=K_i+1 are located at -6, -4, -4, -7, -13 meV. K_i=1,2,3... give values like (41), but times 0.89 P_1, 0.80 P_2, 0.6P_3,... giving, with (41):

$$\lambda(\text{negative } E_r) = 4 \times 10^8 s^{-1} (\rho/\rho_o)^2 \qquad (42)$$

The main contributions to the rate from positive energy resonances are from those at 11,15 meV which allow L=1 terms in (12) and give:

$$\lambda(\text{positive } E_r) = 0.8 \times 10^8 s^{-1}(\rho/\rho_o) \qquad (43)$$

where we have ignored collision broadening and used (32) along with widths from my talk on Decay Modes ($\Gamma \hbar^{-1}=0.5, 0.9 \times 10^{10} s^{-1}$ for $(Ki,K)=(1,3),(2.4)$ states at resonance energies 11,15meV. It is notable that the ρ^2 contribution to the rate (42) is larger than the linear team (43) for densities $\rho > 0.2\rho_9$. (42) and (43) imply that the rate divided by (ρ/ρ_o), increases from 1.2 to $5.7 \times 10^8 s$ between $\rho/\rho_o=0.1$ and 1.2. Idaho-Los Alamos data at $<125°$ shows a range 2.0 to $7.0 \times 10^8 s^{-1}$, which is remarkably similar. The SIN data at $<35°$ shows a similar trend with rather smaller values. For $kT > E_r$ and $\Gamma \propto T$, (39 is $\approx$ constant as T increases, so λ falls as T^{-1}. From (39), the ρ^2 effect (42) must flatten for ρ increasing past the valve where $\Gamma \approx kT$.

Finally let us note that the estimate of collision width may be too small because it identifies spreading width with relaxation to other rotational states of "D_2". In fact, elastic, phase-changing, collisions cause spreading as well as true relaxation processes. Therefore the ρ^2 effects may be larger than indicated by (42).

Acknowledgement

The work described in this paper was carried out for the Euratom/UKAEA Fusion Association.

References

Cohen, J.S. and Leon, M., 1985, Phys.Rev.Lett. 55,52.
Cohen, J.S. and Martin, R.L., 1984, Phys.Rev.Lett. 53,738.
Faifman, M.P., Menshikov, L.I., Ponomarev, L.I. and Strizh T.A., 1986a
        Dubna preprint.
Faifman, M.P., 1986b, Z.Phys.D 2,79.
Frolov A.M. and Efros, V.D., 1985, J.Phys.B. 18, L265.
Gocheva, A.D., Gusev, V.V., Melezhik, V.S., Ponomarev, L.I.,
        Puzynin, I.V., Puzynina, T.P., Somov, L.N. and Vinitsky, S.I.,
        1985, Phys.Lett. 153B,349.
Lane, A.M. and Thomas, R.G., 1958, Rev.Mod.Phys. 30,257.
Lane, A.M., 1987, J.Phys.B. At.Mol.Phys, in press.
Leon, M., 1984, Phys.Rev.Lett. 52,605.
Leon, M., 1986, private communication.
Melezhik, V.S., Ponomarev, L.I. and Faifman, M.P., 1983, Sov.Phys.JETP
        58,254.
Melezhik, V.S. and Wozniak, J., 1986, Phys.Lett. 116A,370.
Menshikov, L.I., 1985, Sov.J.Nucl.Phys. 42,750.
Menshikov, L.I. and Faifman, M.P., 1986, Sov.J.Nucl.Phys. 43,414.
Menshikov, L.I. and Ponomarev, L.I., 1986, Phys.Lett. 167B,141.
Ostrovkii, V.N. and Ustimov, V.I., 1980, JETP 52,620.
Petrov, Y.V., 1985, Phys.Lett. 163B,28.
Ponomarev, L.I. and Faifman, M.P., 1976, Sov.Phys.JETP 44,886.
Ponomarev, L.I., Puzynin, I.V. and Puzynina, T.P., 1974, JETP 38,14.
Vinitsky, S.I., Ponomarev, L.I., Puzynin, I.V., Puzynina, T.P., Somov,
        L.N. and Faifman, M.P., 1978, Sov.Phys. JETP 47,444.

DECAY MODES OF MESO-MOLECULES

A.M. Lane

U K Atomic Energy Authority
T.P. 424.4, Harwell
Oxon OX11 ORA, United Kingdom

GENERAL INTRODUCTION

We discuss the various decay modes of a quasi-molecule, to be written as "D_2" in which a deuteron is replaced by a meso-molecule (mm) which is $(dt\mu)$ or $(dd\mu)$ in its 4th excited state. This state has quantum numbers: rotational A.M.=1, number of vibration quanta=1, and is denoted as $(dt\mu)_{11}$. For definiteness we refer to the $(dt\mu)$ case, but these general remarks apply equally to the $(dd\mu)$ case, unless specified otherwise. The quantum numbers of the D_2 and "D_2" molecules are written as $K_i \nu_i$ and $K\nu$ respectively, and the formation mode of the quasi-molecule is:

$$(t\mu) + (D_2)_{K_i \nu_i} \rightarrow \left[(dt\mu)_{11} dee \right]_{K\nu} \equiv ("D_2")_{K\nu}$$

(At temperatures of interest, T<2000°, the D_2 population is almost entirely ν_i=0).

The decay modes of "D_2" are (Lane 1983):
(0) muon decay; the decay rate is $4.5 \times 10^5 s^{-1}$.
(1) Auger decay of $(dt\mu)_{11}$ to lower states. These are (along with approximate binding energies in eV): (00)320, (10)230, (20)100, (01)35.
(2) E1 radiative decay of $(dt\mu)_{11}$ to (00),(01)
(3) collisional de-excitation of "D_2" with a reduction in the vibration quantum number
(4) radiative E1 de-excitation of "D_2" (with $\Delta K = -1$, $\Delta \nu = -1$)
(5) fusion decay of $(dt\mu)_{11}$:

$$(dt\mu)_{11} \rightarrow \begin{cases} \alpha + \mu + n \\ (\alpha\mu) + n \end{cases}$$

(6) "back decay" which is the reverse of formation, and leads to any rotational or hyperfine states that are energetically accessible. Note that, at 300° and liquid hydrogen density (LHD), rotational transitions of "D_2" due to collisions with D_2 are so rapid ($> 10^{-13} s^{-1}$, Ostrovskii and Ustimov 1981) that, no matter which K formation (1) leads to, before decay occurs, the rotational state will relax to a thermal (Boltzman) distribution, so the relevant back decay width is

$$\Gamma_6 = \sum_{K,K_i} P(K)\Gamma(K,K_i)/\sum_K P(K) \text{ with } P(K) \equiv (2K+1)e^{-cK(K+1)/kT} \tag{1}$$

where c is the rotational energy parameter (2.4 meV for dtμ, 2.2meV for ddμ), and $\Gamma(K,K_i)$ is the back decay width for transition $K \to K_i$. (The estimate of rotational relaxation is compatible with earlier estimates for HD+H$_2$, $0.23 \times 10^{13} s^{-1}$ (Chu 1975) and recent data on HD(ν=1)+HD, $0.14 \times 10^{13} s^{-1}$ (Chandler and Farrow 1986)).

All "downward" transitions (1)-(4) lead to states which lead on to fusion, possibly preceded by further downward transitions. If we ignore the small possibility of muon decay in these intermediate stages, the probability that the decay of "D$_2$" leads to fusion is $1-(\Gamma_6+\Gamma_0) \left(\sum_{i=1}^{6} \Gamma_i \right)^{-1}$, where Γ_i is the width associated with decay process (i) listed above. Note that Γ_3 depends on density (linearly),so, when it is important, it introduces a density dependence into the fusion rate.

ESTIMATION OF DECAY RATES

(1) Auger Decay

Using the Golden Rule Formula, the width for the E0 process is:

$$\Gamma_1 = 2\pi\left(\frac{2mk}{\pi\hbar^2}\right) \; <u_f \phi_{(10)}|V|u_i \phi_{(11)}>^2 \tag{2}$$

where the electron escape energy $\hbar^2 k^2/2m$ is about 220eV, and the initial electron state u_i may be approximated as the 1s state about (dtμ) as centre. For an E0 process, u_f may be taken as $(4\pi)^{-\frac{1}{2}}(kr)^{-1}\sin(kr)$ if we ignore Coulomb distortion at such high energy. The E0 component of the Coulomb interaction V is $2e^2/r$ for 2r>R and $4e^2/R$ for 2r<R where R is the separation of d and t. Integration over r gives:

$$\Gamma_1 = \frac{mke^4}{9\hbar^2 a_o^3} \; <\phi_{(10)}|R^2|\phi_{(11)}>^2 \tag{3}$$

where a_o is the Bohr radius. Graphs of wave-functions ϕ (Ponomarev et al. 1974) give the matrix element as $(550f)^2$, giving

$$\Gamma_1(E0)\hbar^{-1} \approx 3 \times 10^8 s^{-1}$$

for the (11) $\to$ (10) transition. (A separate evaluation, quoted by Bogdanova et al. (1982) gives $2 \times 10^8 s^{-1}$). This estimate applies to both (ddμ) and (dtμ), as does an estimate (Lane 1983) of E2 decay, which is of order 8% of that for E0. There is an E1 part for (dtμ) which has no counterpart for (ddμ) where E1 transitions are almost totally forbidden. For (dtμ), the relevant E1 part of V is $e(\underline{R} \cdot \underline{r})/5r^3$, and one estimates for decays to spin 0 states (using (2) summed on magnetic quantum numbers of u_f)

$$\Gamma_1(E1)\hbar \approx 2 \times 10^9 s^{-1}, \; 10^{11} s^{-1}$$

for final states (00), (01) respectively while the final state (20) has $1.3 \times 10^{11} s^{-1}$.

We note in passing the Auger rates for transitions between lower states. For (ddμ) and (dtμ), Γ_1(E0) for (01)$\to$(00) is $3 \times 10^8 s^{-1}$, while for (dtμ), Γ_1(E1) is $\sim 4 \times 10^{10} s^{-1}$ for (10)$\to$(00) and $4 \times 10^{10} s^{-1}$ for (01)$\to$(10).

(2) El Radiative Decay

As mentioned El effects do not occur in the (ddμ) case. For (dtμ) we have, for A.M. one to zero transitions

$$\Gamma_2 = \frac{4e'^2 E_\gamma^3}{9\hbar^3 c^3} <R>^2 \qquad (4)$$

where (e'/e) is the effective El charge (= 0.2) for dtμ) and <R> is the radial matrix element . For the (01) state E_γ is small (35eV) and <R> is large (~ 1000f), while the (00) state has large E_γ (320eV) and small <R> (~ 60f). In both cases, Γ_2 is small,

$$\Gamma_2 \hbar^{-1} \lesssim 10^6 s^{-1} \ .$$

(3) Collisional De-excitation of "D_2"

At 300°K and LHD, the de-excitation rate for ν=1 of D_2 is $2.7 \times 10^5 s^{-1}$ 2.1×10^6, $4.4 \times 10^6 s^{-1}$ for D_2, HD and H_2 gases respectively (Lukasik and Ducuing 1974). The rate of such processes increases rapidly with ν. For HD-HD collisions, the rate for ν=2,6 is about 3x, 60x larger than for ν=1 (Rohlfing et al. 1984). On the basis of these values, we can tentatively assign the rate for "D_2":

$$\Gamma_3 \approx 0.8 \times 10^6 s^{-1}, \ 1.7 \times 10^7 s^{-1}$$

for the (dtμ), (ddμ) cases respectively. Note that the rates rise rapidly with kT, being about 15x, 700x, 2×10^4x larger at 500°, 1000°, 2000°. This means that, in the (ddμ) case, collisional de-excitation Γ_3 competes with the Auger process Γ_1 for kT $\gtrsim$ 500°, while, for (dtμ), this happens only for kT considerably above 2000°.

The collision width depends on density (linearly at low density), so, when it is significant (i.e. comparable with the largest of the other widths), it gives rise to a density effect.

(4) Radiative De-excitation of "D_2"

The effective dipole charge (e'/e) is 1/3 for (ddμ), 3/7 for (dtμ) in "D_2". If we assume <R> in (4) is 10^{-8}cm, then, for (dtμ), $\Gamma_4 \approx 0.75 \times 10^4 s^{-1}$ which is negligible.

(5) Fusion

The routes to fusion from $(dt\mu)_{11}$ and $(dd\mu)_{11}$ are believed to be quite different as a result of the suppression of El transitions in (ddμ) which means that fusion occurs from states of A.M. one. In (dtμ), rapid El Auger transitions means that states of A.M. zero are populated, and fusion occurs from these.

The theory of fusion requires the Born-Oppenheimer expansion of the (dtμ) wave-function:

$$\Phi_{JM} = \sum_j R^{-1} \chi_j(R) \phi_j(\underline{\rho}, \underline{R}) D_{mM}^J(\Omega) \qquad (5)$$

where Ω are the angles of the d-t axis relative to space-fixed axes. j represents the usual (nℓm) quantum numbers of the muon state $\phi_j(f;\underline{R})$ for fixed R. m is A.M. component along the d-t axis. As R→0, ϕ_j becomes a He-type orbit, and nℓm have their usual meaning: ℓ is orbital A.M.. For all J, the main term, j=0 (say), has (nℓm)=(100); however this need not be the main contribution to fusion. The simplest case is J=0. j=0 is the main contribution and we can assume that χ_o below the barrier is proportional to the usual Coulomb solution $F_o(R)$ for $\ell=0$. For low positive energy, this has a radial dependence independent of energy, although the chosen normalisation makes $F_o \to 0$ as $E \to 0$, viz. $F_o = C_o f_o(R)$ where C_o^2 is the usual $2\pi\eta(e^{2\pi\eta}-1)^{-1}$, independent of R. For small R, $f_o(R) \to kR$. An incident wave of F_o-type in the d+t channel gives rise to an outgoing wave in the α+n channel: $\frac{1}{2}Se^{ik \cdot r}(v/v')^{\frac{1}{2}}$, where S is the S-matrix element, and v,v' the velocities in the incoming, outgoing channels. This means outgoing flux $\frac{1}{4}|S|^2 v$. Defining cross-section $\sigma = \pi|S|^2 k^{-2}$, the fusion rate is

$$\Gamma_5 \hbar = \frac{1}{4}\left|\frac{\chi_o}{F_o}\right|^2 |S|^2 v = \left[(4\pi)^{-\frac{1}{2}} R^{-1}\bar{\chi}_o(R)\right]^2_{R \to 0}(\sigma v C_o^{-2}) \qquad (6)$$

The first factor is the density at R=0 obtained from $\bar{\chi}_o$, which is χ_o modified by suppressing the nuclear potential. The last factor can be obtained from cross-section data, and is independent of energy (excepting resonance effects), so it is reasonable that it applies to the present problem (where the d-t energy is slightly negative).

For J=1, the main contribution to fusion comes from competition from two sources. It may arise from j=0, in which case the relative A.M. of d-t is unity, so $\bar{\chi}_o$ is suppressed near R=0 by the extra A.M. barrier. It may arise from j≠0, specifically j=210, in which case χ_j is small, being a correction to the Born-Oppenheimer leading approximation j=0. For j≠0, (5) applies with χ_o replaced by the s-wave component that χ_j has near R=0. For j=0, (5) needs to be modified for p-waves, which means that C_o^2 is multiplied by $(1+\eta^2)/9R^2$, and $\bar{\chi}_o$ is replaced by $R^{-1}\bar{\chi}_j$.

In the (dtμ) case, j≠0 dominates since the s-wave cross-section is particularly large due to a resonance. This does not hold for (ddμ) since j=210 does not contribute, being ungerade and forbidden by deuteron statistics if the main state (j=0) is gerade. The leading gerade correction is j=211, but this is small, so the dominant contribution is j=0, despite the implied p-wave penetration reduction.

Bogdanova et al. (1981) calculated the rates for (dtμ), using also an improved theory in which the nuclear interaction for a two-channel model is solved explicitly. They found that this confirmed the results of the simple model (5). For A.M. zero states, they found

$$\Gamma_5 \hbar = 10^{12}, \ 0.8\times10^{12} s^{-1}$$

for (00), (01). For A.M. unity states, from j≠0, they found

$$\Gamma_5 \hbar = 1.1\times10^8, \ 4.2\times10^7 s^{-1}$$

for (10), (11) respectively. (The associated level shifts are ~ 1meV and << 1meV for A.M. zero and one).

A recent study (Hu 1986) gives the separate contributions in the case of the (10) state. Hu finds $\Gamma_s \hbar = 1.6\times10^7$ (from j=0) and $10^8 s^{-1}$ (from j≠0, agreeing with the above value). Allowing for non-adiabatic effects increases these by about 70%. Note that the sticking probability is different for the two contributions, being the same (0.89%) as for the (00) state for j=0 (p-wave), but only 0.24% for the j≠0 (s-wave) part.

For (ddμ), Bogdanova et al (1982), 1986) found, for the A.M. unity states from j=0,

$$\Gamma_5\hbar = 1.5 \times 10^9, \ 4.4 \times 10^8 s^{-1}.$$

These are larger by about 100 than the (dtμ) values (from A.M. unity, j=0), so apparently the p-wave cross-sections are larger by this factor in the d+d case relative to d+t.

Finally we mention hyperfine effects. These are not relevant to (ddμ) since orbital A.M. unity states necessarily have fixed value of total spin of the two deuterons, viz. unity. For (dtμ) with orbital A.M. zero, however, spins are relevant since the d+t cross-section is much larger (by order 100) for $(\underline{S}_d + \underline{S}_t)$ = 3/2 than for 1/2. There are four hyperfine states. For the (00) state they are $S(\equiv \underline{S}_d + \underline{S}_t + \underline{S}_\mu)$ = 0,1,2 built on $F \equiv (\underline{S}_t + \underline{S}_\mu)$ = 1 and S=1 built on $(\underline{S}_t + \underline{S}_\mu)$ = 0. The total splitting is 157 meV. The state S = 2 must have $(\underline{S}_d + \underline{S}_t)$ = 3/2, while S=0 must have $(\underline{S}_d + \underline{S}_t)$ = 1/2, while the states of S=1 are a mixture. It follows that the component S=0 has a fusion rate of order 100 times slower than the S=2 (whose rate is of order $10^{12}s^{-1}$). The S=0 state of the (11) resonance is formed only by F=1, which means that it is almost not excited at low T at LHD. The fusion rates of the S=1 states are reduced below $10^{12}s^{-1}$ by their overlap on $(\underline{S}_d + \underline{S}_t)$ = 3/2.

(6) Back Decay

As noted in the Introduction, the effective back-decay width is an average (1) over rotational states K because rotational relaxation is so fast (of order $10^{13}s^{-1}$ at LHD and 300°). Further it is to be summed over energetically-accessible target rotational states K_i. At low and moderate resonance energies, the terms $M_L^{'2}$ controlling the widths (see my talk on Resonance Formation) are much larger for L=0 than L>1 (factor 20 at E_r=25meV, >100 at E_r=4meV). Thus, at low and moderate temperatures, $T \lesssim 300°$, L=0 transitions are dominant if they are allowed. For (ddμ) L=0 transitions occur, but for (dtμ) all the lowest resonances have L=1 or 2.

The (KS)=(1,1/2) state of (ddμ) occurs at 6,54meV in the (K_i,F) = (0,3/2), (0,1/2) channels. The widths are respectively 1.4×10^{-7}eV and 7.3×10^{-7}eV giving a total rate of:

$$\Gamma_6\hbar = 1.3 \times 10^9 s^{-1}$$

(See my talk on Resonance Formation, for details). The (KS) = (2,1/2), (3,1/2) states occur at 8meV, 6meV in the (K_i,F) = (K-1,3/2) channel. The total rates are $\Gamma_6\hbar$=0.9, $0.8 \times 10^9 s^{-1}$. Higher K states can decay only into the F=1/2 channel and have $\Gamma_6\hbar \approx 0.6 \times 10^9 s^{-1}$. The Boltzman average (1) over K then gives, at 300°

$$\Gamma_6\hbar = 0.8 \times 10^9 s^{-1}$$

Faifman et al. (1986) give a rather larger value ∼ $1.2 \times 10^9 s^{-1}$.

All the lowest resonances in the (dtμ) case have L⩾1. Thus their widths are considerably less than for L=0, for which a typical value (for a hypothetical resonance with K_i=0,K=1 at 6meV) is 10^{-4}eV. Some positive energy resonances with L=1 are (K_i,K)=(0,2) at 3meV, (1,3) at 11meV, (2,3) at 15meV. The widths are 0.6×10^{-7}eV, 3.2×10^{-6}eV, 6×10^{-6}eV, i.e. down by a factor of 50 on more on the L=0 value, and giving sample rates:

$$\Gamma_6\hbar = 0.9 \times 10^8, \ 0.5 \times 10^{10}, \ 0.9 \times 10^{10} s^{-1}$$

SUMMARY

$dd\mu$: The main decay is back decay (8×10^8 s^{-1}), followed by fusion (4.4×10^8), then E0 Auger (3×10^8). The latter leads to the (10) state which has a fusion rate 1.5×10^9s^{-1}.

$dt\mu$: The main decay is E1 Auger (10^{11}s^{-1}), followed by back decay ($\lesssim 10^{10}$). The former leads to the (01) state which has a fusion rate $\sim 10^{12}$s^{-1}. Note that back decay may be enhanced by collision broadening effects which broaden the negative energy resonance (KS)=(1,1/2) at -6meV in the (K_iF)=(00) channel. (See my talk on Resonance Formation). This raises part of the state above threshold where its decay rate is $\sim 2\times10^{11}$s^{-1}.

Acknowledgement

The work described in this paper was caried out for the Euratom/UKAEA Fusion Association.

References

Bogdanova, L.N., Markushin, V.E., Melezhik, V.S. and Ponomarev, L.I., 1981, Sov.J.Nucl.Phys. 34,662.

Bogdanova, L.N., Markushin, V.E., Melezhik, V.S. and Ponomarev, L.I., 1982, Phys.Lett. 115B, 171; also correction 1986 Phys.Lett. 167B,485; also 1982 Sov. Phys. JETP 56, 931.

Chandler, D.W. and Farrow, R.L., 1986, J.Chem.Phys. 85,810.

Chu, S-I, 1975, J.Chem.Phys. 62,4089.

Faifman, M.P., Menshikov, L.I., Ponomarev, L.I. and Strizh, T.A., 1986, Dubna preprint.

Hu, C-Y, 1986, Phys.Rev. 34A,2536.

Lane, A.M., 1983, Phys.Lett. 98A,337.

Lukasik, J. and Ducuing, J., 1974, Chem.Phys.Lett. 27,203, J.Chem.Phys. 60,331.

Ostrovskii, V.N. and Ustimov, V.I., 1981, Sov.Phys.JETP 52,620.

Ponomarev, L.I., Puzynin, I.V. and Puzynina, T.P., 1974, Sov.Phys.JETP 38,14.

Rohlfing, E.A., Rabitz, H., Gelfand, J. and Miles, R.B., 1984, J.Chem.Phys. 81, 320.

CAN 250⁺ FUSIONS PER MUON BE ACHIEVED?

CAN 250^+ FUSIONS PER MUON BE ACHIEVED?

S.E. Jones

Brigham Young University
Department of Physics and Astronomy
Provo, UT 84602, USA

INTRODUCTION

Nuclear fusion of hydrogen isotopes can be induced by negative muons (μ) in reactions such as:

$$\mu^- + d + t \rightarrow \alpha + n + \mu^-$$

$$\underset{N}{\underline{\hspace{4cm}}}$$

(1)

This reaction is analagous to the nuclear fusion reaction achieved in stars in which hydrogen isotopes (such as deuterium, d, and tritium, t) at very high temperatures first penetrate the Coulomb repulsive barrier and then fuse together to produce an alpha particle (α) and a neutron (n), releasing energy which reaches the earth as light and heat. Life in the universe depends on fusion energy.

In the case of reaction (1), the muon in general reappears after inducing fusion so that the reaction can be repeated many (N) times. Thus, the muon may serve as an effective catalyst for nuclear fusion. Muon-catalyzed fusion is unique in that it proceeds rapidly in deuterium-tritium mixtures at relatively cold temperatures, e.g. room temperature. The need for plasma temperatures to initiate fusion is overcome by the presence of the muon. In analogy to an ordinary hydrogen molecule, the muon binds together the deuteron and triton in a very small molecule. Since the muonic mass is so large, the dtμ molecule is tiny, so small that the deuteron and triton are induced to fuse together in about a picosecond - one millionth of the muon lifetime. We could speak here of muonic confinement, in lieu of the gravitational confinement found in stars, or magnetic or inertial confinement of hot plasmas favored in earth-bound attempts at imitating stellar fusion.

Room-temperature fusion is perhaps no more fantastic and no less exciting than room-temperature superconductivity. But, in both cases, questions of practicality arise. The relevant question for reaction (1) is: How many times (N) can the muon repeat or catalyze the fusion reaction during its lifetime? The answer depends on a number of factors identified in Figure 1. The diagram shows the complex chain of reactions which occur spontaneously when negative muons stop in a mixture of the hydrogen isotopes (p, d, and t) and helium (helium-4 is the product of reaction 1 while helium -3 arises from tritium decay). All of these factors have been measured at least once in recent years and are discussed in some detail elsewhere.[1-7]

Here it suffices to express the number of fusions which a muon can catalyze in dense d-t mixtures as:

$$N \simeq \left(\frac{\lambda_o}{\lambda_c^{obs}} + W\right)^{-1} \qquad (2)$$

where λ_o = muon-decay rate = 0.455/μsec; λ_c^{obs} = observed muon-catalysis cycling rate, evaluated empirically as 1/(time between fusion neutrons); and W = fraction of muons lost each cycle due to capture or scavenging by helium-3 or helium-4. Equation (2) can be understood as a sum of probabilities:

1/Yield = Probablility of muon decay + Probability of muon capture (3)
           during catalysis cycle   by ^{3}He or ^{4}He

In order to make N as large as possible, it is clear from the yield equation (2) that one must maximize λ_c and minimize W. It is the purpose of this paper to describe what we have learned about achieving these ends, and then to estimate yields anticipated during forthcoming LAMPF experiments. It will be shown that based on previous LAMPF results, we expect to exceed 250 fusions per muon, although such high yields are impossible if results from SIN experiments are taken at face value. The significance of achieving 250 fusions per muon for possible energy applications of muon-catalyzed fusion will also be explored.

74

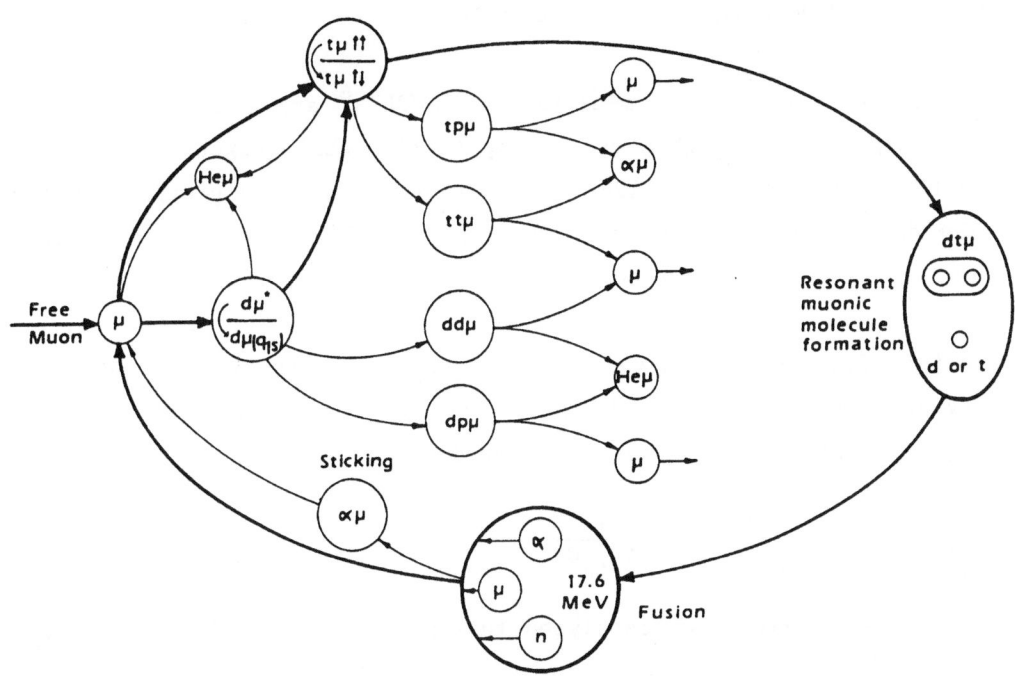

Figure 1. Muon catalysis cycle, showing reactions which occur when negative muons(μ) stop in a mixture of the hydrogen isotopes (p,d,t) and helium (He or α).

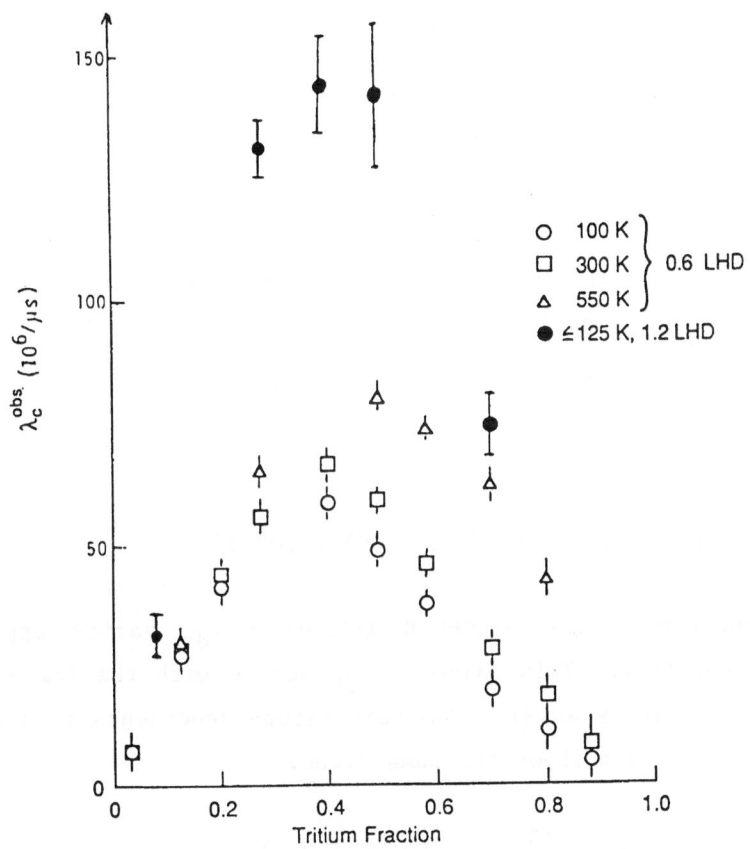

Figure 2. Dependence of the observed muon catalysis cycling rate on tritum fraction, temperature and density.

Observed cycling rates measured at LAMPF[4,6] are summarized in Figure 2. An example of the high-pressure (up to 1000 atm) target flasks used for the LAMPF experiments since 1982 to obtain these data is provided in Figure 3. The data demonstrate clearly that muon catalysis proceeds more rapidly as one increases either the temperature or density of the deuterium-tritium target mixture. In order to assess the physical processes which are responsible for these observed effects, we must sort out and measure the underlying reactions depicted in Figure 1. Salient features of the reaction kinetics can be succinctly stated:

$$\frac{\phi}{\lambda_c^{obs}} \simeq \frac{C_d}{(\lambda_{dt}/q_{1s})C_t} + \frac{1}{\lambda_{dt\mu}C_d} \tag{4}$$

where λ_c^{obs} = observed muon-catalyzed fusion cycling rate (see equation 2);

ϕ = deuterium-tritium mixture density relative to liquid-hydrogen density (LHD = 4.25 x 10^{22} atoms/cm^3);

C_d, C_t = fractions of deuterium and tritium, respectively: $C_d + C_t = 1$;

λ_{dt}/q_{1s} = rate of the muon-transfer reaction $d\mu + t \to t\mu + d$ (The factor q_{1s} will deviate from unity both because of muon transfer to t from excited states of $d\mu$,[8] and because the initial atomic capture ratio might differ somewhat from the ratio C_d/C_t.)

$\lambda_{dt\mu}$ = the rate of resonant formation of $dt\mu$ molecules in collisions of $t\mu$ atoms with D_2 or DT molecules (see Figure 1).

It is worthwhile to summarize what we have learned about the rates which determine the overall cycling rate:[6]

$$\lambda_{dt} \simeq [1 + (6 \pm 1) \times 10^{-4}T] \, (280 \pm 40) \times 10^6 \, s^{-1} \tag{5}$$

for the temperature range 20-500 K (observed λ_{dt} varies approximately linearly with density). This value of λ_{dt} agrees with the low - ϕ, low-C_t experiment of Bystritsky et al.[3] The temperature dependence is not quite as strong as predicted[9] but shows the same trend.

$$q_{1s} \simeq [1- (1 - \gamma) \, C_t] \, \gamma^{2C_t}, \, \gamma = (0.75 \pm 0.2). \tag{6}$$

The q_{1s} factor has been the subject of extensive experimental[6,7] and theoretical[8] study, and we are currently analyzing new data in which we

Figure 3. Target capsule built at the Idaho National Engineering laboratory for muon-catalyzed fusion research at LAMPF. This second-generation capsule is used for temperatures in the range 18K < T < 800 K and pressures up to 1000 atm.

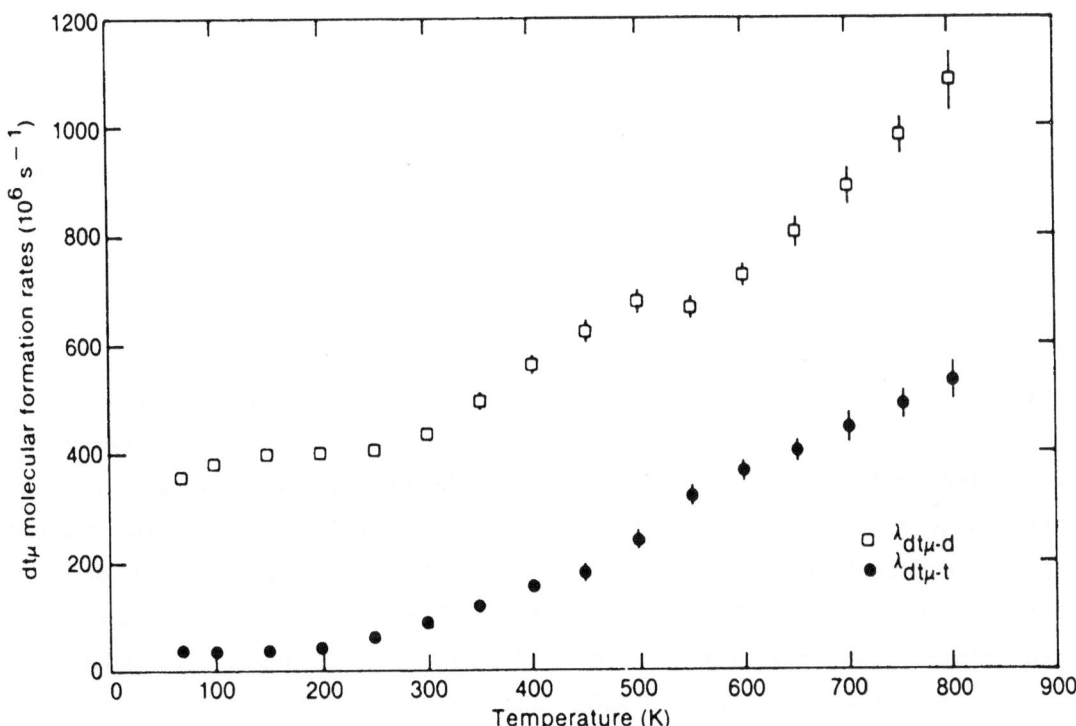

Figure 4. Resonant dtμ-molecular formation rates $g_{dt\mu-d}$ and $\lambda_{dt\mu-t}$ for (tμ + D_2) collisions and (tμ + DT) collisions, respectively.

explore the ϕ - and C_t - dependencies of q_{1s} in detail. But the empirical expression above (from Ref. 6) is adequate for present purposes.

The empirical relations (5) and (6) show that the $d\mu$ to $t\mu$ muon transfer rate $(\lambda_{dt/q_{1s}})$ is relatively insensitive to variations in density or temperature. Yet the observed cycling rate clearly depends strongly on these experimental conditions (see Fig. 2). Thus, from equation (4) we must conclude that it is the $dt\mu$-formation rate that increases strongly with increasing temperature and density. These results are displayed quantitatively in Figures 4 and 5.

Figure 4 displays a number of surprises uncovered by the LAMPF experiments. First, there are two distinct $dt\mu$-formation ratios, $\lambda_{dt\mu-d}$ and $\lambda_{dt\mu-t}$, corresponding to collisions of $t\mu$ atoms with D_2 and DT molecules, respectively, so that

$$\lambda_{dt\mu} = C_d \, \lambda_{dt\mu-d} + C_t \, \lambda_{dt\mu-t} \tag{7}$$

Both of these $dt\mu$ - formation rates are seen to be resonant,[10] in that the rates are very large ($\sim$ 3 orders of magnitude larger than expected for non-resonant Auger reactions) and show a strong dependence on temperature. It is interesting to note that our observation of a strong-temperature dependence of $dt\mu$-formation rates was initially challenged. At an earlier school held at Erice in 1984, W. Breunlich et al. reported[11] that "$d\mu t$ formation rates, on the other hand, only show weak temperature dependence," in direct contradiction of our published results [see also ref. 5]. However, a recent paper by W. Breunlich et al.[12] now reports a strong temperature dependence, at least for $\lambda_{dt\mu-t}$, confirming the LAMPF result. Both groups now concur that $\lambda_{dt\mu-t}$ has the expected property of approaching zero as $T \to 0$ (Figure 4).

On the other hand, $\lambda_{dt\mu-d}$ is rather constant for $T < 300K$. This striking behavior was disturbing when first discovered in the research at LAMPF, but since has been explained in context of the resonant $dt\mu$ formation model by Yuri Petrov.[13] A beneficial consequence of this unexpected effect is that high cycling rates (λ_c^{obs}) can be realized even at low temperatures. This explains in part why yields of well over 100 funsions per muon could be achieved at LAMPF in 1984[6] even in liquid deuterium-tritium mixtures at 20K. This result has since been reproduced in experiments at SIN (Switzerland)[7] and KEK (Japan).[14]

That $\lambda_{dt\mu-d} >> \lambda_{dt\mu-t}$ for $T < 100K$ was proven directly at LAMPF in 1984 [see footnote 16, ref. 6]. This was accomplished by keeping a $C_t =$

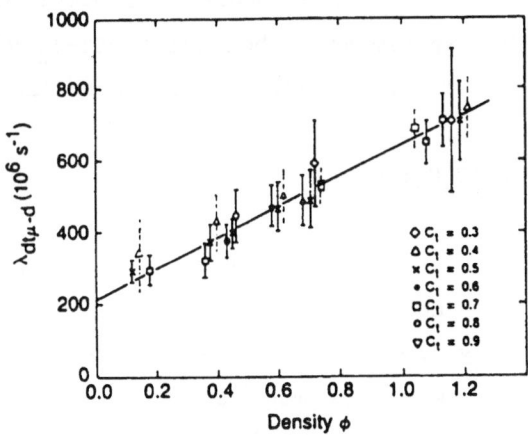

Figure 5. Density dependence of the normalized dtµ-molecular formation rate for (tµ + D_2) collisions observed at LAMPF. Preliminary results from August 1986 runs are displayed with dashed error bars.

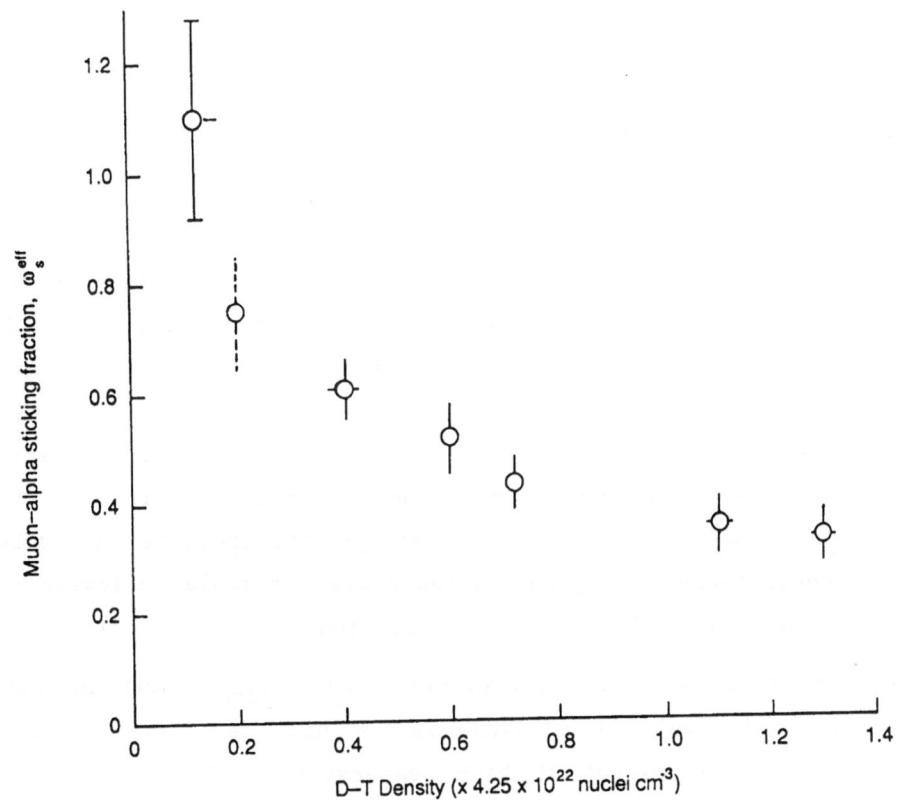

Figure 6. Dependence of ω_s^{eff} on deuterium-tritium density based on data acquired at LAMPF, 1982-1986. Preliminary result from August 1986 run is shown with dashed error bars.

0.7 mixture liquefied at T < 30K for 30 hours. We observed a slow rise in λ_c over the course of the experiment. When the D_2-DT-T_2 mixture was first liquefied, relative molecular concentrations were approximately:

$$C_{D_2} : C_{DT} : C_{T_2} = c_d^2 : 2c_dc_t : c_t^2, \qquad (8)$$

representing high-temperature equilibrium ratios. As expected, the equilibrium slowly evolved at cold temperature to favor (D_2 + T_2) at the expense of DT. Then the cycling rate increased, showing directly that (tμ + D_2) collisions yield dtμ-molecules much more rapidly than do (tμ + DT) collisions; i.e., $\lambda_{dt\mu-d} \gg \lambda_{dt\mu-t}$ at low temperatures in keeping with Figure 4. This result has also been confirmed at SIN.[7] The results imply that a liquid (D_2 + T_2) mixture can be prepared and maintained at cold temperatures, and that cycling rates will be much higher than in (D_2 + DT + T_2) mixtures. This means of enhancing λ_c is more difficult to accomplish at higher temperatures since then the (D_2 + T_2) mixture would quickly equilibrate according to relation (8), evolving DT at the expense of D_2.

Figure 5 suggests yet another reason for preferring cold (D_2 + T_2) mixtures at high densities: $\lambda_{dt\mu-d}$ increases rapidly with increasing density, thus enhancing λ_c. Interestingly we find no such strong density dependence for $\lambda_{dt\mu-t}$.[6] These LAMPF results have been confirmed qualitatively at SIN, but the density-dependence of $\lambda_{dt\mu-d}$ is weaker.[7] However, data taken at LAMPF in August 1986 (dashed bars in Figure 4) agree very well with our previously published results.[6]

The striking density-dependence of $\lambda_{dt\mu-d}$ [Figure 5] provides evidence of significant resonant dtμ formation via three-body collisions. Menshikov and Ponomarev[15] have discussed a mechanism for three-body resonant molecular formation, e.g., tμ + D_2 + D_2 → [(dtμ) d2e]* + D_2 + ΔE. The singlet (tμ+ D_2) collisions are special in having their strongest resonances just below threshold, where they are not accessible in two-body collisions. By absorbing some kinetic energy, the spectator molecule (D_2, DT, or T_2) moves these strong resonances above threshold, allowing them to contribute to very rapid dtμ molecular formation.

There exist theoretical predictions[13] that $\lambda_{dt\mu-d}$ and therefore λ_c must saturate and level off at sufficiently high densities. Indeed, we have reported such an effect at high temperatures (T ≳ 450K).[16] But no saturation in $\lambda_{dt\mu-d}$ (or λ_c) is yet seen for densities up to 1.3 liquid-hydrogen density (LHD) for T < 200K.

In summary, a cold (D_2 + T_2) target at high density will allow us to achieve very fast muon-catalysis cycling rates. Since $\lambda_{dt\mu-d}$ is large for these conditions, such a target will provide much larger cycling rates than, for instance, a high-temperature (T > 800K) target, although high temperatures are very interesting in their own right. Engineering constraints also limit the densities which can be achieved at high temperatures.

Consider, then a ($D_2 + T_2$) mixture at ~30K, compressed to a density of approximately 2.3 LHD (nearly twice the density now achieved at LAMPF, SIN or KEK). The optimum tritium fraction for D_2-DT-T_2 at high temperature equilibrium is $C_t \sim$ 30-40% (Figure 2). Using ($D_2 + T_2$), we can increase λ_c by moving to higher C_t (~ 50%) so as to increase λ_{dt}/q_{1s}. Based on an extrapolation of the LAMPF results described above (assuming no saturation of $\lambda_{dt\mu-d}$ at high density), we would then expect for these conditions:

$$\lambda_c^{obs} \approx 7 \times 10^{-4} \text{ s}^{-1} \text{ (extrapolation)} \tag{9}$$

It is useful to look at the fusion yield which this rate would produce if W were zero (see equation 1):

$$N_{W=0} = \frac{\lambda_c^{obs}}{\lambda_o} \approx 1500 \text{ fusions/}\mu\text{-} \text{ (extrapolation)} \tag{10}$$

Thus, it is clearly possible to achieve well over 250 fusions per muon based on what we have learned about increasing the muon-catalysis reaction rates. This result is a clear departure from theoretical predictions of slow cycling rates expected just ten years ago, and is an exciting confirmation of the resonance model of dtμ-formation developed by S.S. Gershtein, L.I. Ponomarev and their co-workers.[10,15,19]

We are in fact preparing to perform a set of experiments with these ultra-high-density conditions at LAMPF in 1987. Before further discussing the new experiments, however, it is important to turn our attention to the muon-loss factor W which is now expected to limit the achievable number of fusions per muon.

MINIMIZING HELIUM CAPTURE LOSSES

Figure 1 illustrates many ways in which the muon may be lost from the muon catalysis cycle. Each of these results in muon capture by a helium nucleus and contributes to the muon-loss probability W (see equations 1,2). The muon may be captured and retained by a helium nucleus synthesized during dtμ, ddμ, or ttμ fusion, with sticking probabilities ω_s, ω_d, and ω_t, respectively. In addition, small amounts of protium may be present resulting in pdµ and ptµ fusion, with sticking probabilities ω_{pd} and ω_{pt}. The muon may also be scavenged by ^{3}He introduced by tritium decay (normally $C_{He} < < 1\%$). The total muon loss probability per cycle can be written as[4,6,7]

$$W \simeq \frac{q_{1s}C_d}{\lambda_{dt}C_t + \lambda_{dd\mu}C_d}(0.58\lambda_{dd\mu}C_d\omega_d + \lambda_{pd\mu}C_p\omega_{pd} + \lambda_{dHe}C_{He}) \tag{11}$$

$$+ \frac{1}{\lambda_{dt\mu}C_d}(\lambda_{tt\mu}C_t\omega_t + \lambda_{pt\mu}C_p\omega_{pt} + \lambda_{tHe}C_{He}) + C_{He}\omega_{He} + \omega_s^{eff},$$

where $\lambda_{dt\mu}$, $\lambda_{dd\mu}$, $\lambda_{tt\mu}$, $\lambda_{pd\mu}$, $\lambda_{pt\mu}$ are the rates for dtµ, ddµ, ttµ, pdµ and ptµ molecular formation, λ_{dHe} and λ_{tHe} are the rates for transfer to ^{3}He from the dµ and tµ ground states, and ω_{He} is the probability for initial capture by ^{3}He. ω_s^{eff} is then the effective sticking probability following dtµ-fusion. Note that it is possible that some other correction term remains to be included in equation (11), in which case $\omega_s < \omega_s^{eff}$.

Now, for the proposed high-density experiment, we will assure very small protium and helium contaminations. Thus, all terms involving C_p and C_{He} will become negligible. Most other terms also become small since (λ_{dt}/q_{1s}) and $\lambda_{dt\mu}$ become large. For example, while $\lambda_{tt\mu}$ scales linearly with density, $\lambda_{dt\mu}$ increases much faster than this (see Figure 5), so that losses following ttµ-fusion become smaller as one elevates the target density. Consequently, for the conditions of our new experiment, although small correction terms will not be neglected, $W \simeq \omega_s^{eff}$, and equation 11 reduces to:

$$N \cong (\frac{\lambda_o}{\lambda_c} + \omega_s^{eff})^{-1} . \tag{12}$$

We have shown that λ_c can become very large at high densities. What happens to ω_s^{eff}?

A strong dependence of ω_s^{eff} on target density is suggested by all the data acquired at LAMPF since 1982, shown in Figure 6. We have acquired data over a broad density range, $0.12 < \phi < 1.4$ in small steps, and find that ω_s^{eff} decreases with increasing density over this range.

At first, we thought that this effect might be due to impurities (methane, water, etc.) in the D-T mixture, so we did the following test. We filled carefully cleaned target flasks at low density ($\phi < 0.4$) and very low impurity concentration and measured sticking losses in the gas at ~ 50 K. At this temperature, only hydrogen (and any helium) should remain in the gas phase. Then we cooled the same targets to ~ 30 K, so as to liquefy the hydrogen, and re-measured the sticking friction. Each time we found a consistent drop in ω_s^{eff} as the target gas liquefied and the density increased.

The LAMPF experiments point to a significant decrease of ω_s^{eff} with increasing density (Figure 6). A recent paper by Breunlich et al. strongly contradicts this observation.[7] But one cannot be terribly disturbed by this latest disagreement. After all, the LAMPF result that $\omega_s^{eff} \approx 0.4\%$ at high densities was also controversial when first presented at the Jackson Hole Workshop in 1984.[17] However, experiments at both SIN and KEK have now quantitatively confirmed (within errors) this surprising result.[7,14]

It is only at low densities that the LAMPF and SIN results diverge. But for these conditions, cycling rates are small making an accurate evaluation of ω_s^{eff} difficult.[16] We can hope to resolve this sticky issue by doing experiments at very high densities, up to 2.3 LHD, for then the sticking fraction becomes quite easy to measure (equation 12). We have seen too many surprises in this research to suppose that we can resolve the issue by theoretical discussion alone, without dedicated experiments. Both the magnitude and density dependence of ω_s^{eff} must be settled experimentally.

Therefore, we welcome the many experiments in the United Kingdom, Japan and the Soviet Union as well as in Switzerland and the United States, which seek to provide further insight regarding the $\alpha-\mu$ sticking fraction. After all this quantity represents the primary bottleneck to achieving high yields with muon-catalyzed fusion.

Figure 7 displays the approximate number of d-t fusions per muon which might be expected in the new LAMPF experiment at 2.3 LHD, based on equations (9) and (12), and boldly extrapolating the LAMPF results for ω_s^{eff} (Figure 6) to $\phi = 2.3$. In conclusion, we expect to exceed 250 fusions/muon in LAMPF experiments later this year. Note that if $\omega_s \approx 0.45\%$, as might be expected from SIN results,[7] then the maximum yield according to equation 12 would be $N < \frac{1}{\omega_s} \approx 220$ fusion/muons. "Se son rose fioriranno" ("if it is a rose, it will bloom").[1]

THE SIGNIFICANCE OF ACHIEVING 250+ FUSIONS PER MUON

For many years, it was thought that muon-catalyzed fusion would be limited to about one fusion per muon. Slow mesomolecular (i.e., dtμ) formation was seen as the limiting factor:

"The present analysis shows that capture of muons by helium nuclei has no relevance to the question of achieving chain reactions since mesomolecular formation rates limit the number of catalyzed reactions to an average one per muon. Since only

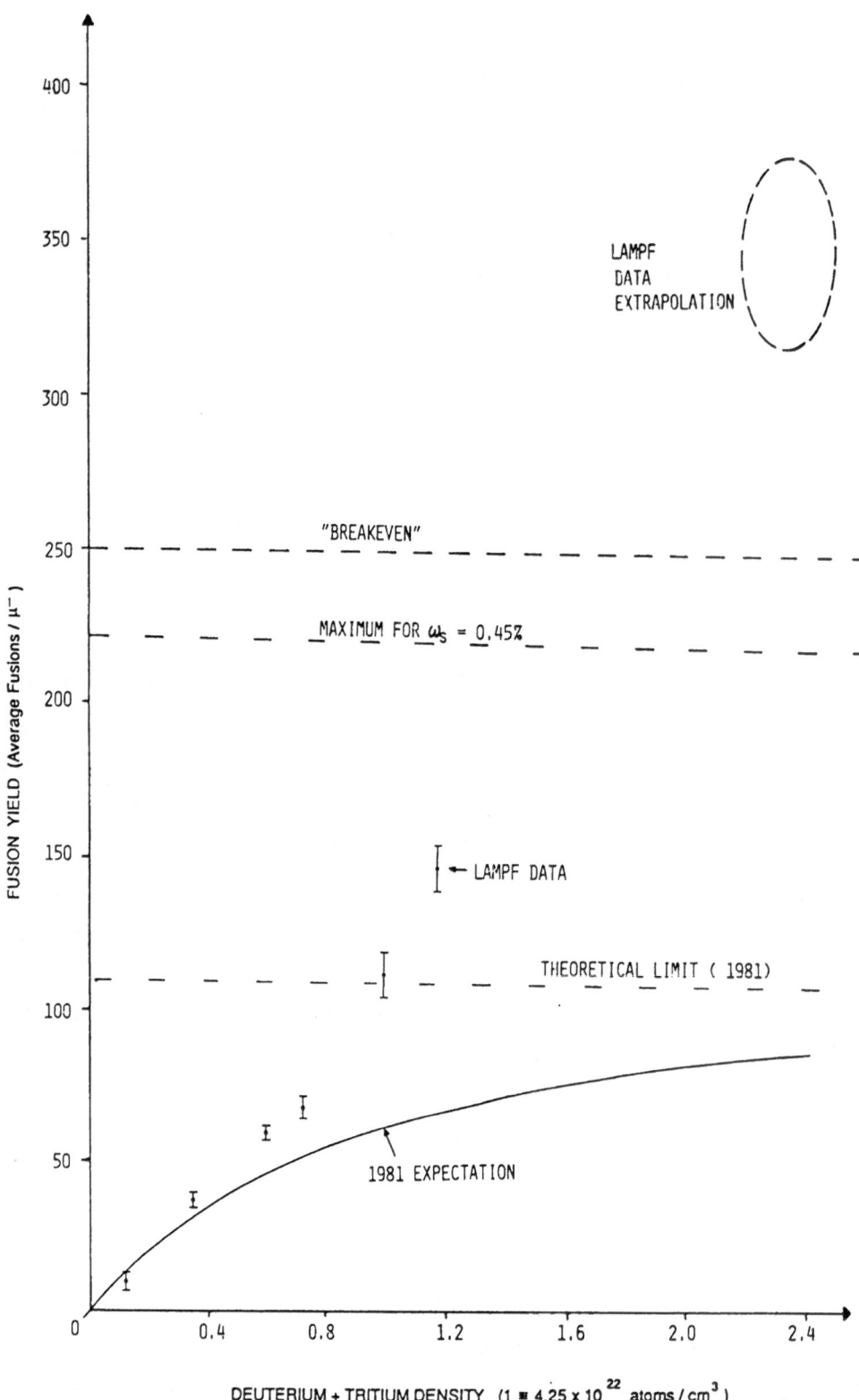

Figure 7. Observed and anticipated muon-catalyzed fusion yields as a function of density. Data acquired at LAMPF for cold (T < 100K) equimolar D_2-DT-T_2 mixtures exceed 1981 predictions. Extrapolating from LAMPF results, a new experiment for cold D_2 + T_2 mixtures at 2.3 LHD should exceed even the "energy breakeven" level described in the text.

one catalyzed reaction per muon can be expected, muon catalysis of fusion reactions falls far short of energy balance."[18]

The subject was revived by a series of important Soviet studies beginning in 1977.[10,19] In their 1977 paper outlining resonant $dt\mu$-molecule formation, S.S. Gerstein and L.I. Ponomarev revive also the possibility of energy production by muon-catalysis:

"It means that every μ- meson in the mixture [of] deuterium and tritium can produce $\sim 10^2$ of the nuclear fusion reaction and release ~ 2 GeV which is 20 times the rest mass of [the] muon."[a)]

"We leave the question open about the possibilities of practical applications of the phenomenon discovered (e.g., the production of energy in fusion reactors, the breeding of tritium by ejected neutrons in the mixture of ^{6}Li and ^{7}Li, the ignition of the thermonuclear fusion reaction provided by lasers, etc.), Nevertheless, we would like to call attention to it."[10]

Then in 1980, Yu. Petrov presented a clever scheme for a power reactor based on muon-catalyzed fusion combined with nuclear fusion processes.[20] He showed that a commercial hybrid reactor would require only about 100 fusions per muon. Subsequent studies have also concluded that 100-200 fusions per muon would suffice for a fusion/fission hybrid power plant.[21,22] It is interesting to note that in the 1979 school on exotic atoms at Erice, J. Rafelski predicted that it might be possible to achieve 100-200 fusions per muon.[23]

In 1984, yields of 150 fusions per muon were in fact realized at LAMPF[6] invigorating speculations regarding pratical applications for muon-catalyzed fusion. Later experiments at SIN[7] and KEK[14] have achieved similar yields.

Are we approaching a limit, or can significantly higher fusion yields be expected? As shown in this paper, muon catalysis cycling rates can be made extremely fast (equations 9, 10), thanks to resonant $dt\mu$-formation. Thus, we find that the limiting factor to muon-catalyzed fusion in fact

---

a). This provides a reasonable definition of "scientific breakeven" for muon-catalyzed fusion. In experiments at LAMPF, SIN and KEK, the fusion yields now exceed the energy of the fusion reaction initiator, ignoring muon production inefficiencies, by an order of magnitude.

stems from muon losses due to alpha-sticking (ω_s).

Prior to the LAMPF experiments, it was calculated that $\omega_s \approx 0.9\%$, so that the yield would be limited to ~110 fusions per muon.[1] Now that this "barrier" has been surpassed (see Fig. 6), it becomes an open question how many fusions per muon can be achieved. The new LAMPF experiment addresses this question, using the best information we now have regarding how to make λ_c large and W small.

I might draw a parallel with recent breakthroughs in superconductivity research. Now that the barrier of attaining superconductivity at 77K (boiling temperature of liquid nitrogen) has been surpassed,[24] room-temperature superconductivity is easily imaginable, and possible applications become enthusiastically discussed.

Similarly, 250 fusions per muon would be significant because it represents an "energy breakeven" level for muon-catalyzed fusion. A number of studies suggest that approximately 5000 MeV of energy must be invested to produce one negative muon by state-of-the-art techniques.[20-22,24] Each muon-induced fusion reaction would produce about 20 MeV, taking credit for the energy released in producing tritium from lithium. Then "breakeven" would require approximately:

$$N_b = \frac{E_{in}}{E_{out}} \approx \frac{5000 \text{ MeV/muon}}{20 \text{ MeV/fusion}} = 250 \text{ fusions/muon}. \tag{11}$$

Thus, reaching 250 fusions/muon would be interesting from a practical standpoint.

Finally, I would submit that the feasibility of producing electrical power by means of muon-catalyzed fusion ought to be re-examined in view of the recent discovery[24] of high-temperature superconductivity. A scheme like Petrov's[20] is actually a power multiplier: input power is fed to an accelerator which generates muons which in turn induce fusion reactions. Fusion energy is then converted to electrical power, much of which must be recirculated to drive the accelerator. But with inexpensive superconductivity all energy conversion processes become much more efficient making a power multiplier more attractive. A muon-catalyzed fusion reactor with superconducting rf sources, accelerator wave-guides and generators would be fascinating to study. In addition to having high efficiency, the accelerator ought also to be shorter and involve less capital expenditures. These factors are particularly relevant to muon-catalyzed fusion schemes where considerable recirculating power is required.

We have considered multiple motivations for studying muon-catalyzed fusions using (D_2 + T_2) mixtures at high densities. We are limited to cold temperatures and ~2.3 LHD for practical reasons: static pressures for these conditions approach 10 kbars. Our goal is to explore (static) high-density limits of muon-catalyzed fusion. An increased understanding of the process in this regime may enhance our ability to achieve high fusion yields with lesser engineering challenges.

The target flasks for this new experiment are being designed and built at the Idaho National Engineering Laboratory under the direction of Dr. A.J. Caffrey. Figure 8 shows a preliminary design of our new target capsule which is actually the fifth generation of target flasks used at LAMPF since 1982. The recessed ports facilitate muon entry and muon-decay electron egress from the D_2-T_2 bearing chamber where fusion occurs. We expect to perform the experiments at LAMPF later in 1987. Thus, we will soon find out whether 250[+] fusions per muon can be achieved.

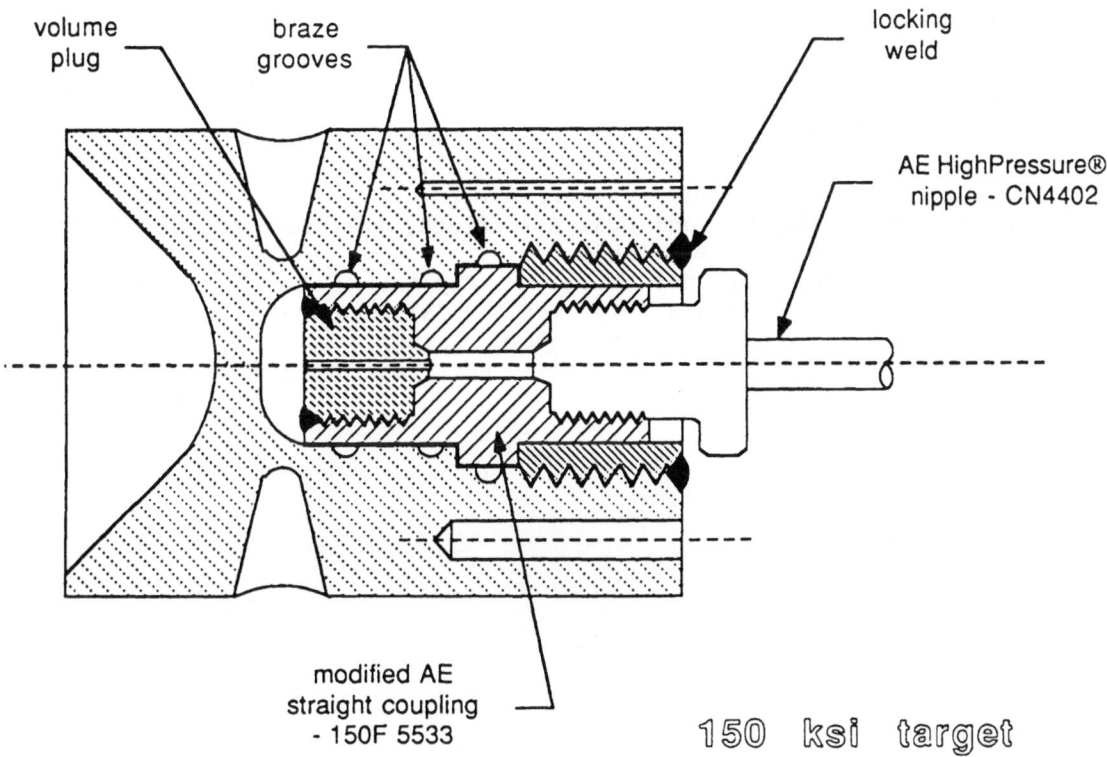

Figure 8. Target capsule design for a new LAMPF experiment, to achieve densities up to 2.3 LHD at cryogenic temperatures. Like its predecessors, this fifth-generation target flask is cylindrically symmetric.

REFERENCES

1. L. Bracci and G. Fiorentini, Phys. Rep. 86:169 (1982).
2. S.E. Jones, Nature 321:127 (1985).
3. V.M. Bystritsky et al., Soviet Phys. JETP 53:877 (1981).
4. S.E. Jones et al., Phys. Rev. Lett. 51:1757 (1983).
5. W.H. Breunlich et al., Phys. Rev. Lett. 53:1137 (1984).
6. S.E. Jones et al., Phys. Rev. Lett. 56:588 (1986).
7. W.H. Breunlich et al., Phys. Rev. Lett. 58:329 (1987).
8. L.I. Menshikov and L.I. Ponomarev, JETP Lett. 39:663 (1984), and JETP Lett. 42:13 (1985).
9. A.V. Matveenko and L.I. Ponomarev, Zh. Eksp. Teor. Fiz. 59:1593 (1970) [Sov. Phys. JETP 32:871 (1971)].
10. S.S. Gerstein and L.I. Ponomarev, Phys. Lett. 72B:80 (1977).
11. W.H. Breunlich et al., International School of Physics of Exotic Atoms, 4th Course, Erice, Sicily, March 31 to April 6, 1984.
12. W.H. Breunlich et al., paper presented by P. Kammel, Proc. International Symposium on Muon-Catalyzed Fusion, Tokyo, Japan, Sept. 1-3, 1986.
13. Yu. Petrov, Proc. International Symposium on Muon-Catalyzed Fusion, Tokyo, Japan, Sept. 1-3, 1986.
14. K. Nagamine et al., ibid.
15. L.I. Menshikov and L.I. Ponomarev, Phys. Lett. 167B:141 (1986).
16. S.E. Jones, Proc. International Symposium on Muon-Catalyzed Fusion, Tokyo, Japan, Sept. 1-3, 1986.
17. A.N. Anderson et al., in Proc. Muon-Catalyzed Fusion Workship, Jackson Hole, Wyoming, 1984, ed. S. E. Jones (EG & G, Idaho Falls, Idaho, 1984); A. J. Caffrey et al., ibid.
18. S.O. Dean, "Muon Catalysis of Fusion Reactions," NTIC Conf-661115-1, November 2, 1966.
19. S.I. Vinitskii et al., Sov. Phys. JETP, 47:444 (1978).
20. Yu. Petrov, Nature 285:466 (1980); Atomkernenergie-Kerntechnik 46:25 (1985).
21. R.C. Miller and R.A. Krakowski, Proc. Muon-Catalysed Fusion Workshop, Jackson Hole, Wyoming, 1984, ed. S. E. Jones (EG & G, Idaho Falls, Idaho, 1984); A. J. Caffrey et al., ibid.
22. S. Eliezer, T. Tajima, and M.N. Rosenbluth, Preprint no. DOE-ET-53088-223, Inst. Fusion Studies (1986).
23. J. Rafelski, International School of Physics of Exotic Atoms, Erice, Italy, March 25 to April 5 1979; Ref. TH.2679-CERN.
24. M.K. Wu et al., Phys. Rev. Lett. 58:908 (1987).
25. M. Jandel, M. Danos and J. Rafelski, CERN Preprint (1987).

THE DEVELOPMENT OF µCF STUDIES AT SIN

W.H. Breunlich

Austrian Academy of Sciences (OeAW)
Institute for Medium Energy Physics
1090 Vienna, Austria

INTRODUCTION

When in the seventies "meson factories" like LAMPF, SIN, TRIUMF
and KEK started operations, this gave rise to a considerable impact on
the field of muon catalyzed fusion (µCF). This fertilizing effect came
along with the strong impact these facilities made on the field of exotic
atoms as a whole, on muon physics [1] and, especially, muonic atoms and
molecules in hydrogen. A good deal of fundamental questions had been
attacked before (for a review, see ref. [2]). A number of problems had been
solved, a number of serious problems was left open or was not answered with
the accuracy desired. They could now be attacked with a reasonable chance
for success due to the strongly improved beam quality. This was partly due
to the strongly enhanced intensity of extracted beams, but also due to new
ideas, e.g. the superconducting muon channel at the Swiss Institute for
Nuclear Research (SIN)[3].

Like many other groups before who piled information about the rich
phenomenology of the µCF reaction chains, we started out with experiments
devoted to weak interaction physics, specifically to nuclear muon capture
in deuterium. It seems to be typical for the early part of µCF history
that most of the surprising observations had been made in experiments that
did not at all aim for µCF (see refs.[4,5]). When the exotic possibility of
spontaneous fusion in muonic molecular hydrogen ions was discovered
experimentally by Professor Alvarez's group in Berkeley [4] in 1956, they
were really engaged in investigating "strange" physics. After they had
found the explanation of all the exotic processes causing the tracks in the
bubble chamber, they found out that Frank [6] with impressive inspiration
had already in 1947 created the idea of µCF. [The motivation came from the
search for alternative exlanations for the tracks indicating the sensational
observation of the π → µ decay (the hypothesis of µCF as an explanation was
correctly rejected by Frank).] Soon after Alvarez's observation, the
excitement about exuberant energy production was superseded by disappointment
[5,7] due to the insufficient efficiency. In the following years, the
variety of atomic and molecular processes (for a review, see [8,9]) already
seen in the bubble chamber experiments [4,10,11] was accompanying muon
capture experiments in hydrogen and deuterium in pioneering counter experiments by the Columbia group [12,13], the CERN-Bologna group [14-16],
and the Dubna group [17]. The high chemical activity of the muonic hydrogen
isotopes is relevant to nuclear muon capture studies in several ways:

(1) Due to the minus sign in the (V - A) theory, weak interaction rates strongly depend on the initial spin state of the muonic atom (or molecule), in which the nuclear capture process takes place (see e.g. [2,18,19]). Thus, reliable knowledge of the population of the hyperfine states with total spin $F = I \pm 1/2$ is necessary for the interpretation of the results (I is the nuclear spin).

(2) The balance of the population of the hf components, which split by an amount ΔE_{hf} (see table), is governed by scattering processes. For low temperatures, the kinetic energy of the muonic atoms is smaller than ΔE_{hf}, and thus the state with $F = I + 1/2$ is depopulated into the lower state with $F = I - 1/2$.

(3) Fusion processes with neutron emission form a serious source of background, e.g. in pure deuterium, HD or DT mixtures, since hydrogen capture (unlike capture in high Z elements) takes place only with a rate of about $10^{-3}\lambda_0$ (λ_0 is the free muon decay rate $0.455 \cdot 10^6 s^{-1}$)[20,21].

(4) Appreciable sticking of the muon to the fusion product He nucleus results in competing nuclear capture which, for example in a liquid HD mixture, is a serious source of background [22].

Table 1

Ground-state properties of muonic hydrogen isotopes

	Ground-state binding energy (eV)	Hyperfine splitting (eV)
μp	2528.4	0.183
μd	2663.1	0.0485
μt	2711.2	0.241

The goal of the first part of the experiments at SIN was an investigation of the question whether the HD mixture (acting as an effective deuterium target due to isotopic transfer) or pure deuterium were more favourable for nuclear muon capture studies in deuterium. Thus, to answer this question we had to consider details of the fusion cycle, and last, but not least, the collision process governing the hyperfine populations.

Hyperfine effects were already known, of course, and the first one reported experimentally in 1962 was the Wolfenstein-Gershtein effect [17,21] in the formation of $p\mu d$ molecules [12,23]. Compared to this rather modest 17% effect, we observed a dramatic effect of about a factor of 100 [24,25] in the formation of $d\mu d$ molecules at 34 K in 1979 at SIN (see chapter 2). The explanation of this experimentally observed effect [19,23,24] came naturally within the framework of the "resonant formation of muonic molecules" (for a review, see [8]), as first developed by Vesman [26]. Experimental results from Dubna [27] had given the first clear-cut indication of the importance of this model (see e.g. [28,29]) before. By including hyperfine effects, this appealing picture was well rounded off experimentally (see refs. [24,25,31]); the theoretical completion followed soon after [30].

μCF had been a phenomenon necessarily to be understood in weak interaction experiments for quite some time, and the "surprising" hf effects had actually helped in weak interaction physics to determine such a hyperfine transition experimentally for the first time (this was important, since the experimental information was quite controversial; for discussion, see [25]). However, with growing confidence in the resonance model, μCF caused excitement for its own sake and called forth more and more dedicated studies. The first one at SIN in 1980 aimed for a determination of the temperature dependence [31] of the molecular formation starting from both the quartet and the doublet state of the μd atoms see chapter 3) to experimentally determine the binding energy of the dμd state ($J = 1, \nu = 1$), which as a freak of nature is so loosely bound that it allows for the resonance mechanism. An even more loosely bound state was predicted theoretically for the dμt molecule, and promised, together with other favourable circumstances, high fusion yields for DT mixtures. Already in the first experiments in Dubna [32,33], it turned out that fusion yields apparently are high. However, together with these favourable findings, new facts indicated that an understanding of the DT system would be more difficult [33]. It remained a characteristic feature of the subsequent experiments at LAMPF [34] and SIN [35] that surprises indicating rising efficiency of μCF along with many more complications rendering the understanding of the DT cycle.

EXPERIMENTS IN DT MIXTURES

A series of experiments (see [36,37] and ref. given there) tried to investigate the energetically relevant parameters governing the number of "fusions per muon". Surprisingly favourable results were found for these parameters:
the cycle rate λ_c (inverse time between successive fusions) showed a maximum value of about 300 times the muon decay rate λ_o (measured in liquid DT) and ω_s, the muon loss in the cycle due to He-sticking, was determined to be substantially lower than theoretically predicted, i. e. 0.45% instead of about 1% [36,38]. The fusion yield is determined to be

$$ Y = \frac{\phi \cdot \lambda_c}{\lambda_o + W \phi \lambda_c} \qquad . \qquad (1) $$

A simplified scheme of the μCF cycle in DT mixture is shown in fig.1. For illustration the SIN results for liquid DT are shown in fig.2.

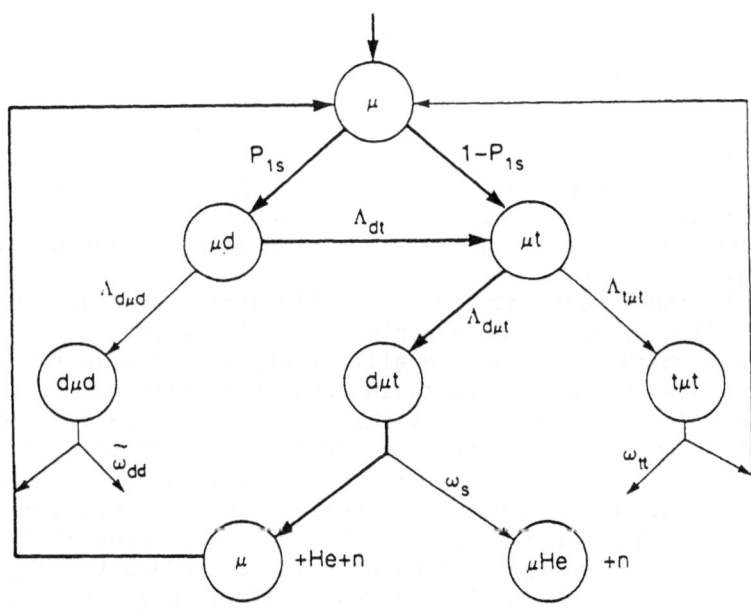

Fig.1: Reaction scheme of μCF in DT mixtures (simplified).

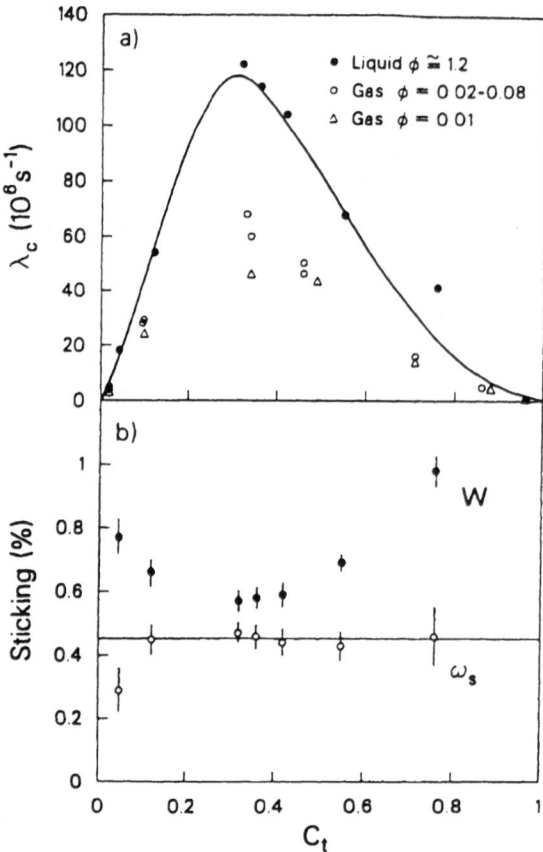

Fig.2: Normalized cycle rates for liquid and gas (a), total muon loss W and sticking probability ω_s for liquid (b), both versus tritium concentration (for details see ref. [36])

A comparison at low temperature between SIN (T=23K (liquid), T=35K (gas)) [36,38] and LAMPF (temperature range <125K not specified) [37] experiments shows agreement in the magnitude of ω_c and W (the total probability of muon loss from the cycle), but discrepancies in the c_t dependence of λ_c and the density dependence of ω_s [38].

Apart from these very interesting results all experiments brought indications that the kinetics of μCF in deuterium tritium mixture is much more involved than anticipated. This conviction grew with increasing insight into of the subject. For instance, even the deduction of ω_s from experimentally determined values for W needs understanding of many details of the kinetics [38,39].

Indeed, the experiments at SIN especially aimed for resolving important kinetic steps. Already the first experiment at SIN [40,41] showed fast decay components which at a glance qualitatively looked like the fast decaying components due to hf-effects in deuterium. Although there was some suspicion before this experiment that it might not be easy to disentangle the hf-problem [42] in the case of deuterium tritium mixture it was necessary and justified to test the hf-hypothesis. As often important facts contradicting this hypothesis stimulated new ideas. Thus, a question rather controversial in the literature, the problem of thermalization, was raised again. Presently there appears to be general agreement that the fast component observed in DT mixtures is most likely to be due to the thermalization of the μt atoms proceeding by far slower than predicted. This explanation was first published by Kammel [43] and

Leon [44|. Any attempt of a "global fit" to experiments covering an
appreciable range of density and tritium concentrations, however, does not
result in a unique solution due to several open questions [36,39]. A survey
of most recent efforts and results is presented in the following.

Density Dependence of λ_c

In a simple picture for the kinetics all relevant rates are proportional
to the density, therefore the cycle rate was expected to scale linearly with
density too. Fig.2 shows the cycle rates λ_c (normalized to liquid hydrogen
density 4.25 10^{22} atoms/cm^3) versus tritium concentration measured in liquid
and gaseous DT mixtures at SIN [36]. This evident and unexpected density
effect was also seen in the LAMPF experiment [37]. A possible explanation of
the density effect was suggested by Petrov [45]: an enhancement of the
molecular formation rate is due to collisional broadening of strong
sub-threshold resonances [46,47,48], thus resulting in a contribution to
the molecular formation rate which is proportional to the density squared.
    Another interesting question is the behaviour of resonant formation
in solid DT mixtures, where structural effects might occur. At present
experiments with cold DT mixtures (minimum target temperature about 10K well
below solidification and scanning the whole tritium concentration range)
are proceeding at SIN.

ω_s – Dependence on c_t and Density

Different methods were used for the determination of the effective
alpha sticking ω_s [38]. From the fusion neutron disappearence rate λ_n

$$\lambda_n = \lambda_o + W \Phi \lambda_c \qquad (2)$$

at steady state condition the raw sticking W (total muon loss per cycle)
which includes contributions from different side channels (dd, pd, pt, tt)
was determined. After correction for these channels the sticking probability
ω_s can be extracted. (Especially valuable for this analysis were the
experimental values of the important parameters of muon catalyzed tt-fusion
determined from SIN data for the first time [39,49].) This method presently
results in the most precise ω_s values for liquid and gaseous DT mixtures
(ω_s (liquid)=0.45 ± 0.05% [36], ω_s (gas)=0.50 ± 0.10% [38]).
    After first presenting a surprising $\Phi\sqrt{c_t}$ dependence of sticking
[37,50,51] the LAMPF group recently considered the c_t-effect not significant
[52] but underscored a density dependence of ω_s [37,52] which has to be
explained by theory yet and is in disagreement to the SIN result [38]
(see fig.3). The latter allows for a weak density effect only which is
compatible to recent theory [53] (however it is remarkable that essentially
only one LAMPF data point shows a significant deviation from the average
value).

Direct Approaches for the Investigation of Sticking

As an extension of former X-rays experiments for the investigation of
pd and dd fusion [54] an experiment in DT mixture was performed at SIN
using very low tritium concentration (0.04%) to reduce tritium Bremsstrahlung
background. A further background reduction was achieved by requiring a 14 MeV
neutron and a delayed muon decay electron, both in coincidence with the
X-ray (fig.4). A quantitative analysis of this data gives a preliminary value
for X(2→1) of 0.21 ± 0.05% [55] in accordance with the results for the
effective sticking determined from the fusion neutron disappearence rate.
    This type of experiment was suggested by Bracci and Fiorentini as
giving complementary information on sticking. While the SIN result is in
perfect agreement with the latest theoretical result (X(2→1) = 0.23 ± 0.05
by Cohen [53]), an experiment at KEK at c_t=30% [56] finding a much smaller

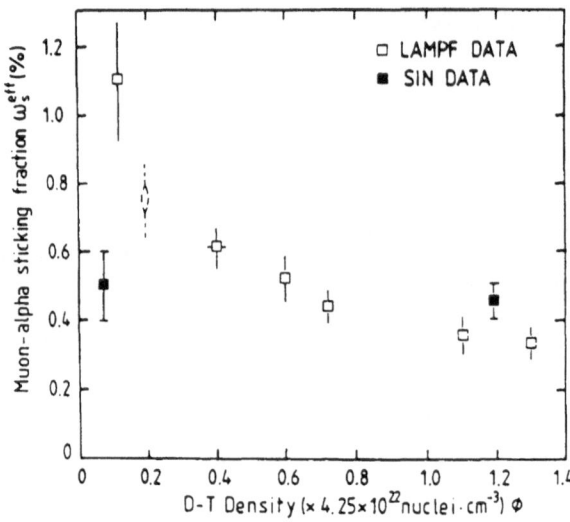

Fig.3: Comparison between LAMPF and SIN results for ω_s as a function of density.

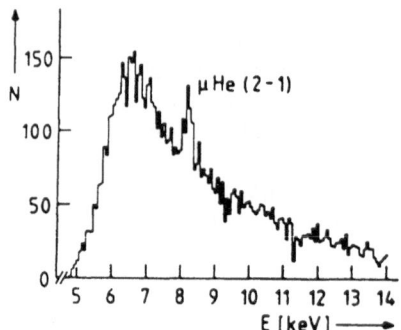

Fig.4: He (2→1) X-ray spectrum measured in the SIN experiment for He sticking after dt-fusion [55].

X-ray yield concludes disagreement between sticking results from X-ray (ω_s <0.36%) and neutron detection, respectively.

Thus, another type of experiments dealing with the direct measurement of the sticking probability attracts interest. This very appealing method is based on the discrimination of $(\mu He)^+$ against He^{++} ions and this was successfully applied at Gatchina in pure deuterium for the determination of the sticking probability after dμd fusion [57]. Although difficult (due to the small sticking and the radioactivity of tritium) the Russian group tries to extend their technique to DT mixtures [58]. Also at LAMPF an interesting experiment of this kind is under way [59]. For both experiments no results were reported so far.

Isotopic Transfer Rate

Experiments at low tritium concentrations offer new possibilities to disentangle kinetic problems like the transfer

$$\mu d + t \rightarrow \mu t + d. \qquad (3)$$

At SIN in parallel to the X-ray measurement the cycle rate at $c_t = 0.04\%$ was determined and consequently a preliminary value for the transfer rate ($\lambda_{dt} \doteq 375 \pm 50$ per μs, normalized to liquid hydrogen density) was found. This rate is in agreement with previous experiments [36] and thus indicates that one has a reasonable handle on the transfer in excited states.

Moreover informations about possibly existent hyperfine effects can be extracted from studies of the time distributions before reaching running equilibrium.

Thermalization of Muonic Tritium Atoms

In low density DT mixtures fast time components with extremely high molecular formation rates were observed in the SIN experiment [40] (the fast component corresponds to a formation rate of about 10^9. According to the "thermalization" model [43,44] the short lived time components are due to nonthermalized μt atoms which pass through energy regions with considerably higher cross sections for $d\mu t$ formation.

On the experimental side a direct measurement of the temperature dependence of $d\mu t$ formation with temperature variation up to 1500K is planned for the near future. A specially designed DT target system for these high temperatures and pressures up to 2000 bar will be developed at Lawrence Livermore National Laboratory [60] for experiments at SIN.

SUMMARY

In experiments at Dubna, KEK, LAMPF and SIN surprising effects in the μCF kinetics showed up, which enhance considerably the fusion yield compared to former theoretical expectations. At present the number of fusion reactions to be catalyzed by a single muon appears to be limited by He-sticking (ω_s) rather than by the speed of the fusion cycle (λ_c). Nevertheless, a detailed understanding of the μCF cycle in DT mixtures has not been reached. Further experiments and theoretical efforts are necessary to bring the understanding of μCF kinetics to the point desirable for judgement upon eventual technical applications.

ACKNOWLEDGEMENTS   My gratitude is due to Prof. Brunelli and Euratom for the splendid hospitality extended to me.

Support by the following institutions is gratefully acknowledged: the Austrian Science Foundation, the Swiss Institute for Nuclear Research, the German Federal Ministry for Science and Technology, and the U.S. Department of Energy.

REFERENCES

[1] Muon Physics, ed. C.S.Wu and V.W.Hughes (Academic Press, New York, 1975).
[2] E.Zavattini, in ref. [1], Vol.II, p.219
[3] C.Petitjean and G.Vecsey, IEEE Trans.Nucl.Sci.NS 18(1971) p.723, SIN-TM 0914 (1970)
[4] L.W.Alvarez et al., Phys.Rev. 105 (1957) 1127.

[5] J.D.Jackson, Phys.Rev. 106 (1957) 330; see also J.D.Jackson, LBL
     Report 18266 (1984).
[6] F.C.Frank, Nature 160 (1947) 525.
[7] L.W.Alvarez, Nobel Prize Lecture.
[8] S.S.Gershtein and L.I.Ponomarev, in ref.[1], Vol.III, p.141.
[9] A.Bertin, A.Vitale and A.Placci, Rev. di Nuovo Cimento 5 (1975) 423.
[10] J.G.Fetkovich et al., Phys.Rev.Lett.4 (1960) 570.
[11] J.H.Doede, Phys.Rev. 132 (1963) 1795.
[12] E.J.Bleser et al., Phys.Rev. 132 (1963) 2679.
[13] J.E.Rothberg et al., Phys.Rev. 132 (1963) 2664.
[14] E.Bertolini et al., Proc.Int.Conf. on High Energy Physics, CERN,
     Geneva (1962) p.421.
[15] A.Alberigi Quaranta et al., Phys.Rev. 177 (1969) 2118.
[16] A.Bertini et al., Phys.Rev. D8 (1973) 3774.
[17] V.M.Bystritskii et al., Dubna Preprint DI-7300 (1973).
[18] W.H.Breunlich, Nucl.Phys.A335 (1980) 137.
[19] W.H.Breunlich, Nucl.Phys.A353 (1981) 201.
[20] M.Cargnelli et al., Part. and Nucl. Tenth Int. Conf., Heidelberg (1984),
     Book of Abstracts Vol.II, K 6; Workshop on Fund.Muon Physics,
     Los Alamos (1986); and to be published.
[21] S.S.Gershtein, Sov.Phys.JETP 13 (1961) 488.
[22] I.T.Wang et al., Phys.Rev.139 (1965) 1528.
[23] W.Bertl et al., Atomkernenergie-Kerntechnik 43,3 (1983) 184.
[24] P.Kammel et al., Phys.Lett. 112B (1982) 319.
[25] P.Kammel et al., Phys.Rev.A28 (1983) 2611.
[26] A.Vesman, Sov.Phys.JETP Lett.5 (1967) 91.
[27] V.M.Bystritskii et al., Sov.Phys.JETP 49 (1979) 232.
[28] L.I.Ponomarev, SIN Preprint Pr-77-011 (1977).
[29] J.Rafelski, Exotic Atoms '79, ed. K.Crowe and E.Duclos (1980).
[30] D.D.Bakhalov, S.I.Vinitskii and V.S.Melezhik, Sov.Phys.JETP 52
     (1980) 820.
[31] J.Zmeskal et al., Atomkernenergie-Kerntechnik 43,3 (1983) 193.
[32] V.M.Bystritskii, V.P.Dzhelepov et al., Phys.Lett.94B (1980) 476.
[33] V.M.Bystritskii, V.P.Dzhelepov et al., Sov.Phys.JETP 53 (1981) 877.
[34] S.E.Jones et al., LAMPF proposal # 727 (1981).
[35] W.H.Breunlich et al., SIN Projekt R-81-05 (1981), R-81-05-1 (1982).
[36] W.H.Breunlich et al., Phys.Rev.Lett.58, no.4 (1987) 329.
[37] S.E.Jones et al., Phys.Rev.Lett.56 (1986) 588.
[38] C.Petitjean et al., Proc.Int.Symp. on yCF, Tokyo (1986).
[39] W.H.Breunlich et al., Proc.Int.Symp. on μCF, Tokyo (1986).
[40] W.H.Breunlich et al., Phys.Rev.Lett.53 (1984) 1137.
[41] W.H.Breunlich et al., Fundamental Interactions in Low Energy Systems,
     1985; Eds. P.Dalpiaz, G.Fiorentini and G.Torelli, p.449.
[42] P.Kammel et al., Atomkernenergie-Kerntechnik 42, no.3 (1983) 195.
[43] P.Kammel, Lett.Nuovo Cimento 43 (1985) 349
[44] J.S.Cohen, M.Leon, Phys.Rev.Lett.55, no.1 (1985) 52.
[45] Yu.V.Petrov, Phys.Lett. 163B (1985) 28.
[46] M.Leon, Proc.Int.Symp.on μCF, Toky (1986).
[47] A.M.Lane, Proc.Intern.School of Fusion Reactor Technology,
     Erice (1987).
[48] Yu.V.Petrov, Proc.Int.Symp. on μCF, Tokyo (1986)
[49] Muon Physics, ed.C.S.Wu and V.W.Hughes (Academic Press, New York,1975).
[50] S.E.Jones, Los Alamos Preprint LA-10714-C (1986) 157.
[51] J.S.Cohen, M.Leon, Phys.Rev.A33 (1986) 1437.
[52] S.E.Jones, Proc.Int.Symp.on μCF, Tokyo (1986).
[53] J.Cohen, Proc.Int.Symp.on μCF, Tokyo (1986).
[54] H.Bossy et al., Phys.Rev.Lett.55 (1985) 1870.
[55] H.Bossy et al., Proc.Int.Symp. on μCF, Tokyo (1986).
[56] K.Nagamine et al., Proc.Int.Symp.on μCF, Tokyo (1986), and this
     conference.
[57] D.V.Balin et al., Phys.Lett.141B (1984) 173

[58] G.G.Semenchuk, Proc.Int.Symp. on μCF, Tokyo (1986).
[59] A.N. Anderson et al., Los Alamos Research Proposal (1985).
[60] H.Heard, private communication.

STICKING IN MUON-CATALYZED D-T FUSION*

C. Petitjean[1]
Swiss Institute for Nuclear Research
5234 **Villigen**, Switzerland

* 1987 Updated version of the SIN Report No. Pr-86-10 (1986)

ABSTRACT

The issue of $\mu\alpha$ sticking after muon catalyzed DT fusion is controversial, since a number of theoretical and experimental results came out recently with sticking values ω_s varying over a large range. After a review of this situation, our measurements at SIN and methods of sticking analysis from neutron time structures are presented in detail. The important point is the correct understanding of the experimentally observed time distributions. At high density (liquid DT) we find, after correction for other fusion channels, for DT sticking $\omega_s = (0.45 \pm 0.05)\%$, not dependent on tritium concentration c_t and in accordance with our X-ray observations. At low density (DT gas, $\phi = 3\%$-8%) our preliminary result is $0.50 \pm 0.10\%$, giving a ratio 1.1 ± 0.2 in agreement with conventional theories, but strongly disagreeing with the LAMPF experiment of S.E. Jones et al. Our result sets the maximum fusion output per muon to less than 220 ± 20. Our newest (1987) data confirm this result.

INTRODUCTION

Muon catalyzed fusion (MCF) of deuterium - tritium (DT) isotopes has become a highly discussed issue, since very large rates of the fusion reaction

$$\mu t + d \rightarrow d\mu t \rightarrow \alpha + n + \mu^- + 17.6\,MeV \tag{1}$$

were observed in recent experiments[1-8]. The reaction kinetics in a DT mixture is shown in Fig. 1. Very rapid DT cycles are induced, due to a resonance mechanism in the formation of mesic molecules[9-11]. This phenomenon was experimentally first observed in pure deuterium (resonant production of $d\mu d$ molecules[12-14]), but surprisingly it leads (as theoretically predicted[9]) in the DT cycle to even much more enhanced $d\mu t$ formation rates. Reported DT cycle rates λ_c

---

[1]Representing P. Ackerbauer, W.H. Breunlich, M. Cargnelli, M. Jeitler, P. Kammel, J. Marton, N. Nägele, A. Scrinzi, J. Werner, J. Zmeskal, (ÖAW, Vienna), J. Bistirlich, K.M. Crowe, M. Justice, (U.C. and LBL, Berkeley), R.H. Sherman, (LANL, Los Alamos), H. Bossy, H. Daniel, F.J. Hartmann, H. Plendl, W. Schott, (T.U. Munich), W. Neumann (ETH Zürich).

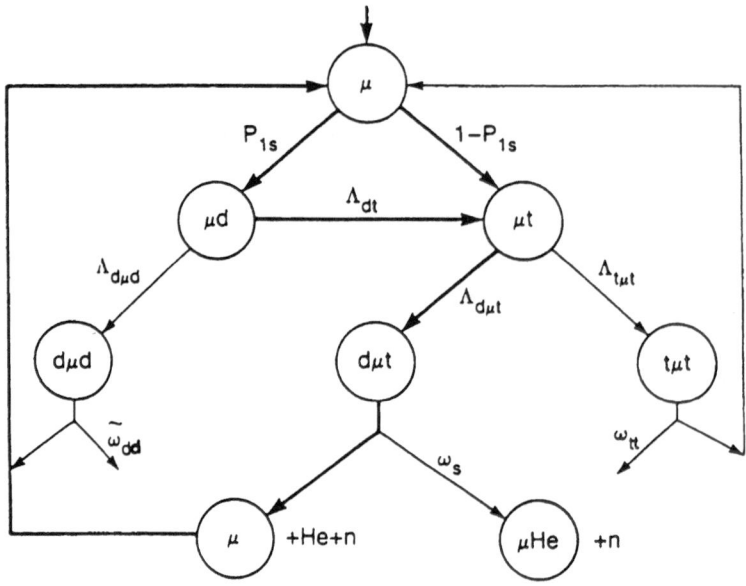

Figure 1   Kinetic cycles of muon catalyzed fusion in DT mixtures (simplified scheme). $P_{1s} = c_d q_{1s}^{11}$ and Λ_{dt} determine the muon transfer. $\Lambda_{d\mu d}$, $\Lambda_{d\mu t}$ and $\Lambda_{t\mu t}$ are the effective rates for mesomolecule formation. $\bar{\omega}_{dd}, \omega_s$ and ω_{tt} are the sticking probabilities after fusion. For a more detailed discussion see Ref. 7.

exceed the muon decay constant λ_o by more than 400[8] and show a rising tendency with target temperature[2,3,4,6]. As a consequence, the discussion of future energy applications of MCF is now lively going on[15].

The attainable fusion yield per muon Y_f, however, cannot grow infinitely with the fusion rate, because it is also limited by the sticking factor ω_s (see Fig. 1) the probability, that the muon gets trapped after fusion by the emerging α particle:

$$d\mu t \xrightarrow{\omega_s} \alpha\mu + n + 17.6\,MeV \tag{2}$$

Muon sticking interrupts the fusion cycles Fig. 1, and if it cannot be overcome (which is difficult to accomplish, one the $\mu\alpha$ system is at rest) the ultimate limit of fusion output is[16]:

$$Y_f \leq \frac{1}{\omega_s} \tag{3}$$

Theoretical values of ω_s range from the original estimate by Jackson[17] of about 1% down to 0.02%[18]. Recent "canonical" calculations yield 0.5% - 0.7% for "final" sticking[19-22].

From our experiments at SIN[7] we have evaluated $\omega_s = (0.45 \pm 0.05)\%$ at high density (liquid) and $\omega_s = (0.50 \pm 0.10)\%$ at low density (gas, preliminary value). The experimental group at Los Alamos by S.E. Jones et al., on the other hand, has reported sticking ω_s to be dependent on tritium density $\phi(c_t)^4$, on $\phi\sqrt{c_t}^{5,23}$ or at least on ϕ^6, with experimental values variing between 0.3% and 1.1%.

Evidently, the issue of sticking is quite controversial, on the theoretical as well as on the experimental side, and needs more clarification. It is the intention of this paper to line out in some detail the present situation concerning expectations, methods of measurement and evaluation of the DT sticking factor, with special emphasis on our experiments, which were performed 1983 - 87 at SIN, Switzerland, in solid, liquid and gaseous DT mixtures.

THEORETICAL ESTIMATES OF DT STICKING

The probability of $\mu\alpha$ sticking after $d\mu t$ fusion depends on initial as well as final configurations. The initial scheme (before fusion takes place) is shown in Fig. 2. In the resonant formation process, the $d\mu t$ system is originally formed in a highly excited and weakly bound state with orbital and rotational quantum numbers $(J,\nu) = (1,1)$[9,10]. It deexcites (mostly by Auger processes) to lower levels, until undergoing fusion from one of the levels. After fusion there is in principle a set of initial sticking probabilities ω_s^n applying to each muonic Helim level n, composed of contribution $\omega_s^{n,i}$ from each $d\mu t$ state i (see Fig. 2):

$$\omega_s^n = \sum_{i=1}^{5} P_i \, \omega_s^{n,i} \tag{4}$$

where P_i are relatively probabilities for fusion ($\sum P_i = 1$) calculated from the transition scheme. The overall "initial" sticking value ω_s^o is thus

$$\omega_s^o = \sum_n \omega_s^n \tag{5}$$

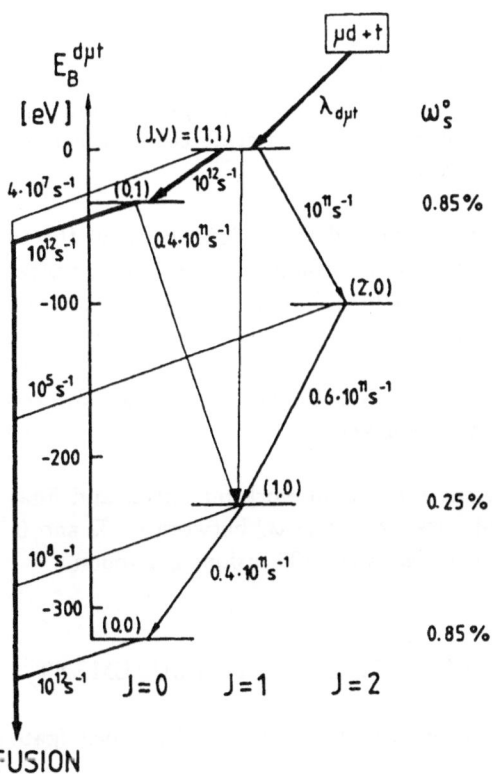

Figure 2   Level scheme and transition rates of the $d\mu t$ mesic molecule (from Ref. 24) leading to fusion. Initial sticking factors ω_s^o are from Ref. 20,27.

The Dubna theory group[24] has calculated the probabilities P_i concluding, that fusion essentially takes place from the J = 0 levels (i = 2,4). So, theoretical work on initial sticking was mainly concentrated on the calculation of ω_s^o from the (J = 0) states[19,20,25,26], see Table 1. Only recently Hu[27] considered also the (1,0) p-state of the $d\mu t$ molecule as starting point and surprisingly obtained a more than three times lower sticking value! Although the contribution from this level to fusion seems to be small, one should keep the scheme of Fig. 2 in mind in case, that some of the transition or fusion rates differ from present theoretical expectations.

It was suggested by S.E. Jones et al.[5], that subdued electron refilling (after 2 Auger transitions) could impede the last transition (1,0) to (0,0), making sticking dependent on the electron density of the target (dependence assumed to be proportional to $\phi\sqrt{c_t}$, see Ref. 5). However, this suggestion is not tenable in view of Hu's result, which leads to a reduced sticking at low $\phi\sqrt{c_t}$, opposite to experimentally observed dependences.

J. Cohen and M. Leon[23] have argued, that a bottleneck state (a side channel, preventing the muon from fast cycling) could lead to a different "effective" sticking ω_s^{eff}, which may then show up in a $\phi\sqrt{c_t}$ dependence of sticking. We clearly reject this argumentation: As long as only one fusion channel is considered open as in Ref. 23, the observable sticking value cannot change due to such a bottleneck state, but rather the cycle rate will be correspondingly reduced. The effect, discussed in Ref. 23, comes from an unphysical definition of ω_s^{eff}, using the cycle rate at time t = 0 to muon stop (before the kinetic system is in equilibrium) instead of using the steady state cycle rate. If the experimental analysis is done correctly as described later in this work, then the effect[23] does <u>not</u> cause any variation of observed sticking.

The "final" sticking value ω_s is defined as the probability, that $\mu\alpha$ system comes to rest in the 1s state. ω_s is significantly lower than ω_s^o, because the $\mu\alpha$'s have an initial kinetic energy of 3.5 MeV and can by inelastic processes shake off the muon. If R is the "reactivation coefficient", we write

$$\omega_s = \omega_s^o (1 - R) \tag{6}$$

Calculations[21,22] have shown, that muon stripping from higher orbits (n > 1) is very likely, while shaking off a muon from the 1s state takes rather place via a multistep process (excitations followed by ionisation).

Although reactivation is expected to be dependent on the target density ϕ (in vacuum R = 0!), Takahashi[22] has found little variability for technically accessible densities, the changes between $\phi = 0.1$ and 1.0 being about 3%, while an approximative formula Ref. 21 predicts an 11% effect. It is not clear, whether the theoretical treatment of multistep activation processes has been advanced enough to fully explain experimental observations (see comments in Ref. 21). Just recently a new calculation by James Cohen[28] became known, resulting now in a significantly larger reactivation coefficient(R $\sim$0.39).

Table 1 summarizes the theoretical work on initial and final sticking. The most accurate calculations yield final sticking factors ω_s between 0.5% and 0.7%, which would limit the maximum DT fusion output to less than 200 fusions per muon.

EXPERIMENTAL METHODS OF STICKING MEASUREMENT

Various methods have been used or are proposed to investigate sticking:

1. direct detection of the $\mu\alpha$ systems

2. measurement of muonic X-rays from excited $\mu\alpha^*$ systems

Table 1   List of theoretical values for "initial" DT sticking ω_s^o, reactivation R and final DT sticking ω_s.

	ω_s^o [%]	R	ω_s [%]	
J.D. Jackson[17], (1957)	~ 1	0.22	~ 0.8	B.O. approx
S.S. Gerstein et al.[25],(1980)	1.12	0.23	0.86	"
L. Bracci and G. Fiorentini[26],(1981)	1.2 ± 0.1	0.24+	0.91	"
S.K. Kauffmann et al.[18],(1981)	0.02			method questioned[21]
D. Ceperly and B. Alder[19],(1985)	0.895 (4)*			MC method
L. Bogdanova et al.[20],(1985)	0.848 (25)*			B.O. S. approx.
			0.58	
L. Menshikov and L. Ponomarev[21],(1985)		0.32+		incl. multistep ionis.
H. Takahashi[22],(1986)		0.24+		incl. multistep ionis.
J. Rafelski and R. Müller[36],(1985)	0.33-0.9			incl. res. doorway state
M. Danos et al.[37],(1986)	0.5-0.9			incl. sens. to dt-res. par.
Chi-Yu Hu[27],(1986)	0.897*			MC method dμt s states
	0.25			MC method dμt (1,0) p state
J.S. Cohen[28],(1986)	(0.87)	0.393	0.53	$\phi = 1.2$

* $\omega_s^{i,n}$ separately listed
+ density dependence given

3. measurement of the yield reduction after first fusion

4. measurement of the fusion disappearance rate.

   We discuss here all methods, but will concentrate on method 4, which has resulted so far in the most precise sticking values.

1. The direct detectionof $\mu\alpha$ systems is a straight foreward method, if proper monitoring and distinction from α's can be made. The discrimination from α's can be achieved by making use of the different range and/or energy loss characteristics (singly versus doubly charged ions!). This method was successfully applied in Gatchina (Leningrad) to determine sticking after $d\mu d$ fusion[29]. For the $d\mu t$ case there is the drawback of much smaller yields and of the enormous beta activity from tritium, inducing large background radiations in detectors placed near or in the tritium target. Nevertheless, attempts are under way to do such measurements also in $d\mu t$ fusion[29,30].

2. The measurement of muonic X-rays from excited $\mu\alpha^*$ atoms is also a direct observation of "sticking" muons, though only of the fraction ending up in orbits above the 1s groundlevel. A full assessment of sticking needs therefore detailed calculations of the initial occupation of $\mu\alpha$ orbits and the successive processes during the $\mu\alpha$ slow down an the muonic cascade. On the other hand, absolute and relative X-ray yields can be compared with theoretical calculations, allowing tests of the validity of theoretical models.

   X-ray measurments have been successfully performed at SIN by our collaboration, first in $p\mu d$ and $d\mu d$ fusion[31], later in $d\mu t$ fusion[32]. An absolute precision of 30% was reached for the $d\mu t$ case. Our results indicate, that higher $\mu\alpha$ orbits (n > 1) get mostly depleted by ionisation during the slow down. The agreement with theoretical expectations[22] is not yet satisfactory, but with the newest calculation[38] is better ($Y_x(2 \cdot 1) = 0.19\%$ per $d\mu t$ fusion).

   X-ray measurements in DT targets also suffer from Bremsstrahlung induced by the tritium radioactivity. The Japaneese group Nagamine et al.[33] has started to perform X-ray experiments at KEK using the pulsed beam technique.

3. The measurement of yield reduction makes use of the fact, that after a fusion only the fraction (1-w) of muons is free to start another fusion cycle:

$$Y_{ff} = Y_{\mu f}(1 - w) \qquad (7)$$

where $Y_{\mu f}$ and Y_{ff} are the fusion yields, measured after a muon stop and after a fusion, respectively. This method was successfully applied in our experiments at high tritium concentrations ($c_t = 96 - 99.5\%$) to determine $t\mu t$ sticking ω_{tt}[34]. It is independent of absolute counter sensitivities, but relies on an accurate measurement of the number of "good" muon stops (muons stopping in the DT mixture are usually less than electronic triggers!). Using the coincidence method[13] (measurement of single and coincident time spectra of fusion events and of electrons from muon decay), we have achieved experimental accuracies better than 2%. For the determination of ω_s this method is not yet precise enough.

The method of determining the fusion parameters from the analysis of experimental time distributions of consecutively detected fusion events has been treated in a general way in a paper by Filchenkov et al.[35], claiming also independence from absolute calibrations.

4. The measurement of fusion disappearance rates

Fusion events disappear due to sticking faster than λ_o, the muon decay constant. For a cycling system (Fig. 1), that has reached steady state equilibrium, one can describe the time distribution of fusion events simply by

$$N_f(t) = \phi\lambda_c e^{-\lambda_n t} \qquad (8)$$

where $\phi\lambda_c$ is the cycle rate (see Ref. 7).

$$\lambda_n = \lambda_o + w\,\phi\lambda_d \qquad (9)$$

is the fusion or (in our case) the neutron disappearance rate and w the "raw" sticking factor. w is not identical with ω_s, because several fusion channels are present in DT, each contributing to the observed disappearance rate. If impurities x are present, there may be more loss terms in Eq. 9 describing the muon transfer to x, e.g. $\Lambda_x = c_x\phi\lambda_x$. The measurement of λ_n in $d\mu t$ fusion is extremely sensitive to the value of w because of the large cycle rates (we have observed $\lambda_n \sim 3.5\,\lambda_o$ in solid DT at $c_t = 0.4$[8]). It has to be pointed out, that a proper evaluation of DT sticking ω_s from λ_n requires:

- knowledge of the absolute cycle rate (and as an ingredient knowledge of the absolute detector efficiency),
- correction for other fusion channels such as $d\mu d$, $t\mu t$ with sticking $\bar\omega_{dd}$, ω_{tt} and (in the presence of protium) $p\mu d$, $p\mu t$ (with ω_{pd}, ω_{pt}) and for transfers to impurities. Such corrections may be model dependent, since some kinetic scheme has to be assumed (see e.g. Fig. 1).

If the cycle rate λ_c is split up into rates for each fusion channel:

$$\lambda_c = \lambda_c^{dt} + \lambda_c^{dd} + \lambda_c^{tt} + \lambda_c^{pd} + \lambda_c^{pt} \qquad (10)$$

then one gets for raw sticking

$$w\lambda_c = \omega_s\lambda_c^{dt} + \bar\omega_{dd}\lambda_c^{dd} + \omega_{tt}\lambda_c^{tt} + \omega_{pd}\lambda_c^{pd} + \omega_{pt}\lambda_c^{pt} \qquad (11)$$

Since sticking values of other channels than $d\mu t$ are much larger, corrections to w are more significant than to λ_c, and an accurate knowledge of the kinetic system in its steady state becomes essential. Denoting $a_{\mu d}$ and $a_{\mu t}$ the (normalized) steady state populations, the cycle rates can be expressed in terms of mesomolecular formation rates[7], e.g

$$\phi\lambda_c^{dt} = a_{\mu t}\Lambda_{d\mu t} = a_{\mu t}\phi[2c_{D_2}\lambda_{d\mu t}^{D_2} + c_{DT}\lambda_{d\mu t}^{DT}] \tag{12}$$

(note, that the hyperfine structure in principle splits up resonant rates into contributions from each hyperfine level. Such effects have so far been experimentally verified only in resonant $d\mu d$ formation[13,14]. In this context, we can define, that a system has reached a steady state, when the relative populations $a_{\mu t}$ and $a_{\mu d}$ remain constant.

On the other hand, the DT cycle rate can also be expressed in terms of rates and parameters of the kinetic scheme given in Fig. 1[7]:

$$\frac{1}{\phi\lambda_c^{dt}} = \frac{P_{1s}}{\Lambda_{dt} + (1 - P_{1s})\Lambda_{d\mu d}} + \frac{1}{\Lambda_{d\mu t}} \tag{13}$$

For the evaluation of DT sticking from w we finaly derive:

$$\omega_s = w - \omega_{tt}\frac{\Lambda_{t\mu t}\lambda_f^{tt}}{\Lambda_{d\mu t}(\lambda_f^{tt} + \Lambda_{t\mu t})} - \frac{P_{1s}(\bar\omega_{dd}\Lambda_{d\mu d} + \omega_{pd}\Lambda_{p\mu d})}{\Lambda_{dt} + (1 - P_{1s})\Lambda_{d\mu d}} - \omega_{pt}\frac{\Lambda_{p\mu t}}{\Lambda_{d\mu t}} \tag{14}$$

In conclusion, we have described several (independent) methods of how to investigate sticking. For DT sticking (ω_s) method 4, the analysis of fusion disappearance rates, presents a very promising approach, but requires good calibration and knowledge of the complete cycling system including side channels.

SETUP OF THE SIN EXPERIMENT

Fig. 3 shows the experimental setup in our DT fusion experiments at SIN 1983/84 and previously described in Ref. 3 and 7. Fig. 3.1 shows a photograph of the tritium target, used at SIN in the 1987 experiments[8]. High density measurements were done with liquid DT mixtures ($V = 20$ cm^3, $T = 10$ K - 30 K) at densities $\phi = 0.9$ - 1.5 (ϕ is always given in units of liquid protium $4.25 \cdot 10^{22}$/cm^3). At low density ($\phi = 0.005$ - 0.08) gas targets with $V = 100$ - 1000 cm^3 were used at temperatures from 30 K to 300 K. Tritium concentrations c_t ranged from 0.3% to 99.5%.

The main features of event detection were:

- A muon beam telescope ($M_1, M_2, \bar{3}, \bar{4}, \bar{E}\bar{T}_1, \bar{E}\bar{T}_2$) defining electronic muon stops. The ratio ϵ_r of real muon stops (in DT) to electronic stops was 64% to 68% in liquid an 6% - 40% in gas. It was evaluated using the time spectra of electrons from muon decay. A pileup circuitry rejected events with a second particle hitting counter $M_1 \pm 9\,\mu sec$ with respect to the muon stop.

- 2 electron telescopes ET_1, ET_2 detecting electrons from muon decay. The overall sensitivity was $\epsilon_e = 38\%$, but also a small, but significant sensitivity for fusion neutrons of $\epsilon^n = 0.08\%$ was encountered. The time distributions were measured in an 8 μsec wide time range (see Figs. 4 - 6).

- A NE 213 liquid scintillation detector with pulse shaping circuit for n-γ discrimination (PSD), positioned far enough from the target (up to 113 cm) to keep the probability of double neutron hits within the integration time (0.5 μsec) at a negligible level (few percent). The proton recoil energy, the PSD signal and the time spectrum were recorded. This detector was previously carefully calibrated in an absolutely monitored 15 MeV neu-

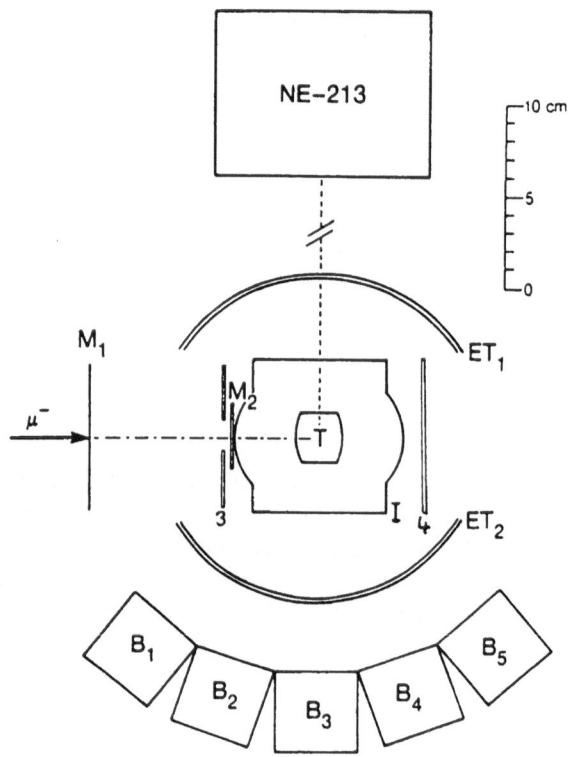

Figure 3   Experimental setup used in liquid and gaseous DT measurements (1984): Target (T),
insulation vacuum (I), muon telescope ($M_1, M_2, 3, 4$), neutron detectors $B_1 - B_5$,
(plastic, sizes 25 cm × 5 cm × 5 cm) and NE 213 (liquid 0 × d = 12.7 cm × 10 cm),
electron telescopes (ET_1, ET_2).

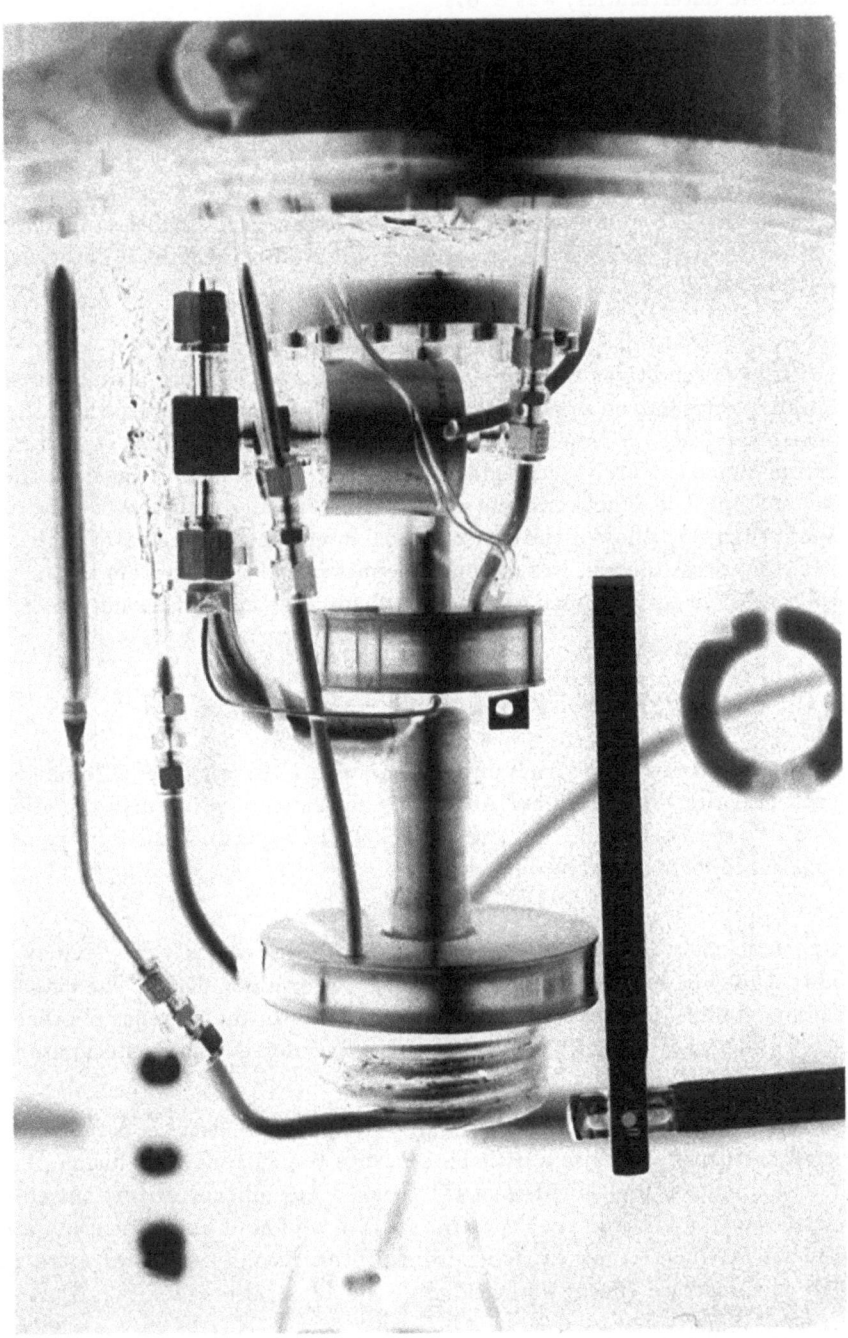

Figure 3.1   Tritium target (in copper) developed at Lawrence Berkeley Laboratory used in the
1987 μCF experiments with gaseous liquid and solid DT mixtures at densities from
3% to 150% of liquid H_2 and pressures up to 20 bar. Below the cylindrical volume,
the liquid helium refrigeration system with heat switches, above, the main filling
line and sniffing lines for the mass spectrometer are visible.

tron beam at GSF (Gesellschaft für Strahlenforschung) Munich and acted as our main calibration instrument. The effects of neutron absorptions and scatterings were carefully studied with Monte Carlo simulations. The absolute precision (including all experimental and electronic uncertainties) was $\pm$ 8%.

- 5 plastic detectors equipped with a fast routing electronics to handle multiple signals acted as fast neutron detectors (dead-time 50 nsec). The time and proton recoil spectra of up to 4 succeeding hits were recorded (see spectrum Fig. 7). The efficiency ϵ_n for detection of a DT fusion neutron was 0.4% per detector, small enough to prevent significant distortions due to dead-time or pileup effects.

- A Ge(Li) diode (not shown in Fig. 3) detected X-rays from muons stopped in the target walls or transferred from the DT mixture to higher Z impurities. The energy and time spectra were registered.

Various trigger conditions were applied for each target mixture, using simple conditions (e.g. an electron or any neutron signal within an 8 μsec time gate with respect to the muon stop) as well as more complicated schemes (e.g. neutron events delayed to muon stop or in coincidence with an electron signal). Purely accidental events were collected by triggering the electronics without a muon stop. The simple triggers yielded electronically undistorted time distributions, while the special triggers allowed the collection of "precleaned" spectra at high statistics, by taking some electronic distortions into account (see next chap.). For certain trigger schemes and experimental conditions an appropriate reduction factor was applied for data recording.

ANALYSIS OF TIME DISTRIBUTIONS

The investigation of sticking from disappearance rates necessitates a detailed understanding of experimental time distributions. Although the real time structures are given by simple exponentials $e^{-\lambda_n t}$, see Eq. 8, and $e^{-\lambda_o t}$ (muon decay), the experimentally observed spectra may be more complicated for several reasons:

- After a particle has hit a detector and triggered its electronics, the system is "dead" for a period that may be long on the scale of event rates or muon decay. The detector thus kills itself from detecting further events with a rate $\epsilon_n \phi \lambda_c$. In the presence of event routing this effect is avoided, but a dead-time Δt has to be considered. These effects are illustrated in Fig. 4.

- The electron telescopes - although primarily designed for electron detection as a charged particle coincidence - are also a little bit sensitive to neutrons from fusion. This leads due to the large neutron multiplicities in DT fusion to significant distortions of electron and correlated neutron-electron (ne) spectra. In fact we found experimentally a much larger neutron sensitivity ϵ^n than expected from proton recoils. Only for cylce rates $\phi \lambda_c <$ 1 μsec^{-1} the observed effects were negligible (for illustrations, see Fig. 5).

- Neutron time distributions generated in correlation with an electron signal are usually distorted, because the probability of a (ne) coincidence varies as function of the time after muon stop. A method (adopted by us) to avoid this problem is to collect fusion events by individually requesting for each event a delayed (ne) coincidence in a constant time delay bin with respect to the fusion event (not the muon stop!). Still an additional term $e^{-\epsilon^n \phi \lambda_c t}$ remains due to the "dying out" of the electron detector (unless it is also routed for multiple events). Fig. 6 shows some of the distortion effects due to limited time bins.

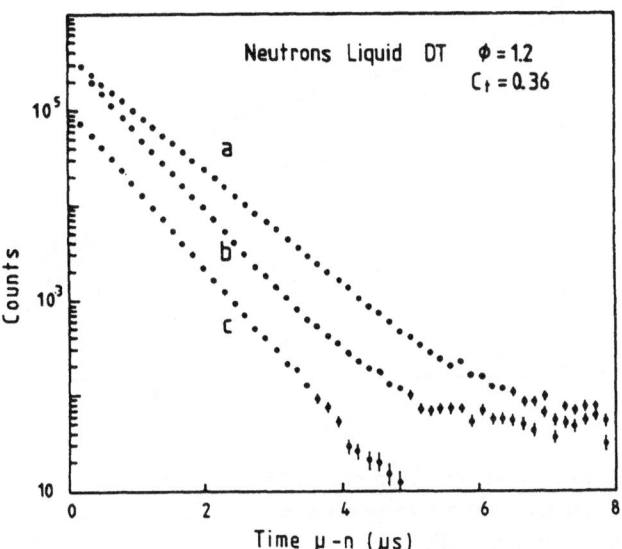

Figure 4   Time spectra of neutron disappearance from one of the fast plastic detectors, measured with liquid DT.
a) sum of first 4 neutron hits ($\lambda \sim \lambda_n = \lambda_o + w\phi\lambda_c$, see Eqs. 17 and 18)
b) first neutron hit only ($\lambda = \lambda_n + \epsilon_n\phi\lambda_c$, see Eq. 17)
c) first neutron hit with delayed (ne) correlation (Eq. 19).

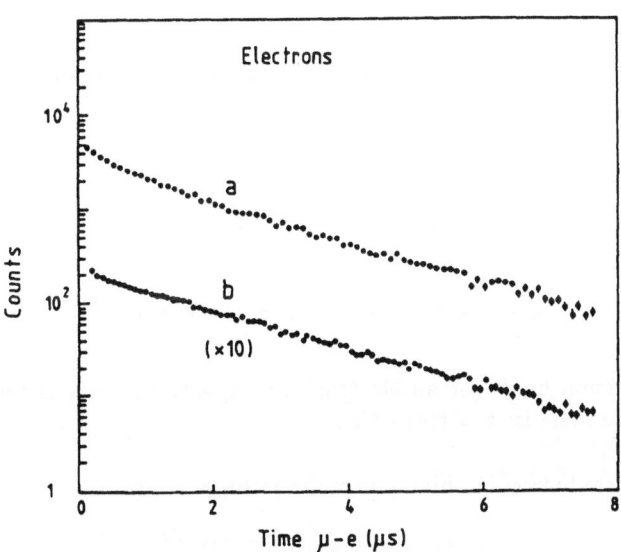

Figure 5   Time spectra of electrons from muon decay, measured
a) in liquid DT ($\phi = 1.2, c_t = 36\%$)
b) in gaseous DT ($\phi = 7.8\%, c_t = 34\%$)

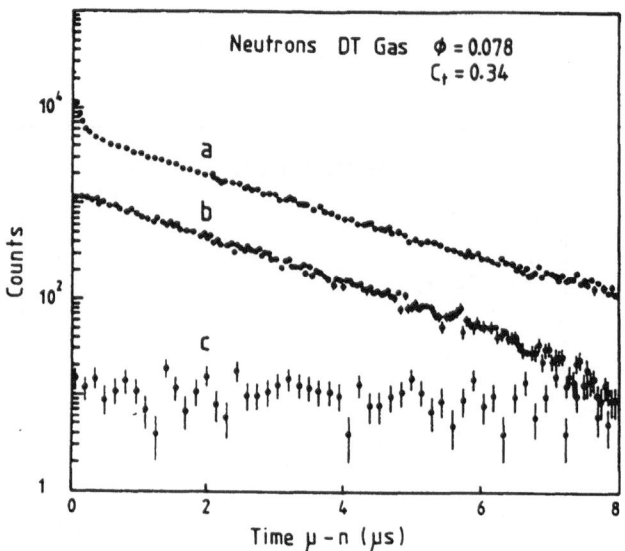

Figure 6   Time spectra of neutrons from one of the gas runs
   a) all neutron hits, simple trigger ($\lambda = \lambda_n$). The spike at $t = 0$ is due to muon
      stops in the target wall.
   b) neutrons with delayed (ne) coincidence [0.3 - 3 μsec] ($\lambda = \lambda_n + \epsilon^n \lambda_c$). Distortions
      after 5 μsec are due to the limited e time window. Events from μ stops in the
      target wall are now eliminated.
   c) accidental background, measured with a fake trigger.

Taking into account these effects we derived the following expressions for observable time spectra (which also were carefully checked with Monte Carlo simulations):

Electron time distribution (1. hit), per muon (simple trigger):

$$e_1(t) = \epsilon_r \epsilon_e \lambda_o \left[\frac{w}{\epsilon^n + w} e^{-\lambda_o t} + \frac{\epsilon^n}{\epsilon^n + w} e^{-\lambda'_n t} \right] + \epsilon_r \epsilon^n \phi \lambda_c e^{-\lambda'_n t} \tag{15}$$

with

$$\lambda'_n = \lambda_n + \epsilon^n \phi \lambda_c = \lambda_o + (w + \epsilon^n)\phi \lambda_c \tag{16}$$

The same expression holds for an electron time spectrum, that is triggered by a fusion neutron (ne). The time scale is: $t = t(e) - t(n)$.

Neutron time distribution (1. hit), per muon (simple trigger):

$$n_1(t) = \epsilon_r \epsilon_n \phi \lambda_c e^{-(\lambda_n + \epsilon_n \phi \lambda_c)t} \tag{17}$$

Neutron time distribution (k-th hit), per muon (simple trigger):

$$n_k(t) = n_1(t) \frac{[(1 - w)\epsilon_n \phi \lambda_c (t - [k - 1]\Delta t)]^{k-1}}{(k - 1)!} e^{+\epsilon_n \phi \lambda_c \Delta t(k-1)} \tag{18}$$

Neutron time distribution per muon (1. hit) with delayed (ne) correlation:

$$n_1^c(t) = [\int_{\Delta t_{nc}} e_1(t)dt]\, \epsilon_n \phi \lambda_c e^{-[\lambda_n + (\epsilon_n + \epsilon^n)\phi\lambda_c]t} \tag{19}$$

In the presence of accidental background some small additional terms are necessary to describe the observations. Fig. 5 shows some observed electron time spectra. While in the liquid case (a) all 3 terms of Eq. 15 are significant, its in gas (b) the first 2 terms (due to the smaller cycle rates). At t = 0 there is an additional component due to muon stops in the target walls, however, dying out quickly. This component can be easily eliminated by using the (ne) time distributions.

Figures 4,6 and 7 show neutron time spectra applying to expressions (17) to (19). For multiple hit time distributions see also Refs. 7 and 8.

RESULTS AND DISCUSSION

Normalized DT cycle rates λ_c have been evaluated for all experimental conditions, see Refs. 7 and 8. From the observed neutron disappearance rates λ_n a set of raw sticking values was extracted by fitting the observed time distributions to the formulas Eq. 15-19. Different analysis methods were applied, (a) using non coincident spectra (1. hit), (b) coincident spectra, (c) fitting the full set of spectra, separated according to hit number (see Fig. 7) and (d) using high discrimination levels (to check for effects of accidental events). All methods yielded for our liquid data (1984) raw sticking factors w with a relative consistency of better than 2%. The resulting averaged values for w are displayed in Figure 8 (open circles).

The sticking analysis of gas data, on the other hand, is much more difficult for reasons of smaller cycle rates (ϕ dependence, enhanced by density effect!). The effect of sticking $w\phi\lambda_c/\lambda_o$ is reduced from typically 100% (liquid data) to 5% or less for gas data. Distortions of the type discussed above or due to backgrounds become large. Also the time after muon stop, until the kinetic system reaches equilibrium, becomes significantly enlarged (e.g. more than 1 μsec transient times were observed at $\phi = 1\%$), which narrows the time bin of meaningful analysis. At too low density the system may even never reach a steady state!

For these reasons we have restricted the sticking analysis to the cases of gas data with highest cycle rates and density ($\lambda_c \sim 2 - 5/\mu sec$, $\phi = 3\% - 8\%$). Preliminary results are shown in Figure 9.

From the w values DT sticking ω_s was obtained by correcting for other fusion channels $d\mu d$, $t\mu t$, $p\mu d$ and $p\mu t$ (the $p\mu d$ and $p\mu t$ channels are present, because the mass spectrometer determined about 0.9% protium content in our samples). Expression (14) was taken using kinetic parameters, determined from a fit to the cycle rate distribution[7]. Different fit assumptions in Ref. 7 did not alter the corrections significantly. No corrections were necessary for muon transfers to impurities, because the Ge(Li) spectra yielded upper limits $\delta w < 0.02\%$. The results for ω_s are also given in Figures 8 and 9 (closed circles).

As can be seen from the figures, we find no significant dependence of sticking ω_s neither on concentration c_t nor on density ϕ. For the liquid data ($\phi \cong 1.2$) we obtain an average value.

$$\omega_s^{LQ} = (0.45 \pm 0.05)\% \tag{20}$$

and for gas ($\phi = 3\% - 8\%$)

$$\omega_s^{GAS} = (0.50 \pm 0.10)\% \tag{21}$$

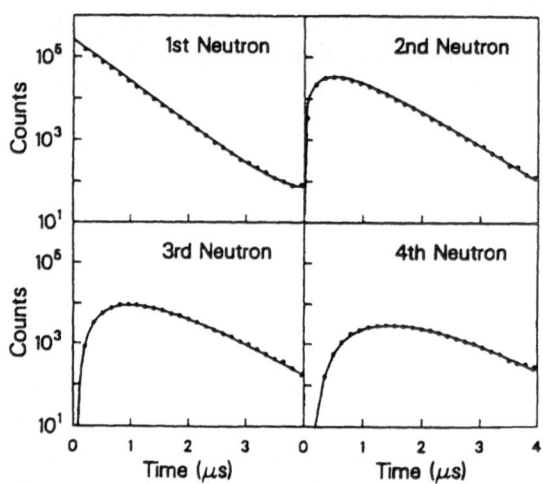

Figure 7   Time spectra of fusion neutrons, observed subsequently in one of the plastic detectors (1.-4. hit) and fitted curves using Eqs. 17 and 18.

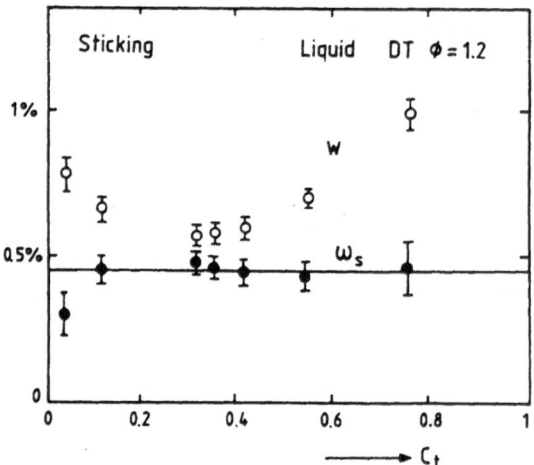

Figure 8   Experimental results (1984) of observed "raw" sticking w (open circles) and corrected DT sticking ω_s (closed circles) in liquid DT mixtures at densities $\phi \cong 1.2$.

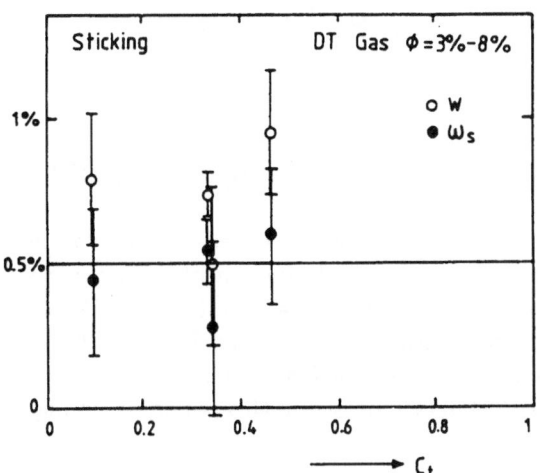

Figure 9 Preliminary experimental results (1984) for raw sticking w (open circles) and corrected DT sticking ω_s (closed circles) obtained in gaseous DT at densities $\phi = 3\%$ - 8%.

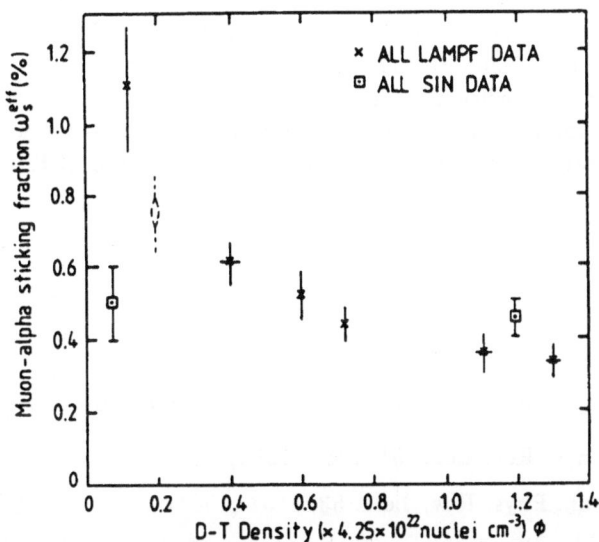

Figure 10 Muon-alpha sticking factors ω_s after DT fusion versus DT density ϕ. The data show experimental results from LAMPF and SIN, after corrections for other fusion channels and for contributions from impurities (LAMPF), and are averaged over different runs, temperatures and tritium concentrations.

The given errors include uncertainties from statistics, absolute calibration and from corrections to w. In our liquid data the systematical uncertainties (calibration and corrections) dominate, while in the gas data it is statistics.

From the results (20) and (21) we can form the ratio, to study the ϕ dependence:

$$r = \frac{\omega_s^{GAS}}{\omega_s^{LQ}} = 1.1 \pm 0.2 \qquad (22)$$

This expression is independent of most systematical errors, because all data were measured with the same detectors, geometry and method. It agrees well with theoretical expectations[21,22,28]

The dependencies of sticking ω_s on density ϕ and/or concentration c_t, previously reported from the LAMPF experiment[4-6,23] are in disagreement with our results. Especially there is no $\phi\sqrt{c_t}$ dependence in our data. Also the ratio $\omega_s(\phi = 0.12)/\omega(\phi = 1.2) = 3.1 \pm 0.5$ extracted from Ref. 6 disagrees with our observation Eq. (22) by many standard deviations. This controversial situation is illustrated in Fig. 10 (from Ref. 38). Our new 1987 results[8] on raw sticking w confirm the 1984 data reported in this paper, extending the range of densities up to $\phi = 1.5$.

ACKNOWLEDGMENTS

Support by the following institutions is gratefully acknowledged: the Austrian Academy of Science, the Austrian Science Foundation, the Swiss Institute for Nuclear Research, the German Federal Ministry for Science and Technology, and the U.S. Department of Energy under Contract No DE-AC03–76SF00098 and AT03-81ER40004. We are indebted to Professors J.P. Blaser and K. Lintner for their continuous support and encouragement. We thank Dr. L. Hansen for the calculation of neutron detector efficiencies at the Lawrence Livermore Laboratory. We thank Drs Hietl, Schulz and Schraube for providing us an absolutely calibrated neutron beam at GSF Munich. We especially thank the SIN technical staff and the workshop of H. Oschwald for their excellent work and expert assistance.

REFERENCES

1. V.M. Bystritsky et al., Phys. Lett. 94B, 476 (1980) and JETP 53, 877 (1981).

2. S.E. Jones et al., Phys. Rev. Lett. 51, 1757 (1983).

3. W.H. Breunlich et al., Phys. Rev. Lett. 53, 1137 (1984).

4. S.E. Jones, "Some surprises in muon-catalyzed fusion", presented at the Ninth Int. Conference on Atomic Physics, Seattle, Washington, USA, 23 - 27 July 1984.

5. S.E. Jones et al., Phys. Rev. Lett. 56, 588 (1986).

6. S.E. Jones, "Muon-induced fusion: experiments at LAMPF", proc. Los Alamos Workshop on Fundamental Muon Physics, Jan. 20 - 22, 1986, LA-10714-C, p.157 and Muon Cat. Fusion 1, 21 (1987), A.J. Caffrey et al., Muon Cat. Fusion 1, 53 (1987).

7. W.H. Breunlich et al., LBL reports 21'174 (1986), 21366 (1986), Phys. Rev. Lett. 58, 329 (1987) and Muon Cat. Fusion 1, 67 (1987),
C. Petitjean et al., Muon Cat. Fusion 1, 89 (1987).

8. C. Petitjean et al., SIN preprint PR-87-07 (1987), to be published in Muon Cat. Fuson $\underline{2}$ (1988).

9. S.S. Gerstein, L.I. Ponomarev, Phys. Lett. $\underline{72B}$, 80 (1977).

10. S.I. Vinitskii et al., JETP $\underline{47}$, 444 (1978).

11. L.I. Ponomarev, Atomkernenergie $\underline{43}$, 175 (1983),
    L.I. Ponomarev and G. Fiorentini, Muon Cat. Fusion $\underline{1}$, 3(1987).

12. V.M. Bystritskii et al., JETP $\underline{49}$, 232 (1979).

13. P. Kammel et al., Phys. Lett. $\underline{112B}$, 319 (1982) and Phys. Rev. $\underline{A28}$, 2611 (1983).

14. J. Zmeskal et al., Atomkernenergie $\underline{43}$, 193 (1983) and Muon Cat. Fusion $\underline{1}$, 109 (1987).

15. Yu.V. Petrov, Nature $\underline{285}$, 466 (1980) and Muon Cat. Fusion $\underline{1}$, 351 (1987).

16. S.S. Gershtein et al., JETP $\underline{51}$, 1053 (1980).

17. J.D. Jackson, Phys. Rev. $\underline{106}$, 330 (1957).

18. S.K. Kauffmann et al., UCT-TP 16, (1984) (unpublished).

19. D. Ceperly and B.J. Alder, Phys. Rev. $\underline{A31}$, 1999 (1985).

20. L.N. Bogdanova et al., Nucl. Phys. $\underline{A454}$, 653 (1986),
    V.E. Markushin, Muon Cat. Fusion $\underline{1}$, 297(1987).

21. L.I. Menshikov and L.I. Ponomarev, Sov. Phys. JETP Lett. $\underline{41}$, 623 (1985).

22. H. Takahashi, BNL Preprint BNL 37714 (1986), Fusion Technology $\underline{9}$, 328 (1986), Phys. Lett $\underline{B174}$, 133 (1986) and Muon Catalyzed Fusion $\underline{1}$, 375 (1987).

23. J.S. Cohen, M. Leon, Phys. Rev. $\underline{A33}$, 1437 (1986).

24. L.N. Bogdanova et al., Sov. Phys. JETP $\underline{56}$, 931 (1982).

25. S.S. Gerstein et al., Sov. Phys. JETP $\underline{53}$, 872 (1981).

26. L. Bracci and G. Fiorentini, Nucl. Phys. $\underline{A364}$, 383 (1981).

27. Chi-Yu Hu, UCRL preprint 94504, Livermore, April 1986.

28. J.S. Cohen, Muon Cat. Fusion $\underline{1}$, 179 (1987).

29. D. Balin et al., Leningrad preprint 895 (1983) and Muon Cat. Fusion $\underline{1}$, 127 (1987).

30. A.N. Anderson et al., Los Alamos Research Proposal, (1985).

31. H. Bossy et al., Phys. Rev. Lett. $\underline{55}$, 1870 (1985).

32. H. Bossy et al., Muon Cat. Fusion $\underline{1}$, 115 (1987) and submitted to P. R. Lett. (1988).

33. K. Nagamine et al., Muon Cat. Fusion $\underline{1}$, 137 (1987).

34. W.H. Breunlich et al., Muon Cat. Fusion $\underline{1}$, 121 (1987).

35. V.V. Filchenkov et al., Nucl. Instr. and Meth. $\underline{228}$, 174 (1984).

36. J. Rafelski and B. Müller, Phys. Lett. $\underline{164B}$, 223 (1985).

37. M. Danos et al., ANR 302L-3 (1986), submitted to Phys. Rev. A.
    J. Rafelski et al., Muon Cat. Fusion $\underline{1}$, 315 (1987).

38. C. Petitjean, Panel Discussion on the future of Muon Catalyzed Fusion, Muon Cat. Fusion $\underline{1}$, 391(1987).

PRESENT ACTIVITY ON MUON-CATALYZED FUSION AT KEK AND FUTURE DEVELOPMENT

K. Nagamine

Meson Science Laboratory, Faculty of Science
University of Tokyo, Hongo, Bunkyo-ku,
Tokyo, Japan
and
Metal Physics Laboratory, Institute of Physical
and Chemical Research (RIKEN),
Wakoh, Saitama, Japan

1. INTRODUCTION

In 1980, the Meson Science Laboratory of the University of Tokyo (UT-MSL) has established a new type of muon experimental facility at National Laboratory for High Energy Physics (KEK)[1]. There, a sharply pulsed muon beam can be used for different types of experiments which are mostly unaccessible for the conventional continuous beam. By utilizing a single bunch extraction of the 500 MeV proton from the booster synchrotron, the instantaneously intense (10^4 μ^-/pulse in 5 x 5 cm^2) muon beam can be available at the experimental target with a unique time structure of 50 ns pulse width and 50 ms pulse separation (20 Hz). The layout of experimental facility is seen in Fig. 1, where pulsed μ^- source can be obtained at the superconducting muon channel.

To our knowledge, the following excellent features exist in the pulsed muon beam compared to the conventional continuous beam: a) a weak signal measurement of a muon-associated event can be done against a huge white-noise type background if it exists; b) long time constant phenomena compared to muon life-time can be easily measured with the help of substantially reduced background; c) extreme experimental conditions like high power laser or rf can be combined to realize various resonance conditions. By using these features, we have been enjoying various new types of muon experiments, in particular, pulsed μSR experiments [2,3]. Recently, it was noticed that there are many interesting muon catalyzed fusion (μCF) experiments realized only by using this unique pulsed muon beam [4,5]. At Rutherford Appleton Laboratory, the same type of pulsed μ^- beam is now available at the commissioning stage [6].

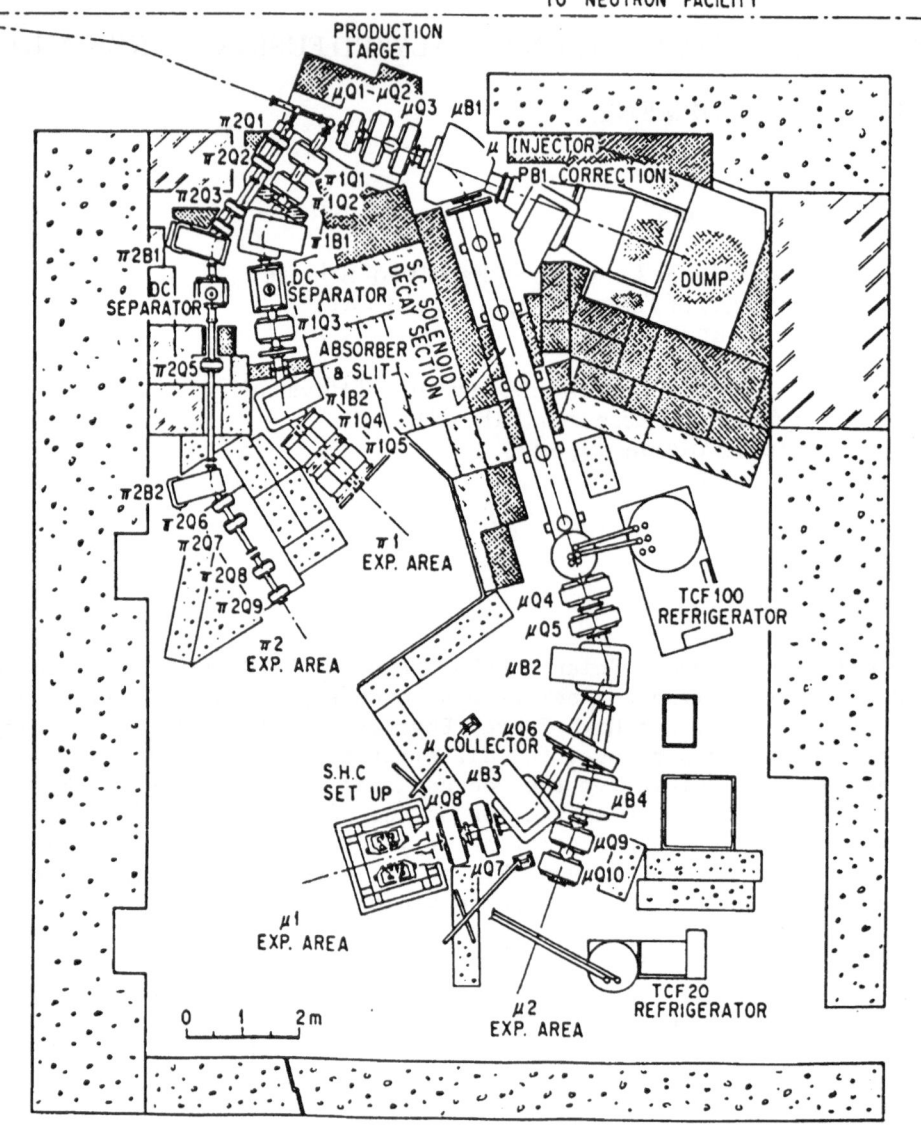

Fig. 1. Layout of the UT-MSL BOOM pulsed muon facility where superconducting muon channel as well as two surface muon channels are installed.

In this report, we are describing two experimental results on μCF carried out at UT-MSL and the one experiment which is now under considerations: 1) muonic X-ray measurements on μ^--α sticking probability in (dtμ) fusion for high-tritium-concentration liquid D_2/T_2 mixture carried out by UT/MSL-RIKEN-BYU-LANL collaboration [4]; 2) observation of radiative transition photons in the muon transfer to He mixture, in liquid D_2 exhibiting the first direct observation of mesomolecule formation carried out by UTMSL-RIKEN collaboration [5]; 3) the future laser resonance experiment to clarity the mesomolecular process in the μCF phenomena. The first and second experiment were realized by the help of the feature a) and b), respectively. The third experiment will be ralized by the help of the feature c).

2. MUONIC X-RAY MEASUREMENT IN (dtμ) FUSION FOR HIGH T_2-CONCENTRATION LIQUID D_2/T_2: DIRECT OBSERVATION OF μ-α STICKING PROBABILITY

Progress in muon catalyzed fusion (μCF) research up to 1985 has been reviewed in several articles [7]. The concept of resonant mesomolecule formation in dtμ-fusion and the dependence of the formation rate on the density of D_2/T_2 mixture were experimentally established in recent experiment [8,9]. So far, most of the important observables have been obtained by detecting fusion neutrons in either absolute number per μ^- or its time distribution after μ^- stopping. In the high density D_2/T_2 target, the absolute neutron yield per muon, Y_n is expressed in the following formula;

$$Y_n = (\lambda_0/\phi\lambda_c + W)^{-1} \qquad (1)$$

where λ_0 is the muon decay constant, ϕ the target density normalized to liquid hydrogen density $\phi_0 = 4.25 \times 10^{22}$ cm^{-3}, λ_c is the normalized fusion cycle rate obtained from the average time interval between the successive fusion neutrons and W is the loss probability whose main term is expected to be ω_s (μ-α sticking probability. Actually the value of ω_s can be obtained by a subtraction of the similar numbers of Y_n^{-1} and $\lambda_0/\phi\lambda_c$, and with further corrections for additional other loss processes. Two neutron experiments have shown a somewhat contradicting behavior; in the LAMPF experiment, ω_s depends strongly on the density of the mixture in the density range from 0.1 ϕ_0 to 1.3 ϕ_0 becoming 0.35 % at liquid hydrogen density [8], while in the SIN experiment it stays at around 0.5 % irrespective of the density in the range from 0.1 ϕ_0 to 1.2 ϕ_0 [9].

One of the major interests in the sticking probability is the fact it is theoretically a solution of Coulomb and nuclear few body problems. There have been many theoretical papers based upon various assumptions and treatments [10-12]. The predicted values for the α-sticking probability in (dtμ)-fusion varies from 0.9 % down to 0.2 % or smaller. At the same time, when the λ_c is large enough, the sticking probability is a measure of the efficiency for energy production; it has been argued that, if the ω_s is below 0.2 %, the energy production exceeds a breakeven.

As a new method to determine the ω_s, we have employed the muonic X-ray measurement method from the μ^4He atom. Compared with the neutron method, the X-ray method provides us more direct knowledge regarding ω_s, when the value of a formation probability of excited states in the total (μα) atom formation (κ) is known beforehand. More strictly, κ should be considered to be a probability for X-ray emission per μ-α sticking. At least, the X-ray method places a lower limit for the ω_s, which is in other words an upper limit for the energy production.

$$Y_x = \kappa \frac{\omega_s \phi \lambda_c}{\lambda_0 + W\phi\lambda_c} \qquad (2)$$

$$Y_x(t) = \kappa\omega_s\phi\lambda_c \; e^{-(\lambda_0 + W\phi\lambda_c)t} \qquad (3)$$

$$Y_n(t) = \phi\lambda_c \; e^{-(\lambda_0 + W\phi\lambda_c)t} \qquad (4)$$

where W represents a total loss probability,

$$W = \omega_s + \text{others} \qquad (5)$$

Thus, there are several ways to determine $\kappa\omega_s$; a comparison between absolute X-ray yield and neutron yield ($Y_x/Y_n^s = \kappa\omega_s$); by the measurements of the $Y_n(t)$ and Y_n to provide W and λ_c and then the measurement of Y_x provides $\kappa\omega_s$.

The experimental method of the X-ray measurement, particularly for (dtμ) system with high density and high C_T (C_T: tritium concentration), has serious difficulties due to an existence of a huge radiation background around the X-ray peak of μ^4He atom (8.2 keV). The most serious one comes from bremsstrahlung γ rays associated with tritium beta decay; it becomes easily several MH$_z$ per keV at around 8.2 keV even for a 1 cc liquid containing 0.3 litre (STP) T_2. Because of this bremsstrahlung background, the only existing experiment at SIN [13] was done for the low T_2 concentration (C_T) of 0.024 to 0.048 %, which is a thousand times smaller than the C_T of 30 % employed in our experiment.

In addition, since the injected muons stop at the surrounding materials other than the d-t mixture, we have an additional radiation background due to the high energy decay-electrons which may also occupy the X-ray detector before the real signal can be detected.

Experimental method

In order to overcome the problem related to the bremsstrahlung background, we applied the pulsed muons at UT-MSL located at KEK. Because of a high instantaneous intensity, a signal to noise ratio is tremendously increased by operating all the detection systems to be in phase with the beam pulse (a phase sensitive detection method). At the same time, we employed 35 kG superconducting Helemholz coil (SHC) along the muon beam axis for a source of a strong confinement field which keeps decay electrons away from the X-ray detector.

In the present experiment, our main concern was focussed on the determination of the α-sticking probability in liquid and high C_T d-t mixture. The C_T was taken as 30 % which is known to have the maximum λ_c at low temperatures [8,9]. The basic structure of D_2/T_2 target chamber is shown in Fig. 2. The D_2/T_2 gas mixture is contained in

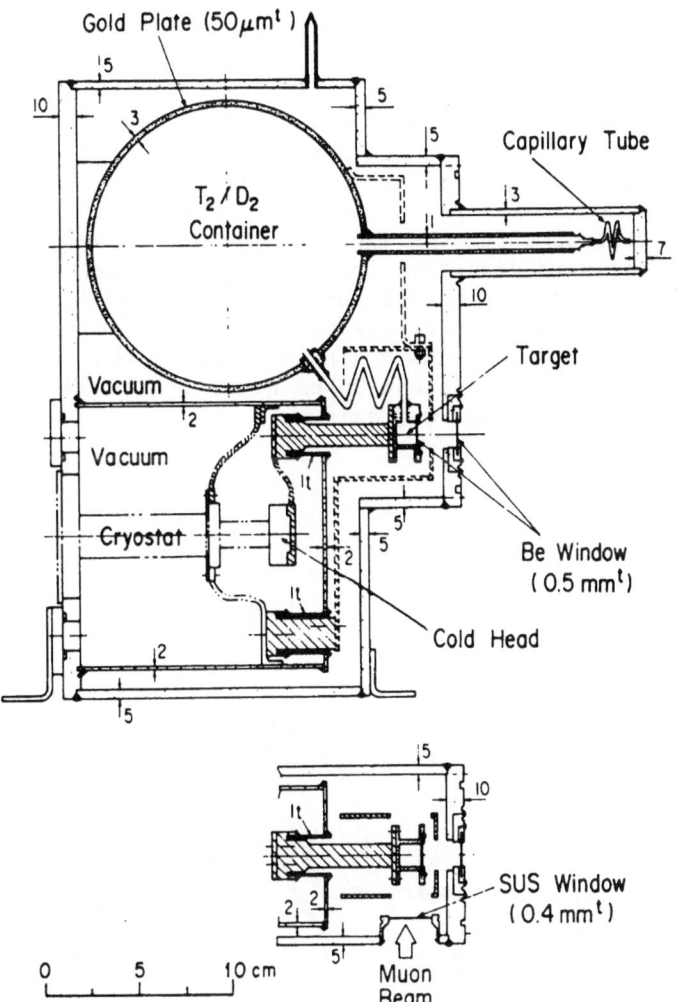

Fig. 2. Schematic view of the doubly sealed tritium-deuterium target
chamber used in the present experiment; vertical view (above)
horizontal cut-view along the muon beam plane (below).

the doubly closed container, an inner part of which is a gas container (1 litre) which is connected to a liquid container (1 cc) made of cupro-nickel. The liquid container is indirectly cooled by the He flow cryostat through a thermal conductive copper rod. The X-ray from the liquid D_2/T_2 can penetrate towards the Si(Li) detector through two 0.5 mm thick Be windows.

During our experiment, the liquid container was maintained at the temperature between liquefaction temperature T_L and T_L-2 K. Although a small amount of ^{3}He is accumulated, its solubility in liquid D_2/T_2 mixture is so small at this temperature that we assumed we can consider ^{3}He contribution being at the level below 0.1 % in W.

The experiment was carried out at the μl port of the superconducting muon channel of the UT-MSL BOOM facility, by using the 60 MeV/c backward μ$^-$ beam. The μ$^-$ beam is confined to a spot of 2 cm diameter with the help of the longitudinal field of the SHC. The essential part of the experimental arrangement is shown in Fig. 3. The 16 channel plastic counter telescopes for the digital type of the μe detections are installed at the backward direction from the D_2/T_2 target chamber. The Si(Li) detector (5 mm thickness) for the low energy X-ray measurement as well as NE213 liquid scintillation counter (2" dia. and 2" length) for 14 MeV fusion neutrons are placed inside the room-temperature gap (10 cm wide and 90 cm high) of the SHC in the direction perpendicular to the beam axis.

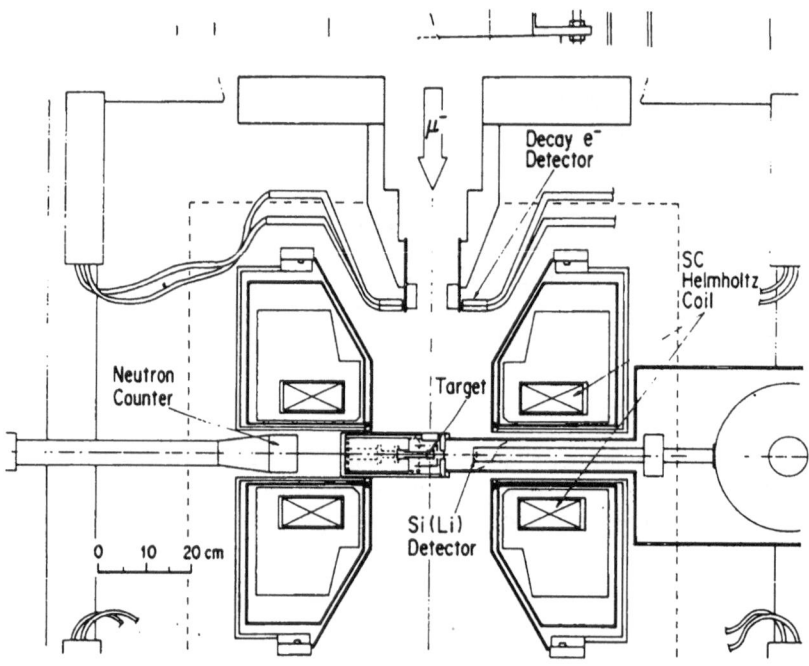

Fig. 3. Experimental arrangement for the X-ray and neutron measurement for liquid d-t target in the μCF experiment.

After formation of liquid D_2/T_2, we have measured
time-sequentially the following spectra; (a) time spectrum of decay
electrons, (b) two parameter (energy and time) correlation spectrum for
the X-ray signals and (c) three parameter (energy, timing and n-γ
descrimination) correlation spectrum for the neutron signals.
Establishment of liquid formation was known by observing the intensity
variation of the bremsstrahlung as a function of the temperature of the
cold head of the cryostat where the complete liquefaction temperature
corresponds to a saturation of the intensity.

Experimental result

Typical example of the observed spectra from the X-ray detector are
summarized in Fig. 4 and 5; integrated energy and time spectrum (Fig. 4)
and time-divided energy spectra (Fig. 5a-5c). The following features
can be readily seen in the time-divided energy spectra: a) the "early"
spectrum is essentially the bremsstrahlung observed at the off-beam
condition, where the 8.0 keV peak is an electronic Cu K_α X-ray due to
tritium beta-rays hitting the cupro-nickel wall of the liquid container;
b) the "prompt" spectrum has on additional contribution from·a low
energy tail due to γ-rays from the multiple scattering of decay
electrons; c) the "delayed" part, which is the most essential part for
the α-sticking effect, is made of a featureless spectrum with the
bremsstrahlung background, which can be removed by using the "early" and
"very delayed" spectrum. The corrected "delayed" spectrum is shown in
Fig. 5d.

The corrected delayed spectrum was used as an object for the
fitting analysis to find out the peak associated with the delayed X-ray
due to the α-sticking. The fitting was made by assuming a Gaussian
shaped peak superposed on a linear background structure. Here, the FWHM
of the Gaussian peak was taken to be either 0.25 keV corresponding to
the detector resolution seen in the observed FWHM for the Cu K_α X-ray
peak or 0.70 keV corresponding to the FWHM to be expected when a Doppler
effect contributes fully to the muonic K_α line. The results are
summarized in Table 1. Hereafter, we will use both of these two fitting
values corresponding to the two assumptions for the X-ray line-width.

In the neutron energy spectrum as well as in the n-γ discrimination
spectrum, we noticed a significant pile-up effect of the photon backgrounds around t=0. Thus, only $Y_n(t)$ data with a n-γ discrimination
gate was used for further discussions. The obtained neutron time
spectrum is shown in Fig. 6. By fitting the time constant of the
neutron intensity $\lambda_n = \lambda_0 + W\phi\lambda_c$ is obtained to be 1.20(9)
$(\mu s)^{-1}$.

Discussions

By using the obtained experimental results, we can discuss the
parameters related to the α-sticking as well as loss probabilities of
various other sources. At this moment, we adopt the experimental value
of the cycling rate $\lambda_c = 112 \times 10^6$ (s^{-1}) obtained by the neutron
experiment [8,9] for the liquid d-t mixture of $C_T = 0.30$ and $\phi = 1.2$.
With thus adopted λ_c, our result of neutron life time (λ_n) gives a
total loss rate probability (W) of 0.55 ± 0.07 %. There are other loss
processes usually considered in the neutron method, e.g. $\lambda_{dd\mu}$, λ_{tHe},
λ_{dHe}, etc. Applying these corrections obtained in the recent neutron
experiments (0.09 %) [8], we obtain an effective sticking probability
seen in the neutron method: $\omega_s(n) = 0.46 \pm 0.07$ %, almost consistent
with the other neutron data.

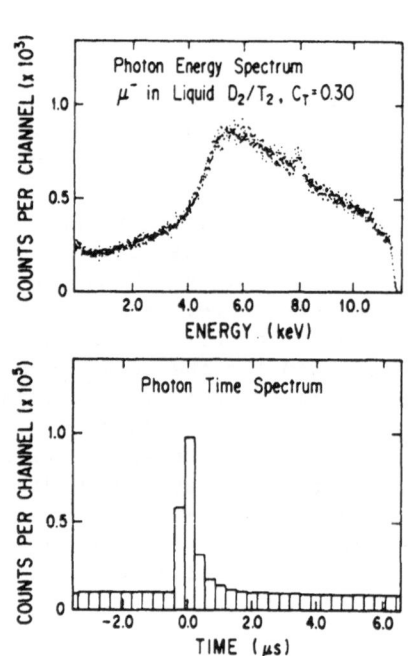

Fig. 4. Integrated energy (above) and time spectrum (below) observed by the X-ray detector for μ⁻ in the liquid d-t mixture.

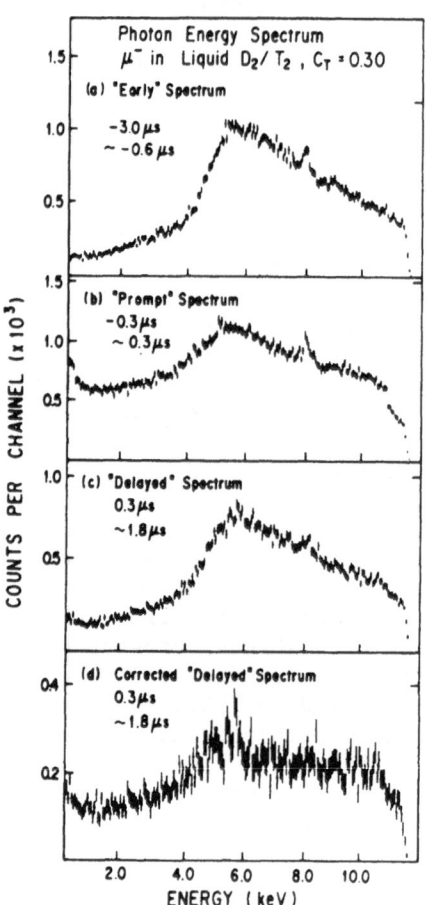

Fig. 5. Time-divided energy spectrum observed by the X-ray detector for μ⁻ in the liquid d-t mixture(a-c) and the "delayed" spectrum after bremsstrahlung background is removed (d).

Table 1.   Summary of the chi-square fitting to the X-ray
data for liquid d-t μCF

Assumed γ-ray width (keV)	E_γ (keV)	X-ray yield per μ^-	Chi-square/DOF	X-ray yield per fusion (x 10^{-2})
0.25	8.30(10)	0.035(22)	131/176	0.031(20)
0.70	8.30(13)	0.055(45)	133/176	0.049(40)

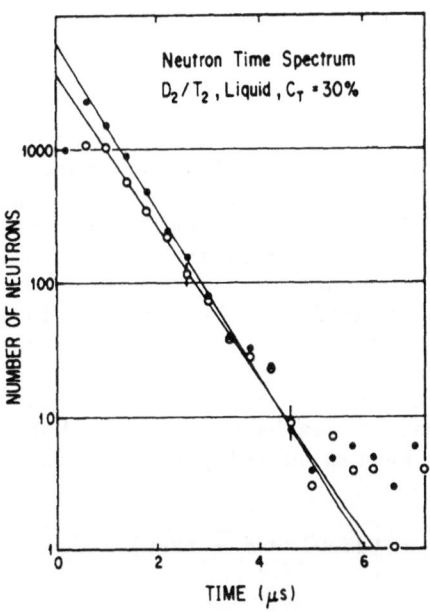

Fig. 6. Neutron time spectrum observed in the neutron counter for the
μ⁻ in the liquid d-t mixture gated by the neutron signal
from the n-γ discriminator.   Filled circle corresponds to the
experimental raw data and open circle corresponds to the data
taken with a prompt pulse rejection for a different run.

Our data on both X-ray intensity and neutron life time and the adopted λ_c with eq. (2) gives us the value of $\kappa\omega_s$ (= Y_x/Y_n, X-ray yield per fusion) to be 0.031(20)% and 0.049(40)% for the two values of FWHM. The experimental ω_s can be compared to the various theoretical works which predict various combinations of the numbers of κ and ω_s [4]. There, we found most of the theories [10-12] predict κ = 0.25 and ω_s being larger than 0.5 %. Assuming κ = 0.25, the sticking probability obtained in the X-ray method $\omega_s(X)$ becomes 0.12(8) and 0.20(16)% corresponding to the two FWHM of the X-ray line-width. At the present stage, the number should be considered as an upper limit for $\omega_s(X)$; it is clear that $\omega_s(X)$ is lower than 0.36 % when κ is 0.25.

Therefore, we can draw the following conclusions: a) the sticking probability for the X-ray method $\omega_s(X)$ (below 0.36 %) is smaller than that for the neutron method $\omega_s(n)$ (0.46 ± 0.07%); b) the experimentally obtained $\omega_s(x)$ is far smaller than any of the theoretical predictions.

In order to explain these discrepancies, let us consider the further corrections in the obtained ω_s such as hidden loss process which produces neither fusion neutron nor ($\mu\alpha$) X-ray, and then:
1) the κ should decrease when the μ^- transfer from the excited states in collision process e.g. $\mu\alpha + t \rightarrow \mu t + \alpha$ is effective [14]. On the other hand, it should increase when the atomic excitation like (1s) $\rightarrow$ (2s,2p) is effective[15];
2) the ω_s is also subject to a correction due to the reactivation process where the μ^- in not only $(\mu\alpha)_{1s}$ but also $(\mu\alpha)_{2s\cdots}$ will be detached.

3. OBSERVATION OF RADIATIVE TRANSITION PHOTONS IN μ^- TRANSFER FOR LIQUID D_2 WITH ^{4}He IMPURITY

Motivation

In the μCF of D_2/T_2 system, there might be an inherent limiting factor for the energy production, namely, an accumulation of the ^{3}He, a decay-product of tritium. Therefore, μ^- transfer mechanism to He impurity in high density D_2/T_2 mixture should be throughly studied. Related to this subject, we have recently studied the μ^- transfer phenomena in liquid D_2 with controlled He impurity [5]. It is well known that the rate of μ^- transfer to He impurity in a gaseous hydrogen isotope atoms is extremely high [16,17]; experimentally observed rate is the order of 10^8 s^{-1} compared to the theoretical value of 10^6 s^{-1} expected for a direct exchange reaction. In order to explain this phenomena, it has been proposed that the charge exchange of mesonic hydrogen with helium nuclei proceeds via the mesic molecular formation [16],

$$(d\mu) + He \rightarrow (dHe\mu)^* \rightarrow (dHe\mu) \rightarrow d + (He\mu) \qquad (6)$$

Or more correctly,

$$(d\mu) + He \rightarrow [(dHe\mu)^* e^-]^+ + e^-$$
$$[(dHe\mu)^* e^-]^+ \rightarrow [(dHe\mu)^{++} e^-] + \gamma \qquad (7)$$
$$(dHe\mu)^{++} \rightarrow (He\mu) + d$$

Several experimental investigations gave consistent exchange reaction rates with the calculation[17]. However, it seems that no one has ever observed the X-ray emitted by the radiative deexcitation process in the (dHeµ) mesic molecule. We started an experiment to detect the X-ray to understand the basic mechanism for transfer reaction and to obtain a direct evidence for the mesomolecular process.

Experimental method

The X-ray measurement was carried out at the UT-MSL BOOM facility located at KEK by using pulsed backward μ^-. The experimental setup was almost the same as that we used in the X-ray measurement on the α-sticking probability in a liquid D_2/T_2 mixture. The muons were stopped in a separately prepared liquid target chamber of 12 cc. There, the upper surface of the liquid D_2 was in contact with the ^{4}He gas of 1 atm. According to the Henry's law, the concentration of He in liquid D_2 is estimated to be 300 ppm. The energy and the timing of the X-ray were measured by a Si(Li) detector.

3.3 Result

In Fig. 7, we present the X-ray energy spectra for the "prompt" part (0.0 to 0.16 µs) and the "delayed" part (0.16 to 7.5 µs). The "prompt" part shows a sharp peak at 8 keV, which is again electonic Cu K_α X-ray from the target chamber hitted by the electrons contaminating the muon beam. For the "delayed" part we can see a bump around 7 keV (6.85 keV) with a FWHM of around 1 keV (0.85 keV). As shown in Fig. 8., the observed photon spectra was consistent with the transition from the mesic molecule excited level to the ground state of (dHeµ); a radiative transition [16]. A more detailed analysis gives a life of around 2.2 µs for the bump, while the background level decays much faster.

Discussions

About 2 % of the muons were found to contribute to the X-ray emission. Using the helium concentration of 300 ppm expected from the Henry's law, the transfer rate was obtained to be around 4×10^7 s^{-1}. The obtained rate is much smaller than either that previously obtained for gaseous target at 40 K [17] or theoretical predictions [16]. The difference may be due to a) the smaller velocity of the atoms in the liquid than in the gas, b) many-body effects on the mesomolecule formation rate, c) other decay branches opened at high density, etc. Also, the effect of inhomogeneous distribution of dissolved He atom should be taken into account. Refined experiments are now in progress.

Admitting the present argument is correct, the effect of He impurity in high density D_2/T_2 mixture does not contribute much to the loss process in (dtµ)-fusion; below 0.1 % loss rate even at 1 % He concentration. Thus, the existence of small amount of ^{3}He impurity could not be a limiting factor for energy production.

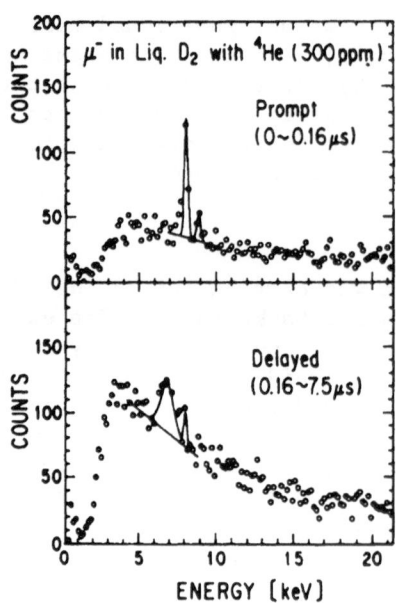

Fig.7. Observed photon spectrum from the μ^- in liquid D_2 with
He impurity (300 ppm); prompt (0.0-0.16 μs) above and delayed
(0.16-7.5 μs) below.

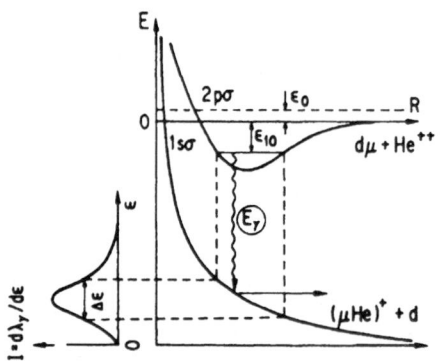

Fig. 8. Schematic picture for the radiative decay photons in the μ^-
transfer process from (dHeμ) to He propsed by Aristov et. al.

Before going into real future experiments, we would like to mention the subjects which are currently scheduled in coming few months from now: a) precise measurement of the ($\mu\alpha$) X-ray in the (dtμ)-μCF for liquid D_2/T_2 mixture which is essentially a repeat of the experiment reported in the ealier section in a 10 times longer measurement period, aiming to know a spectrum shape of the X-ray, whether a Dopper-broandening is effective or not; b) precise measurement of the radiative photon spectrum associated with the μ^- transfer to He in liquid D_2, which is again a high-statistics version of the experiment reported here aiming to learn more detailed character of the radiative photon such as asymmetric energy spectrum, etc.

In the previous section, we have demonstrated the <u>direct</u> observation of a role of the mesomolecule in μ^- transfer reaction which involves the mesomolecular processes related to the μCF process. Considering the discrepancies between many experimental observations and theoretical predictions for the (dtμ)-μCF process, it is highly demanded to observe a role of the mesomolecule in the high density d-t mixture. For this purpose, atomic or molecular resonance experiment is quite a powerful method. With the help of sharply pulsed beam structure, laser experiment is now under preparation at UT-MSL. A part of these projects are described in the followings.

Hyperfine structure resonance in (tμ) atoms

The application of a pulsed laser with the pulsed muon beam should bring a change in the mesomolecule formation cross sections. Typical example is to pump the hyperfine levels of the (tμ) states by an M1-transition. The required laser power was estimated under the following assumption: the hf splitting is 0.241 eV (5.0 μm laser wave length); the laser power is adjusted to make the induced transition rate close to the experimental λ_{10} at 30 K; the other broadenings of each level are given by the experimental $\lambda^1_{dt\mu}$ and $\lambda^0_{dt\mu}$ [18]. The required laser power becomes 70 kW in a 1 cm^2 cross section during 2 μs. Then, the result of the rate equation calculation gives the change in the reaction paths along the hyperfine levels. A major part of this argument was based upon the interpretation shown in the publication of the SIN experiment[18], which was reconsidered in the later publication [9].

Electronic molecular level resonance: [(dtμ)d2e] molecules

The laser resonance could be much easily carried out for the electronic molecular state. If the present understanding is correct for the resonant formation process of the (dtμ) molecule, a substantial effect might be expected when the electrons in ordinary electronic molecule are excited or ionized.

Concerning laser-technology for this type of experiments, there is essentially no difficulty to produce a sufficiently strong pulsed laser. This point has been demonstrated by the recent experiments at UT-MSL for 2 photon-absorption laser resonance experiments in thermal muonium performend mainly by A.P. Mills Jr, S. Chu (both AT and T Bell Laboratories) and the present author.

5. CONCLUSION

As described here, important experiments on the μCF phenomena are now in a rapid growth with the help of pulsed muon beam at UT-MSL. Recently we learned that same type of experiments are under planing at RAL. Soon or later, further insights will be obtained for the basic mechanisms in (dtμ)-μCF phenomena. These insights will eventually be a key information for the final decision on the possible energy production from muon catalyzed fusion.

The author acknowledges Drs T. Matsuzaki and K. Ishida for their great contributions to the muon catalyzed fusion experiments at UT-MSL.
He also acknowledges the following persons for their kind helps and encouragements throughout the expriments at UT-MSL/KEK: Prof. T. Yamazaki and Prof. H. Miyazawa, the former and present directors of UT-MSL; Dr T. Nishikawa, Prof. H. Sasaki, Prof. K. Katoh and other related members of KEK; Prof. A. Arima, Prof. T. Tominaga and associated members of University of Tokyo; Dr G.T. Garvey and related persons at Los Alamos National Laboratory. The experiment at UT-MSL/KEK were achieved in collaboration with the following members to whom sincere thanks are given; Mr. Y. Hirata, Mr. Y. Watanabe, Mr. R. Kadono, Dr. Y. Miyake, Dr K. Nishiyama, Prof. S.E. Jones and Dr. H. R. Maltrud, Helpful discussions with Dr. M. Leon and Mr T. Ohi are also acknowledged.
This work was supported in part by the Grant in Aid of the Japanese Ministry of Education, Culture and Science.

REFERENCES

[1] K. Nagamine, Hyperfine Interactions $\underline{8}$ (1981) 787

[2] UT-MSL Newsletter 1-6 (1981-1986), eds., K. Nagamine and T. Yamazaki, unpublished

[3] Collected Papers on Muon Science Research at Meson Science Laboratory, University of Tokyo, (February, 1986), eds. K. Nagamine and T. Yamazaki, unpublished

[4] K. Nagamine, T. Matsuzaki, K. Ishida et al., contribution to μCF86, Muon Catalyzed Fusion 1(1987) Ch.2

[5] T. Matsuzaki, K. Ishida, K. Nagamine et al., contribution to μCF 87 (Leningrad)

[6] G. Stirling and S.F.G. Cox, private communication (March, 1987)

[7] L.I. Ponomarev, Atomkernenergie-Kerntechnik $\underline{43}$(1983)175; L.Bracci and G. Fiorentini, Phys. Report $\underline{86}$(1982)169; S.E. Jones, Nature $\underline{321}$(1986)127

[8] S.E. Jones et al., Phys. Rev. Lett. $\underline{56}$(1986)588

[9] W.H. Breunlich et al., LBL-Report 21174(1986) and contribution to μCF86, Muon Catalyzed Fusion 1(1987) Ch.2.

[10] J.D. Jackson, Phys. Rev. $\underline{106}$(1957)330; S.S. Gerstein, Yu.V. Petrov, L.I. Ponomarev, N.P. Popov, L.P. Presnyakov, and L.N. Somov, Sov. Phys. JETP $\underline{53}$(1981)872

[11] L.Bracci and G. Fiorentini, Nucl. Phys. $\underline{A364}$(1981)383
L.N. Bogdanova, L. Bracci, G. Fiorentini, S.S. Gerstein, V.E. Markushin, V.S. Melezhik, L.I. Menshikov and L.I. Ponomarev, Nucl. Phys. A454(1985)653

[12] M. Kamimura, contribution to μCF86, Muon Catalyzed Fusion 1(1987) Ch.5.

[13] F.J. Hartmann, contribution to μCF86, Muon Catalyzed Fusion 1(1987) Ch.2.

[14] J. Rafelski, contribution to μCF86, Muon Catalyzed Fusion 1(1987) Ch.5.

[15] J.S. Cohen, contribution to μCF86, Muon Catalyzed Fusion 1(1987) Ch.3.

[16] Yu. A. Aristov et al., Yad. Fiz. $\underline{33,}$ 1066 (1981);
A.V. Kravtsov et al., Phys. Lett. $\underline{83A.}$ 379 (1981)

[17] S.E. Jones, in "Atomic Physics " R.S. VonDyck Jr. and E.N. Fortson e.d., World Scientific, Singrpor, 1984, p99 and references therein.

MUON-CATALYZED X-RAY LASER AND NEUTRON PULSES

S. Eliezer

Plasma Physics Department
Soreq Nuclear Research Center
Yavne 70600, Israel

ABSTRACT

In this paper we discuss[1]a possible scheme to build an X-ray laser by irradiating a thin rod of deuterium-tritium by a μ^- beam. The excited state of dtμ molecules $(J,V) = (1,0)$ (where J and V are the rotational and vibrational quantum numbers of the $(dt\mu)^+$ hydrogen type molecule) can be induced to make radiative chain transitions with photon energy of about 90 eV to the ground state $(0,0)$. The unique feature of this laser system is the fact that each photon is associated by a 14 MeV neutron derived from the fusion of the d-t ground state (also a 3.5 MeV α particle is released during this process). This laser scheme is described in section 1. In section 2 we estimate the interesting possibility to induce Stokes or anti-Stokes transitions in the μ-molecules by an external laser.

1. X-RAY LASER

The lasing conditions occur when (i) the population inversion is sufficient, (ii) the induced transition is much shorter than the spontaneous decay time, and (iii) the gain factor of the laser is sufficiently larger than the opacity of the medium. The gain factor is defined[2][3]

$$\gamma = \frac{\left(N_2 - \frac{g_2}{g_1} N_1\right) \lambda^2 g(\nu)}{8\pi n^2 \tau_{sp}} ,$$

(2.1)

where the intensity of laser as a function of the distance of propagation goes like $I(z) = I(0)e^{\gamma z}$, N_2 and N_1 are the number density of the excited state and the ground state, λ is the wavelength of radiation, g_1 and g_2 are the numbers of states for the given energy, n the index of refraction, $g(\nu)$ is the line shape of the transition. Equation (2.1) reduces to

$$\gamma = \frac{N_\mu \lambda^2}{2\pi^2 \tau_{sp} \Delta\nu} \tag{2.2}$$

where we have used the relation $g(\nu) = 4/(\pi\Delta\nu)$ and we have assumed an index of refraction $n = 1$, and $N_2 - (g_2/g_1)N_1 = N_\mu$, where N_μ is the number density of the muons. The intensity I of the laser obeys

$$\frac{dI}{dz} = \left(\gamma - \frac{1}{\ell_{op}}\right) I \tag{2.3}$$

where the opacity length ℓ_{op} is defined by the Born approximation as

$$\ell_{op}^{-1} = N_{D_2} \sigma = \frac{2^8 \pi e^2 a_B^2 N_{D_2}}{3\hbar c} \left(\frac{Ry}{\hbar\omega}\right)^{7/2}, \tag{2.4}$$

where a_B is the Bohr radius and Ry is the Rydberg energy (≈ 13.6 eV), N_{D_2} is the deuterium density and $\hbar\omega$ the emitted X-ray energy. From Eqs. (2.2) - (2.4) the threshold muon number density N_μ^{cr} for lasing is obtained as

$$N_\mu^{cr} \approx 3\pi^2 \tau_{sp} \Delta\nu/\lambda^2, \tag{2.5}$$

where ℓ_{op} is taken to be 0.7 cm for the present medium. The spontaneous decay time τ_{sp} is determined [3] by

$$\tau_{sp}^{-1} = \frac{32\pi^3 \nu^3}{3hc^3} |M_{ij}|^2. \tag{2.6}$$

By comparing this formula with the appropriate electron transition, ν increases by a factor (m_μ/m_e), while $|M_{ij}|^2$ the dipole transition probability, decreases by a factor $(m_\mu/m_e)^2$, thus we obtain $\tau_{sp}^{-1} \sim 10^8$ sec^{-1}. Asserting that for the lasing amplification $\tilde{g}L=5$ with $\tilde{g}=\gamma - \ell_{op}^{-1}$ then using Eq. (2.2), the lasing medium L should be L ≈ 10 cm. Since the formation of [(dtµ)d2e] molecule is (almost) in resonance, it is expected that the transition rate saturates at a certain low limit temperature due to the intrinsic quantum fluctuations.

The linewidth $\Delta\nu$ is [3]

$$\Delta\nu = \sum_i \Delta\nu_i, \tag{2.7}$$

where the main contributing widths are (i) the Doppler broadening, (ii) the collision between molecules, (iii) the pressure broadening, (iv) the saturation effects, (v) the fusion time, and (vi) the natural width. In our case, the first four are the main contributions, all of which is proportional to the temperature square root, $\sqrt{T}$, except for (iv). For example,

$$\Delta \nu_c \sim N_{D_2} <\sigma_e v> , \tag{2.8}$$

$$\Delta \nu_{Dopp} \sim \frac{\nu}{c} \sqrt{\frac{2kT \ln 2}{m}} . \tag{2.9}$$

N_{D_2} is the deuterium density, σ_e is the appropriate cross-section and v is the particle velocity and ν is the collision frequency. At a temperature of $T \leqslant 1000^\circ K$, $\Delta \nu \approx 10^{12} sec^{-1}$. From Eq. (2.5) we get the critical muon density, $N_\mu^{cr} \simeq 6 \cdot 10^{18} cm^{-3}$ at $T = 1000^\circ K$. Here we have taken a typical temperature of $1000^\circ K$ because the mu - molecule formation time is short around this temperature.

To determine the factors of the heating and thus the temperature of the lasing medium (dt mixture) the following stopping lengths (ranges) must be taken into considerations: the muons, the α-particle at 3.6 MeV due to the fusion of dt, and the electrons due to muon capture by the hydrogen isotope molecules and atoms. The muon capture, which is about 10^{-14} sec at $v_\mu \simeq \alpha c$ (i.e. at an energy of about 4 keV), corresponds to a range of 10^{-5} cm, and the range of α-particle at 3.6 MeV is about 10^{-4} cm. The range of electrons at energies $\varepsilon_e = p_e^2/2m_e$ which equal the muon energy $\varepsilon_\mu = p_\mu^2/2m_\mu + V_\mu \sim 6 keV$ at $v_\mu \simeq c\alpha$ ($p_e = \sqrt{m_e/m_\mu} p_\mu$) is about 10^{-5} cm at liquid hydrogen density. As long as the dimensions of the lasing medium are larger than the stopping length of μ^-, but thinner than the ranges of α and electrons, then the irradiated μ^- and the fusion product do not cause the material to overheat. In this case the specimen of the lasing medium should have a thin rod or wafer type cross-section of the linear dimensions of 10^{-5} cm and the length of 10 cm and thus a volume of about $10^{-9} \div 10^{-6}$ cm^3. The necessary number of muons per pulse in the specimen is $n_\mu \sim 6 \times 10^9 \div 10^{12}$ at $T = 10^3$ $^\circ K$.

Once the induced radiation by a noise takes place, the rest of dtμ-molecules are triggered by spontaneous radiation. This ensures the temporal coherence within 10^{-8} sec, while the spatial coherence is guaranteed because the photons are chain-triggered.

2. LASER INDUCED STOKES AND ANTI-STOKES TRANSITIONS.

We consider a 3 energy-level system of a μ-molecule, such as dtμ, ddμ, dpμ, etc. The quantum levels are denoted by a, b and c where $E_a > E_c > E_b$ are the appropriate energies. The lifetimes satisfy the condition $\tau_a \gg \tau_c$. We calculate explicitly a Stokes transition (the anti-Stokes calculations are analogous). The μ-molecule is initially (t=0) in state a and decays spontaneously to its ground state b with a rate γ_{ab} sec^{-1}. In the presence of a laser with frequency ω (energy $\hbar\omega$) we induce a two step decay of the a level. During the first step a virtual transition to level c is induced by absorption of a single photon while the second step is a spontaneous decay from c to b with a rate γ_{cb}. This process is described by the interaction Hamiltonian, H_{int} = Vcos t, and the equations of motion for the probability amplitudes of levels a and c are given by

$$\frac{da}{dt} = -\frac{1}{2}\gamma_a a + \frac{i}{2\hbar}V_{ca}\,e^{-i\Delta t}c$$

$$\frac{dc}{dt} = -\frac{1}{2}\gamma_c c - \frac{i}{2\hbar}V_{ac}\,e^{-i\Delta t}a \tag{3.1}$$

where $\gamma_a = \gamma_{ac} + \gamma_{ab}$, $\gamma_c = \gamma_{ca} + \gamma_{cb}$ and V_{ac} is the transition matrix element between states a and c. Δ is defined by $\Delta = (E_a - E_c)/\hbar - \omega$. The solution of eqs. (3.1), for the initial conditions a(o) = 1, c(o)=0, is

$$|a(t)|^2 = e^{-\gamma_a t}$$

$$|c(t)|^2 = \left|\frac{V_{ac}}{2\hbar}\right|^2 \left\{\frac{e^{-\gamma_a t} + e^{-\gamma_c t} - 2e^{-\gamma t}\cos\Delta t}{\Delta^2 + \delta^2}\right. \tag{3.2}$$

where $\gamma \equiv (\gamma_a + \gamma_c)/2$ and $\delta \equiv (\gamma_a - \gamma_c)/2$. Since in our case the c level decays much faster than a, i.e $\gamma_c \gg \gamma_a$, one gets $\gamma = -\delta = \gamma_c/2$. One likes to calculate the rate of Stokes transitions up to a time T (T is the laser pulse duration) which is small compared to the natural lifetime of the a level, i.e. T $\ll \gamma_a^{-1}$, while on the other hand T $\gg \gamma_c^{-1}$. In this case the probability for Stokes transitions to state b up to a time T is given by

$$P_{st}(T) = \gamma_{cb}\int_0^T dt\,|c(t)|^2 = \left|\frac{V_{ac}}{2\hbar}\right|\frac{\gamma_{cb}T}{\Delta^2 + (\gamma_c/2)^2} \tag{3.3}$$

The appropriate cross section for this two step process is

$$\sigma_{ab} = \frac{P_{st}(T)}{FT} = \frac{1}{F}\frac{|V_{ac}|^2\gamma_{cb}}{4\hbar^2(\Delta^2 + \gamma_c^2/4)} \tag{3.4}$$

where F is the laser flux ($cm^{-2} sec^{-1}$). The probability of spontaneous decay (P_{sp}) from a to b, during a time T, is

$$P_{sp} = \gamma_a \int_0^T dt |a(t)|^2 = \gamma_a T \quad .$$

(3.5)

From eqs. (3.3) and (3.5) one gets efficiency η, defined as the ratio between Stokes induced transitions and the spontaneous decay,

$$\eta \equiv \frac{P_{st}(T)}{P_{sp}(T)} = \frac{|V_{ac}|^2 \gamma_{cb}}{4\hbar^2 \gamma_a (\Delta^2 + \gamma_c^2/4)}$$

(3.6)

This result is correct for laser pulse duration which satisfies $\gamma_c^{-1} \ll T \ll \gamma_a^{-1}$. In order to calculate the interaction matrix element we assume a dipole transition between quantum levels a and c.

$$|V_{ac}|^2 = \frac{1}{3} (eE_1)^2 |< c|\vec{r}|a>|^2$$

(3.7)

where the electric field of the laser, E_1, is related to the flux F, $|E_1|^2 = (4\pi)^2 \hbar F / \lambda_1$, λ_1 is the laser wavelength. Whenever the masses of the hydrogen isotopes inside the μ molecule differ from each other, the above dipole approximation is appropriate. The above electric dipole moment in eq. (3.7) is related to γ_{ca}^{-1} (i.e. the c lifetime with respect to the radiative decay to a) using Einstein's "A-coefficients",

$$\gamma_{ac} = \frac{4e^2}{3\hbar^4 c^3} (E_a - E_c)^3 |<c|\vec{r}|a>|^2$$

(3.8)

From (3.8) and (3.7) the transition matrix element is derived,

$$|V_{ac}|^2 = \frac{\hbar}{2\pi} (\frac{\lambda_{ca}}{\lambda_1})^3 \lambda_1^2 F \cdot \Gamma_{ca}$$

(3.9)

where $\Gamma_{ca} = \hbar \gamma_{ca}$ and $\lambda_{ac} \equiv 2 \hbar c/(E_a - E_c)$. Using the last relation into eq. (3.4) we get the well known Breit-Wigner formula for the cross section,

$$\sigma_{ab} = \frac{\lambda_1^2}{8\pi} (\frac{\lambda_{ac}}{\lambda_1})^3 \frac{\Gamma_{ac} \Gamma_{cb}}{(\hbar\Delta)^2 + \Gamma_c^2/4} \quad .$$

(3.10)

The induced Stokes efficiency is given from Eqs. (3.6) and (3.9) by

$$\eta = \frac{\hbar F \lambda_1^2}{8\pi} \left(\frac{\hbar\omega}{E_a - E_c}\right)^3 \frac{\Gamma_{ac}\Gamma_{cb}}{\Gamma_a [(E_a - E_c - \hbar\omega)^2 + \Gamma_c^2/4]} \tag{3.11}$$

For a resonance effect, $\hbar\omega = E_a - E_c$, and assuming $\Gamma_{ac} \simeq \Gamma_a$, $\Gamma_{cb} \simeq \Gamma_c$ one gets

$$\eta_{res} \simeq \frac{F\lambda_1^2 \tau_c}{2\pi} \tag{3.12}$$

where τ_c is the lifetime of the intermediate c level. The interesting case, $\eta \gg 1$ is easily satisfied for the resonance case (Eq. 3.12)); however it is a non trivial task to obtain $\eta \gg 1$ for laser irradiation with a wavelength far away from the resonance condition. In this (second) case the constrain is so severe that $\eta \gg 1$ implies a laser flux high enough to produce a plasma.

A generalization of the above formalism implies for the cross section (Eq. (3.10)) for laser induced transitions with polarities higher than the dipole, the following expression

$$\sigma_{ab} = \frac{\lambda_1^2}{8\pi} \left(\frac{\hbar\omega}{E_a - E_c}\right)^{2J_a + 1} \left(\frac{E_a - E_b - \hbar\omega}{E_c - E_b}\right)^{2J_b + 1} \frac{\Gamma_{ac}\Gamma_{cb}}{(E_a - E_c - \hbar\omega)^2 + \Gamma_c^2/4} \tag{3.13}$$

where the multipolarities from a to c and c to b are denoted by J_a and J_b respectively.

The laser induced transitions calculated in this section are relevant not only for building new schemes for X-ray laser, but also for increasing the probabilities for resonance formation or decreasing the probability of the sticking problem. At the moment, it appears that the main difficulty in using induced Stokes or anti-Stokes physics is the lack of appropriate short wavelength lasers. The development of X-ray lasers is a necessary factor to pave the way for improving the muon catalyzed fusion process.

ACKNOWLEDGEMENT

I would like to thank my colleague Prof. T. Tajima for the ideas summarized in this paper. The useful and enlightening discussions with Prof. M.N. Rosenbluth are very much appreciated.

REFERENCES

[1] T. Tajima and S. Eliezer, Laser and Particle Beams (to be published, 1987).

[2] A. Yariv, Quantum Electronics, (John Wiley and Sons, N.Y. 1967), pp. 237 - 242.

[3] C. H. Townes and A. L. Shalow, Microwave Spectroscopy, (McGraw-Hill, N.Y., 1955) pp. 19 - 371.

ACCELERATORS FOR μ⁻ PRODUCTION

M. Weiss

CERN
1211 Geneva 23, Switzerland

INTRODUCTION

Muon catalysed fusion (MCF) relies on an abundant production of μ^- particles, which, in turn, are essential in the formation of so-called meso-molecules, capable of fusing spontaneously in about a picosecond[1]. The μ^- particles (in fact 'heavy' electrons) are created in the π^- decay and these latter mesons are a product of the bombardment of a target by an energetic particle beam. The target is, at the same time, the fusion 'fuel'; it is composed of a gaseous mixture of deuterium and tritium at high density ($\sim 5 \times 10^{22}$ atoms per cm³), but at low temperature ($\sim$ room temperature). The charged particles composing the beam are deuterons; they are more efficient than protons in π^- production. Tritons would be still more efficient, but being unstable they are not considered here for use in particle accelerators.

The π^- production is efficient at beam energies of $\sim$ 1 GeV per nucleon. Therefore, accelerators capable of accelerating deuterons up to 2 GeV will be considered. These accelerators must operate in the continuous wave (CW) mode and handle intense beams, which is not a trivial task.

In this paper, the principle of operation of linear ion accelerators is reviewed, in particular with respect to their possible application to MCF. Several accelerators are needed to compose the chain which brings the particles to the desired energy; it is hoped that the reason for the choice of specific accelerators is demonstrated with sufficient clarity.

An example of a 1 GeV per nucleon, 100 mA accelerator chain is given. It is composed of 'standard' accelerator types, the only difference being that here they operate in the CW mode. To reduce the power consumption, the use of superconducting accelerators has been imposed.

An attempt has been made to compare the fusion power with the required mains power. A rough estimate of the cost of the accelerator chain is also included.

PRINCIPLE OF OPERATION OF LINEAR ACCELERATORS

The principle of operation of linear accelerators is recalled here in order to help in understanding the choice made for the accelerator chain.

A waveguide, as shown in Fig. 1, can propagate electromagnetic waves above a given (cut-off) frequency. A wave component, either electric or magnetic, must be in the direction of propagation. This is in contrast to free space, where electromagnetic fields are perpendicular to the propagation.

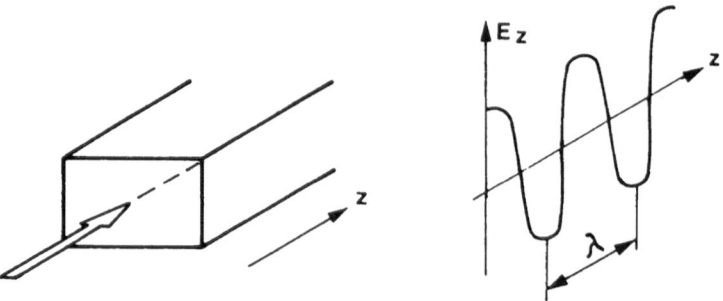

Fig. 1 Longitudinal field along the axis of a rectangular waveguide

If one wishes to accelerate particles, one must have a longitudinal electric-field component and, in addition, the wave and the particle must move synchronously (or quasi-synchronously). Unfortunately, the wave velocity (phase velocity) in waveguides is always larger than the velocity of light. There is a possibility of slowing down the phase velocity by introducing periodic obstacles into the waveguide ('disc-loaded waveguide'), see Fig. 2. The smaller the distance between obstacles (cell length), the lower the velocity. In this way, we can establish the wave-particle synchronism, but we pay for it by complicating the boundary conditions, which now require the presence of a whole spectrum of waves. The point is that the spectrum contains one wave which is synchronous and that the amplitude of this wave is preponderant.

Very often accelerators do not work in this 'travelling-wave' regime, but in the standing-wave one. Closing a waveguide at both ends we get wave reflection and an

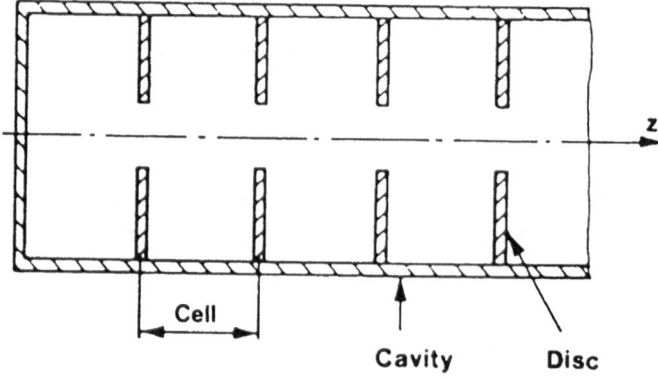

Fig. 2 Circular waveguide with periodic obstacles (discs)

eventual establishment of standing waves. Although apparently different, both regimes are alike in the sense that they require a 'synchronous' speed of the particles in order to accelerate them. We will not go into more details here, but just mention that the accelerators we shall deal with operate in the standing-wave regime.

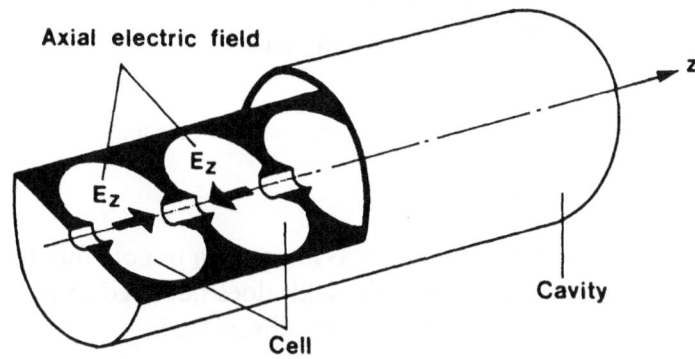

Fig. 3   Accelerating cavity operating in π mode

Figure 3 shows schematically an accelerating cavity with electric fields 180° out of phase in two adjacent cells. The bunch of particles is in the first cell, where the electric field is accelerating. At a moment π/ω later (half-period) the bunch is in the second cell, but during this time the field has reversed its sign and is again accelerating. If the bunch is in the middle of the cell before the field has reached its maximum, the particles will not only be accelerated, but also kept together longitudinally. The phase angle with respect to the top of the field is called the synchronous phase.

Transversely, the situation is less favourable: the electromagnetic field defocuses the particles while, on the contrary, a focusing is needed. The particles have transerse momenta and, left alone, would spread apart. The focusing is usually achieved by magnetic quadrupole lenses, which focus in one of the transverse planes and defocus in the other. A sequence of quadrupoles with alternating polarities has, nevertheless, a focusing effect in both transverse planes (alternating-gradient focusing).

With accelerators as shown in Fig. 3, quadrupoles are placed between cavities. The bigger the transverse momenta in the beam, the shorter the distance between quadrupoles. The beam must, therefore, have small transverse emittances (area in phase space occupied by position and momentum coordinates of particles).

It is customary to divide the transverse particle momenta by the longitudinal one. We get thus the angles that the motion of particles forms with the longitudinal axis. At low energies, these angles are large and require closely-spaced quadrupoles and very-short accelerating cavities, which is impractical. In addition, at low energies the cells in the cavity are short and therefore the power loss in the dividing walls is increased.

There is another ion accelerator, more suitable for lower energies, the Alvarez linac, shown schematically in Fig. 4. The 'periodic obstacles' are created here by hollow cylinders (drift tubes) placed along the longitudinal axis. In each drift tube, there is a quadrupole lens. The particles are accelerated in gaps between drift tubes; when the field reverses its sign, the particles are in the drift tubes and are thus screened

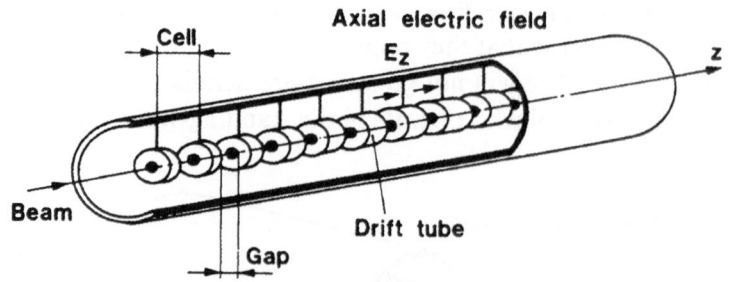

Fig. 4   Section of an Alvarez or drift-tube linac

from the decelerating action of the field. It is interesting to note that the Alvarez linac, owing to the shape of the electromagnetic field, does not need to have dividing walls between cells; the power losses are, in this way, diminished.

The particles travel from a gap to the next in a time $2\pi/\omega$. The required velocity corresponds to kinetic energies of several hundreds of keV per nucleon. This energy was, until recently, provided beforehand by electrostatic acceleration, employing important high-voltage installations. The particles have also to be grouped (bunched) before entering the Alvarez linac, in order to arrive at the right phase of the electromagnetic field; this requires still more equipment.

With the recent advent of a novel accelerator type, the radio-frequency quadrupole (RFQ), the situation at low energies has changed drastically. The RFQ accepts particles at very low energies (only a few tens of keV per nucleon) and bunches, focuses and accelerates them by using electromagnetic fields only. The principle of operation of the RFQ can be understood in the following way: consider a long electric quadrupole with alternating voltages on its electrodes, see Fig. 5. Particles travelling along the axis will experience alternating gradient focusing. Instead of having several quadrupoles with opposite polarities, one has here a single quadrupole with alternating fields, producing the same effect.

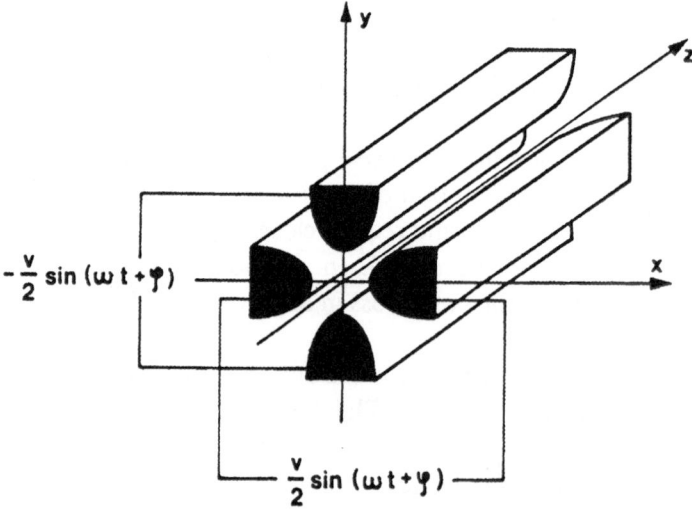

Fig. 5   Electric quadrupole with alternating electrode voltages

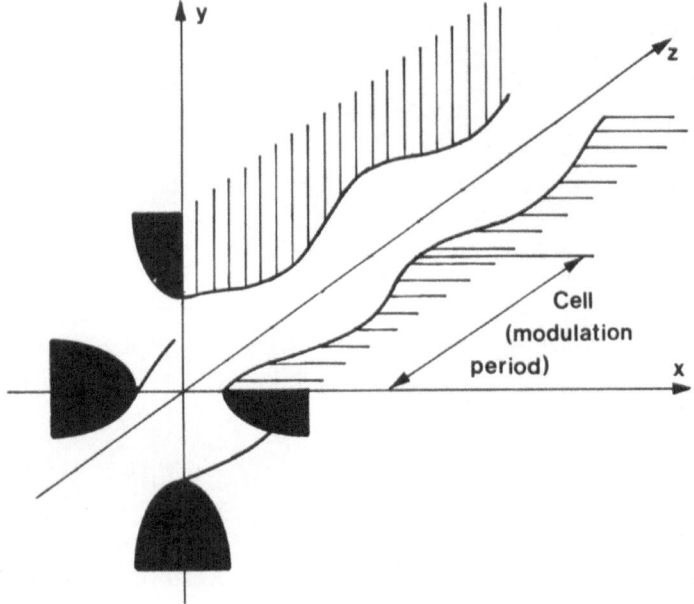

Fig. 6 Creation of longitudinal fields by modulation of electrodes of an electric quadrupole

Bunching and acceleration of particles is achieved by 'perturbing' periodically (modulating) the electrodes so as to create longitudinal electric-field components. The period of the perturbation can be made very small, permitting the use of the RFQs at much lower particle velocities than in the case of an Alvarez linac, see Fig. 6. The electrodes of the RFQ have to be closed in a cavity to limit the electromagnetic field. Figure 7 shows the cavity of the CERN proton RFQ during construction (1982); Figure 8 shows the electrodes placed inside the cavity (input cover removed). The electrode modulation is clearly seen in the middle.

Fig. 7 Cavity of the CERN RFQ

Fig. 8   CERN RFQ: inside view from the input end.

These are the types of accelerators we envisage for use in MCF. They operate, typically, at frequencies of a few hundred MHz. The energy ranges they cover and their approximate lengths are presented in Fig. 9.

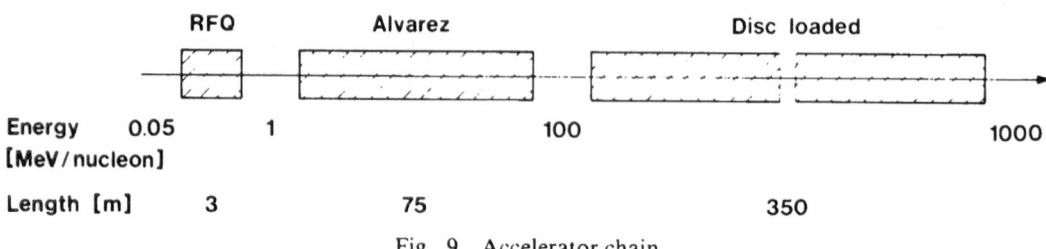

Fig. 9   Accelerator chain

POWER REQUIREMENTS—SUPERCONDUCTING CAVITIES

If accelerators have to become a part of an energy-producing plant, the power needed to operate them is of primary importance. A part of the power is needed to accelerate the beam, i.e. it goes into the beam kinetic energy. This part is fixed (unless we find a better way of producing π^- mesons or μ^- leptons). Another part is dissipated in the walls of the accelerating cavities. This part must be minimized.

An important accelerator parameter is the effective shunt impedance per unit length:

$$R = E_{acc}^2/P_1 \qquad [M\Omega/m] \ ,$$

where E_{acc} is the effective accelerating field in MV/m (averaged over the cavity) and P_1 is the power loss per length in MW/m. The power dissipated in an accelerator of length L is evidently

$$P = P_1L = (E_{acc}^2/R)L \ .$$

146

The value of R varies along the accelerator and depends also on the frequency ($\propto f^{1/2}$). Typical values are (at $f \simeq 300\,\text{MHz}$)

$$R \simeq 35\,\text{M}\Omega/\text{m} .$$

With $E_{acc} = 6\,\text{MV/m}$ and $L = 100\,\text{m}$ we get

$$P = 100\,\text{MW} ,$$

which is certainly prohibitive.

It seems that the only way out is to employ superconducting cavities. In this case use is made of the fact that some materials, when cooled down close to the absolute zero, lose almost all resistance to the current flow. This advantage is paid for by incurring some complications connected with:

– the production of cavities from superconducting materials (usually Nb);
– putting the cavity into a cryostat and cooling it to liquid He temperature (4.2 K);
– bringing the RF power into the cavity via a feeder or coupler, of which one end is at room temperature and the other at 4.2 K;
– keeping the resonant frequency of the cavity (low-loss oscillating circuits have very sharp frequency characteristics);
– damping higher order modes (HOM) in the cavity (unwanted and harmful, beam-deflecting field configurations can be excited in the cavity by the beam itself; in normal cavities the energy dissipation in the walls usually prevents such fields from developing; in superconducting cavities, higher order modes have to be artificially damped via so-called HOM couplers).

Superconducting cavities are nowadays constructed and used in several places (USA, Europe, Japan). Figure 10 shows schematically such a cavity, developed and constructed at CERN, for use in the LEP ring. Each cryostat module contains a 4-cell

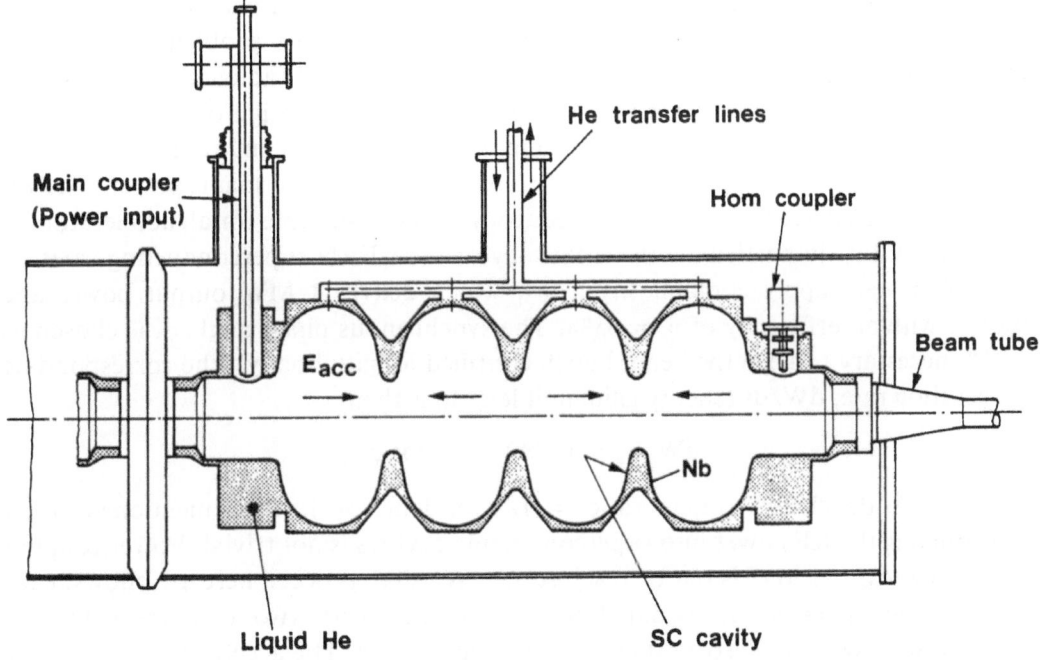

Fig. 10   Simplified sketch of a LEP cryostat module with a 4-cell Nb cavity (f = 350 MHz)

cavity, equipped with a RF feeder and HOM couplers. The liquid He is kept in a stainless-steel container placed around the Nb cavity walls. The cryostat is made of an easily demountable stainless-steel envelope.

Figure 11 shows the Nb cavity; the rounded shape is required to keep the peak electric field within permissible limits (relatively low peak field is a weak point of superconducting cavities).

Fig. 11   4-cell Nb cavity for LEP; f = 350 MHz.

The superconducting cavities are of a type we have analysed previously from the electrodynamics and beam dynamics point of view. It is suited for beam energies ≥ 100 MeV per nucleon.

ACCELERATOR CHAIN FOR MCF

The analysis of various accelerator types carried out so far enables us to specify, tentatively, an accelerator chain for MCF. We are aware of problems that a CW operation poses: power dissipation, beam losses, and induced radioactivity. However, we shall assume that with very good controls and adequate design of components we can master these problems. For the focusing of the beam we intend to use, as much as possible, permanent magnet quadrupoles to reduce the required mains power. The RF frequency is fixed at 350 MHz; it is a frequency which is suitable for all accelerators in the chain and is, in particular, the frequency of the CERN superconducting cavities. The RF power supply contains klystrons, which deliver 1 MW output power and operate with an efficiency of $\eta \simeq 65\%$. The synchronous phase angle ϕ_s is chosen as $-25°$ (necessary to keep the beam bunch confined longitudinally); the corresponding acceleration rate, dW/ds (energy gain/unit length) is then

$$dW/ds = (q/A)e\, E_{acc} \cos \phi_s \,,$$

where q/A is the charge-to-mass ratio (= 1/2 for deuterons). It was mentioned earlier that bringing the RF power into superconducting cavities is not trivial. Power couplers for 200 kW exist; 1 MW is out of reach so far. We shall proceed here as follows: each superconducting cavity is assumed to be equipped with two couplers delivering together 500 kW; one klystron thus supplies two cavities. The accelerating field in the cavity must comply with this power restriction.

To get an approximate idea of the cost of such an accelerator chain, we rely on the CERN list of prices[2,3] extrapolated (maybe somewhat arbitrarily) for the envisaged CW operation.

	RFQ	Alvarez	SC
Cavity [MSF/m]	0.2 → 0.3	0.12 → 0.2	0.25 → 0.4

Note that the cavity prices include vacuum and focusing as well as couplers, tuners, etc. (The sign → stands for extrapolation.)

The cost of RF power is 1.8 MSF/MW; the cost of cryogenics is

$$2.5 \, P^{3/5} \quad [MSF/kW] \, ,$$

where P is the power, in kW, dissipated in the superconducting cavities. The power from the mains for the cryogenic installation is approximately 1 kW for each watt dissipated at 4.2 K.

The price for instrumentation and controls is approximately 10% of the total accelerator cost.

The total length of the accelerator chain comprises the effective accelerator lengths, to which 20% is added to account for the lengths of focusing sections between superconducting cavities, various connections, etc.

Based on the above, the parameters of the MCF accelerator chain are given in Table 1. It should be recalled that a 100 mA deuteron beam is accelerated to 1 GeV per nucleon.

Table 1. Parameters of the MCF accelerator chain

	RFQ	Alvarez	SC
Energy [MeV/nucleon]	0.05 → 1	1 → 100	100 → 1000
Relativistic factor β	0.01 → 0.046	0.046 → 0.43	0.43 → 0.875
E_{acc} [MV/m]		3.2	4.75
Effective length [m]	3	72	405
Shunt impedance R [MΩ/m]		35	8.25×10^5
Power [MW]			
cavity	1.2	20	0.011
beam	0.2	20	180
Cost [MSF]			
cavity	1	15	162
RF power	3	72	324
cryogenics			11

Total power from mains is (note that $\eta = 65\%$):

$$\text{RF power}/\eta + \text{cryogenic power} = 352\,\text{MW} .$$

More than 50% of this power is in the kinetic energy of the beam.

$$\text{Total length is: Effective length } (1 + 20\%) \simeq 575\,\text{m} .$$

$$\text{Total cost is: Accelerator + controls} \simeq 650\,\text{MSF} .$$

CONCLUSION

In this paper some basic facts about accelerators have been reviewed, in order to contribute to the understanding of problems connected with their possible use in MCF. The example of the accelerator chain is to be considered only as the first feasibility analysis. The parameters presented in Table 1 can be used as preliminary guidelines.

This paper would not be complete without a few remarks concerning the energy gained by MCF. Following Ref. 1, we assume that 2.5 deuterons in the beam are needed, on the average, to produce one μ^-. The energy liberated by the fusion of the muon-bound deuteron–triton molecule is 17.6 MeV; after that, the muon is either freed to repeat the cycle or lost by sticking itself on the α particle created in the fusion process. Extrapolating the results of Ref. 1, we assume that the muon may catalyse ~ 300 fusions during its lifetime (2.2 μs).

Putting all the above values together, the power gained by MCF amounts to ~ 200 MW. This equals to the power in the beam, which means that the so-called scientific break—even is feasible. Comparing this figure with the 350 MW from mains, we see that $> 40\%$ of the power is still missing for the technological break-even. It should, however, be mentioned that the beam has lost only a fraction of its energy by impinging on the target. No effort is made, in this paper, to use the remaining beam kinetic energy.

Acknowledgements

U. Amaldi introduced me to the subject of MCF and made several suggestions concerning the layout of this paper. H. Lengeler and Ph. Bernard explained to me the problems connected with superconducting cavities. I would like to express my appreciation to them.

BIBLIOGRAPHY

1. S.E. Jones, Muon catalysed fusion, Nature **321,** 127 (1986).
2. E. Boltezar et al., The new CERN 50 MeV linac, Brookhaven National Laboratory Report, BNL 51134 (1979), p.66.
3. Ph. Bernard, H. Lengeler and E. Picasso, Upgrading of LEP energies by superconducting cavities, Internal Report CERN/LEP-DI/86–29 (1986).

ON THE PRODUCTION OF PIONS AND MUONS FOR THE PURPOSES OF

MUON-CATALYZED FUSION

A. Bertin, M. Capponi, S. De Castro, I. Massa, **M. Piccinini**, M.Poli[+],
N. Semprini-Cesari, A. Vitale, E. Zavattini* and A.Zoccoli

Dipartimento di Fisica dell'Università and
Instituto Nazionale di Fisica Nucleare, Sezione di Bologna,
 40100 **Bologna**, Italy

+ Dipartimento di Energetica dell'Università di Firenze,
and Istituto Nazionale di Fisica Nucleare, Sezione di Bologna,
Italy,

*CERN, 1211 Geneva 23, Switzerland

INTRODUCTION

In order to get the necessary yield of parent particles for muon-catalyzed fusion (μCF) purposes, an adequate production of negative pions represents the most effective mean, provided the pions are released in suitable beam-target collisions. If one takes into consideration the possibility of exploiting μCF for energy applications, the problem of the cost of
the primary pions turns out to be essential.

While direct experimental information on this point is missing, we have performed a set of systematical calculations to investigate the cost of
producing pions with different <u>impinging particle-target</u> configurations.
Following an approach different from those of previous estimates,[1] the present study is based on a Monte Carlo method,[2] the predictions of which were experimentally checked in a recent measurement[3] of the yield of low-energy pions from a high-energy primary proton beam.

The Monte Carlo calculation, which was originally foreseen to study
safety radiation conditions in the neighbourhood of high-energy particle
beams, has an excellent flexibility with respect to different <u>beam-target</u>
configurations, and good reliability as to the production of pions. Nevertheless, the calculation needs an improvement - which is presently under
study - as far as regards the nuclear interactions of pions produced at
low energy until they stop within the target material.

In the present report, we shall discuss the results obtained so far,
pointing out the improvements one should reasonably achieve while developing the described method.

With the aim of optimizing the cost of pion production, the calculations were performed by keeping into account several pieces of information
available from literature.[1] Among these, the most significant ones are that
(a) the primary beam must be neutron-rich (with an energy of few GeV/nuc-
leon); and (b) the targets must also be made of neutron-rich materials.

With this in mind, the systematic study was started with the following
main features:

    i)particle beams considered: neutrons, protons.

    ii)primary beam momentum: from 2 to 8 GeV/c.

    iii)targets considered: mainly liquid deuterium and tritium (unless
otherwise specified). The target length was generally fixed at three nuclear interaction lengths of the impinging nucleons.

    iv)cost of one produced pion:this figure, namely the most significant
parameter, was defined as the ratio $C_{\pi^-} = (T \times N)/N_{\pi^-}$ (where T is the
kinetic energy and N the number of incoming nucleons, and N_{π^-} is the number
of negative pions produced by the incoming nucleons).

    v)high-atomic number targets (always having lengths fixed at three
nuclear interaction lengths of the incoming nucleons) were also taken into
account to study the problem of the negative pion emission outside the production target. In this case, the cost of one emitted pion $C_{\pi^-}^e$ will be defined in analogy to C_{π^-} (see point iv)).

The results obtained following the criteria of points i) to iv) are
presented in Figs. 1 and 2.

For different combinations of incoming nucleons and targets, the variation of C_{π^-} with the target diameter is shown in Fig. 1 (for two typical
values of the momentum of the incoming nucleons). Asymptotically, it is
seen that the lowest cost for pion production is obtained for large values
of the target diameters (namely, greater than about 3 m).

The dependence of this optimal C_{π^-} value on the incoming nucleon momentum is given in Fig. 2 a), from which one can easily see that:

    a) when the primary particles are neutrons, the variation of C_{π^-} over
the whole interval of considered neutron energies is less pronounced than
in the case of primary protons. With these beams, in any case, the optimal
momentum of the primary beam turns out to be at about 3.5 GeV/c.

    b) On the absolute scale, the minimum cost ($C_{\pi^-} \cong 1$ GeV) is obtained
for (3-4) GeV/c neutrons impinging on a tritium target.

    c) Concerning more realistic beam-target configuration estimates,
Fig. 2 a) provides a cost $C_{\pi^-} = 1.7$ GeV, in correspondence to 3.5 GeV/c
protons impinging on a deuterium target, and $C_{\pi^-} \cong 1.5$ GeV for 4 (GeV/c)/nu-
cleon deuterons hitting a deuterium target.

It is interesting to compare these results with the ones obtained (e-
ven with a less detailed scanning of the beam momentum scale) on the cost
C_{π^+} of positive pions (defined in the same way as C_{π^-}). The results obtained on C_{π^+} are presented in Fig. 2 b), where the roles of protons and neu-

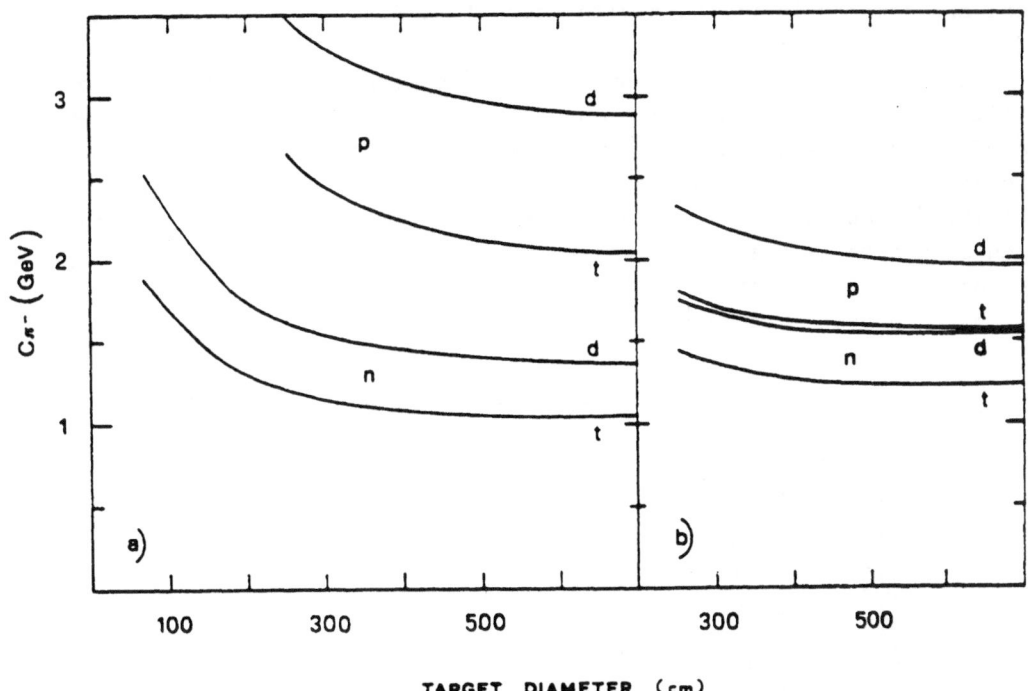

TARGET DIAMETER (cm)

Fig. 1. Calculated cost $C_{\pi-}$ of negative pions produced in neutron
(n, lower curves) and proton (p, upper curves) collisions
on liquid deuterium (d) and tritium (t) targets having different diameters. Impinging particle momentum P_B = 2 GeV/c
(a); P_B= 5 GeV/c (b).

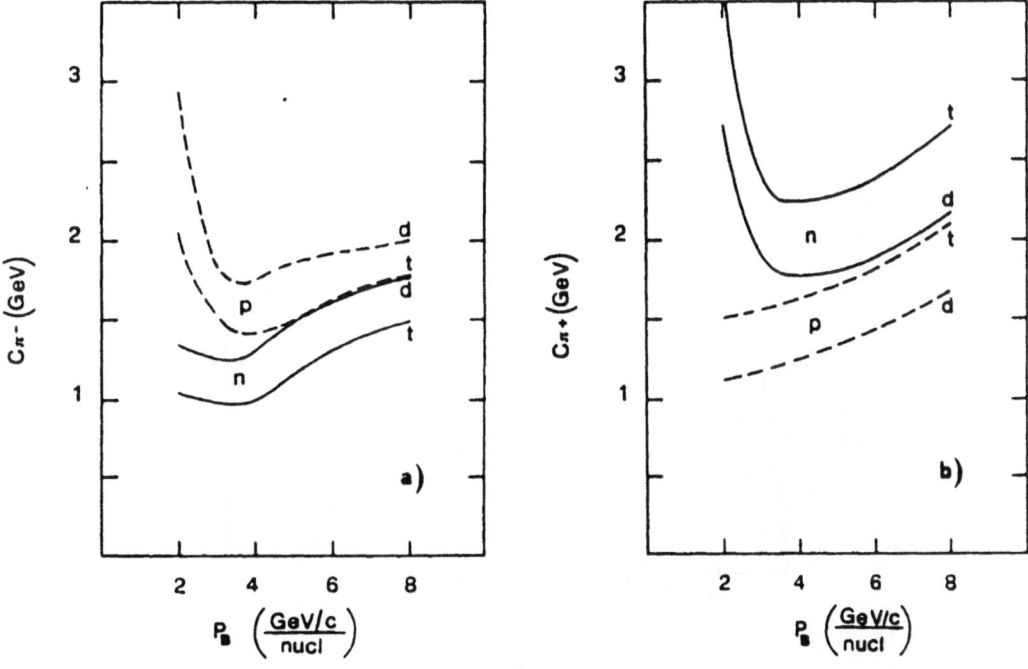

Fig. 2. Calculated cost C_π of negative (a) and positve (b) pions
produced by neutrons (n, continuous curves) and protons
(p, dashed curves) impinging on liquid deuterium (d) and
tritium (t) targets having optimal diameters, as a function of the impinging particle momentum P_B.

153

trons are apparently inverted with respect to Fig. 2 a) (as is to be expected).

In the discussed conditions, due to the target dimensions, one has to recall that a significant fraction of the produced pions (the momentum spectra of which are shown in Fig. 3 for two typical cases) remains inside the production target. As a consequence, it becomes obvious to examine the possibility of considering a suitably dimensioned deuterium-tritium target which also acts as the medium where μCF takes place (once the pions produced inside it decay into muons).

With respect to this target=fuel approach, an alternative possibility

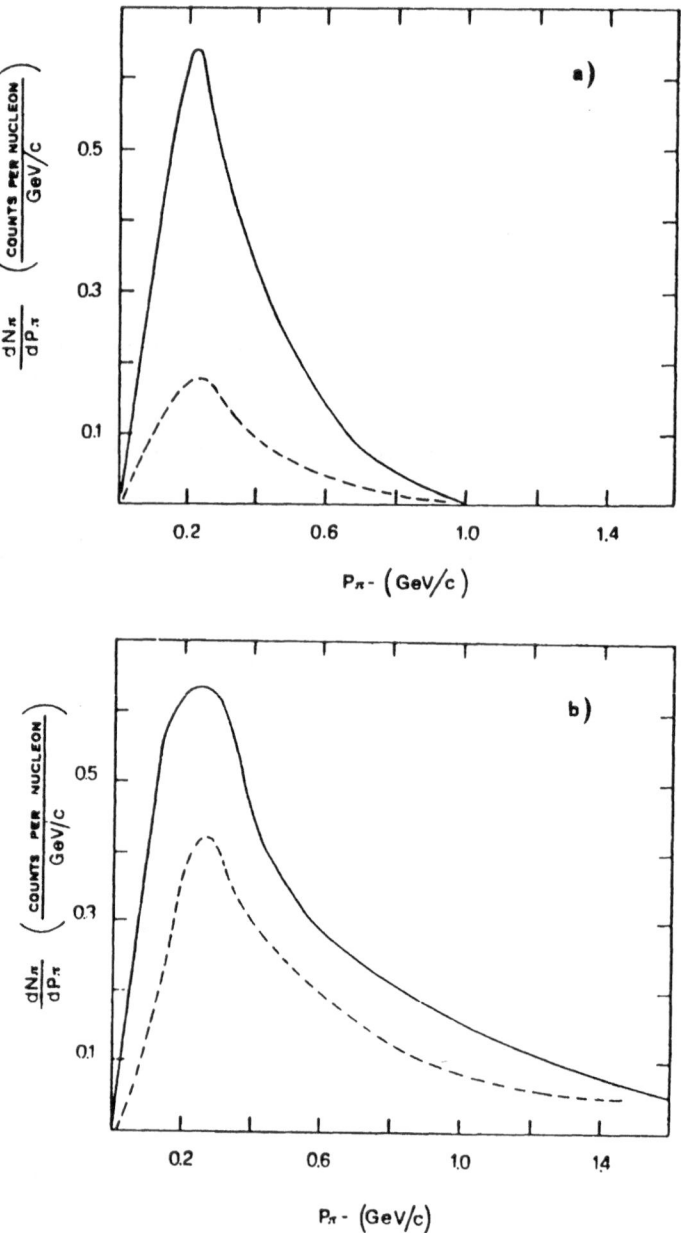

Fig. 3. Momentum distribution of the negative pions produced in p+d (dashed curves) and n+d (continuous curves) collisions at two typical impinging particle momenta P_B: (a): P_B= 2 GeV/c; (b) P_B= 5 GeV/c.

is to consider a device where the three different phases of pion production, pion-muon conversion, and muon catalyzed fusion are distinguished, and occur in different spaces. If this attitude is taken, concerning the first of these three phases, the problem can be examined to find out the optimal target (as far as shape, dimensions and composition are considered) from the standpoint of the negative pion emission <u>outside the target</u> (following its bombardment with nucleons or nuclei having a suitable energy). Some typical results we obtained in this way, both for hydrogen isotopes and for high-atomic number targets (see point v) mentioned above) are reported in Table 1, from which the following pieces of information can be extracted:

i) also in this case, the minimum cost is obtained for neutrons impinging on a tritium target.

ii) The cost of pions produced by medium-atomic number targets exceeds the preceeding lower limit by about 30%, but targets in the solid state compensate this disadvantage with their reduced dimensions, which allow a better control of the flux of released pions.

Table 1.   Cost $C_{\pi^-}^e$ of negative pions released from different targets bombarded by protons (p) and neutrons (n) having different momentum P_B

Target	Beam	P_B (GeV/c)	Diameter (cm)	$C_{\pi^-}^e$ (GeV)
$^{2}_{1}H$	n	2	75	4.1
	p	2	100	7.7
	n	5	150	5.0
	p	5	150	6.3
$^{3}_{1}H$	n	2	60	3.2
	n	5	100	4.1
$^{12}_{6}C$	n	2	5	4.2
	p	2	5	15
	n	5	7	4.5
	p	5	9	6.8
$^{56}_{26}Fe$	n	5	2	5.0
	p	5	2	7.5
$^{184}_{74}W$	n	5	1	6.3
	p	5	1	9.9

For the purposes of μCF, the next point to clarify is how many muons one can obtain from the negative pions produced in the conditions discussed. In what follows, we shall consider the problem of pion-muon conversion, assuming the target=fuel approach which was discussed in the previous Section.

In order to calculate the probability that pions decay into muons inside the medium where they are produced, one has to consider the following significant facts:

(a) once the pions have been taken to rest, after loosing all their energy by ionization, they may undergo nuclear capture. Given the high rate of this process (about 10^9 s^{-1}), one can state in practice that a stopped pion represents a pion lost via the nuclear capture channel. The significant parameter, in this case, is the negative pion range.

(b) The pions produced in the target can disappear also due to inelastic interactions with the surrounding nucleons, since the inelastic cross sections [4] of negative pions against nuclei are not negligible even well below a 1 GeV/c momentum of the incident pions.

(c) In the c.o.m. system, the pions may decay at a rate 3.85×10^7 s^{-1}, providing in this case a negative muon which is effective for the fusion purposes.

We are presently developing the Monte Carlo code in such a way that the negative pion-nucleus interactions at low energies are completely taken into account. In the meantime, and for the case of a deuterium target, we have prepared a simplified Monte Carlo program, which keeps into account points (a) to (c),in order to calculate by numerical methods the fraction of negative pions (produced with a given energy) which succeed in decaying before stopping.

A first set of approximate results were obtained by assuming a constant value of 60 mb for the inelastic cross section concerning the processes mentioned at point (b). These results are shown in Fig. 4, in correspondence to different densities of the deuterium target.

By combining this piece of information with the energy spectrum of the produced pions (see Fig. 3), it is possible to extract out the integral probability of pion decay: as it is seen, at the density of liquid deuterium the decaying pions are only about 15% of the total number. This means that a muon cost of (7-11) GeV would correspond to the typical costs of negative pions mentioned in the previous Section. This situation can be improved by lowering the density of the target where the pions are produced, obtaining for instance a cost of (2-3) GeV per muon produced at a density ten times smaller than the density of liquid deuterium (see again Fig. 4).

CONCLUSIONS

The discussion performed represents an extension of the results of recent measurements[3] on the yield of low-energy pions from a high-energy proton beam, which checked the predictions of the Monte Carlo method. The latter at present is being improved as far as the behaviour of the single particles

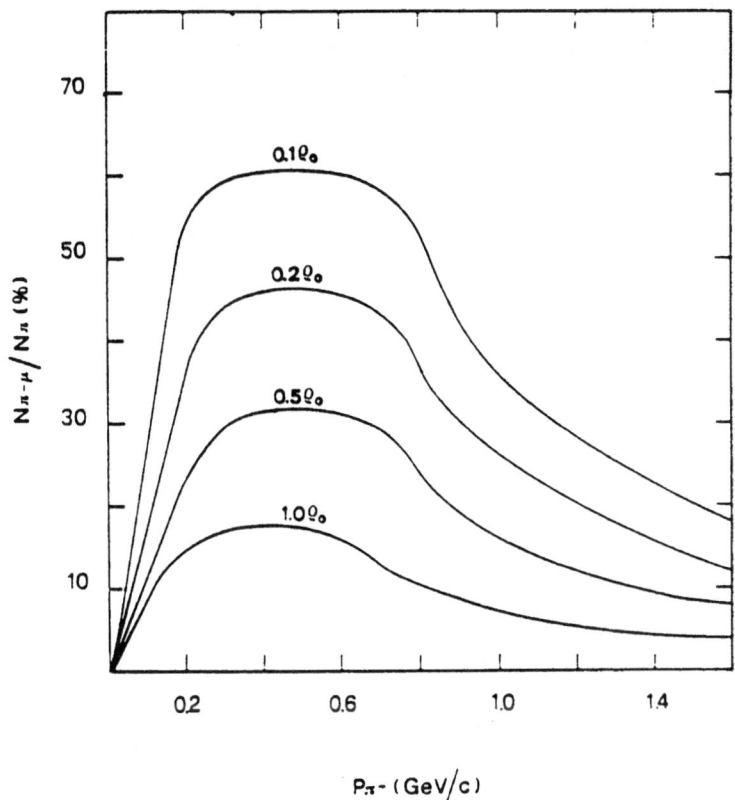

Fig. 4. Fraction of decaying negative pions as a function of their
momentum, when pions are formed within deuterium targets
having different densities. ϱ_o represents the density of
liquid deuterium.

produced at low energies is concerned.

The results obtained so far indicate as a lower limit to the ideal cost
of a negative pion to be used for ~μCF purposes a value of

    (a) C_{π^-} = (1-2) GeV in the target=fuel approach;

    (b) C_{π^-} = (3-5) GeV in the approach where the pions outgoing from
the production target are considered.

As to the corresponding cost of one muon, the calculations are in progress. Preliminary results, referring to the approach of point (a), provide a cost of (7-11) Gev for muons produced in a liquid target, and significantly lower costs for lower-denity targets (see Fig. 4).

REFERENCES

1. Yu. V. Petrov and Yu. M. Shabel'skii, Estimate of the Expenditure of E-
   nergy in Production of π^- Mesons by Nucleons in Light Nuclei, Yad. Fiz.
   30:129 (1979)(English translation: Sov. J. Nucl. Phys. 30:66 (1979)).
   See also S.E. Jones, Muon-catalysed Fusion Revisited, Nature 321:127
   (1986); and references therein.

2. P.A. Aarnio, A. Fasso, H.J. Moehring, J. Ranft and G.R. Stevenson, FLUKA
   85 Code: see FLUKA 86 User's Guide, CERN TIS-RP/168, 21st January 1986.

157

3. A. Bertin, M. Capponi, S. De Castro, I. Massa, M. Piccinini, M. Poli, F. Scuri, N. Semprini-Cesari, A. Vacchi, A. Vitale, E. Zavattini and A. Zoccoli, High Yield of Low-Energy Pions from a High-Energy Primary Proton Beam, Europhysics Letters (to be published).

4. Review of Particle Properties, Phys. Letters 170 B, April 1986. See also: V. Flaminio, W.G. Moorhead, D.R.O. Morrison, N. Rivoire, Compilation of Cross Sections I: π^+ and π^- Induced Reactions, CERN-HERA 83-01, 30 August 1983.

II. FUSION WITH POLARIZED NUCLEI

POLARIZED ADVANCED FUEL REACTORS

R.M. Kulsrud

Princeton Plasma Physics Laboratory
P O Box 451
Princeton, NJ 08544, USA

ABSTRACT

The D-^{3}He reaction has the same spin dependence as the d-t reaction. It produces no neutrons so that if the d-d reactivity could be reduced it would lead to neutron-clean reactor. The current understanding of the possible suppression of the d-d reactivity by spin polarization is discussed. The question as to whether a suppression is possible is still unresolved. Other advanced fuel reactions are briefly discussed.

INTRODUCTION

The two nuclear reactions of most interest for advanced fuel reactors are the d-d reactions d(dp)t and d(dn)^{3}He where d represents a deuteron and the d-^{3}He reaction. As in the d-t reaction, that is the principle reaction of interest in nuclear fusion, these advanced fuel reactions also depend on the spin of the reacting particles.[1]

THE D-^{3}He REACTION

Let me first consider the ^{3}He(dp)He^4 reaction. This reaction is the mirror image of the t(dn)^{4}He reaction since interchanging neutrons and protons in the second reaction changes d to d, t to ^{3}He and ^{4}He to ^{4}He. Like t, ^{3}He has spin 1/2. Therefore, at least to first approximation, one has the same spin dependence of the two reactions.

Both reactions proceed through a resonant state of the corresponding compound nuclei. ^{5}He and ^{5}Li respectively. These states both have 3/2 units of angular momentum and even parity. However, the ^{5}He state contributes more than 99% to the reaction because ^{5}He has no other resonant states close by. Thus, for the d-t reaction the spin dependence is very simple. The cross section can be increased by a factor greater than 1.49. For the ^{5}Li compound nucleus there is another resonant state within 2 MeV of the principal one so that the contribution of the principal resonant state to the d-^{3}He reaction is closer to 95% than to 100% of the total cross section. Thus, the maximum enhancement of the cross section is less than 1.5. (The resonant state is 467 KeV above that of a free ^{3}He, t pair).

Ignoring this small change the d-^{3}He reaction is identical to the d-t reaction in all respects. Further, the depolarization rates of a polarized d-^{3}He plasma are also very similiar because the magnetic moment of ^{3}He is nearly the same as that of t.

The reaction ^{3}He(dp)^{4}He has only charged reaction products so that a thermonuclear reactor based on this reaction alone has no neutrons. This makes it a more desirable reaction than the t(dn)^{4}He reaction because the absence of neutrons makes the reactor much easier to handle and more economical to construct.[2] However, in general, neutrons are not entirely absent because the deuterons are energetic enough to react and the d(dn)^{3}He reaction yields enough neutrons to require some shielding and to produce some activation of the walls surrounding the plasma. If the reactor has both the ^{3}He and the d polarized along the magnetic field, then the spin dependence of the d(dn)^{3}He reaction must be considered.

THE D-D REACTION

Unfortunately the spin dependence of the d-d reaction has not been determined by direct experimental measurements and some controversy surrounds the question. In particular, even the order of magnitude of the d-d cross section for the case in which both spins are aligned is under active debate. For this case the total spin is two and the state is denoted as the quintuplet state. Originally it was argued[3,4] that the contribution of this state to the d-d reaction should be much smaller than other states for two reasons. First, the nucleon spins in each d are parallel and if the d spins are parallel all four spins are lined up. In the final state the spins of nucleons in the ^{3}He nucleus are not lined up so that the spin of one nucleon must flip over. Thus, it was argued that the central nuclear force could not produce a d-d reaction from the quintuplet state. Hence, only the spin dependent nuclear forces could produce such a reaction. Since these forces are

weaker than the central nuclear force, the cross section for the
quintuplet state should be much smaller than the unpolarized cross
section. (The central force can produce reactions out of other spin
states such as the state of antialigned d spins where no spin flip is
required so that the central force produces most of the unpolarized
reaction.) The second reason for the reduction is the Pauli exclusion
principle that makes it difficult for nuclei to approach each other when
the nucleon spins are all parallel.[5]

The argument that the central force plays no role in the reaction
from the quintuplet state is not quite correct.[6] The reason is that
the ^{3}He nucleus has a 4% mixture of a D state in which the nucleons are
all aligned, and the central force can cause a transition from the
quintuplet state to this D state plus a neutron with spin in the same
direction.

I will try to summarize the evidence for and against the
suppression.

The first experimental evidence bearing on the problem can be
abstracted from a paper of Ad'yasevitch and Fomenko[7] in 1969 on the
d(dp)t. They analyzed data, taken at different energies, on the
singular distribution of protons from a polarized beam of deutrons
impinging on a thin unpolarized target. Assuming that the energy
dependence contributed by various channels was only due to variation of
the penetrability of the Coulomb barrier they were able to determine the
various matrix elements for the reaction at 290 KeV bombarding energy.
These matrix elements depended on the initial and final spins, on the
orbital angular momenta L and on the total angular momentum J. The
largest contributions came from initial J,L,S = 1,1,1 and 0,0,0. There
was a small contribution from the initial state 2,0,2 corresponding to
the quintuplet state of interest. Making use of these matrix elements
it is straightforward to calculate relative cross section of the various
spin states and therefore the suppression of the d(dp)t cross section
achieved by polarizing the spins of the d along the magnetic field. The
relative cross section for the quintuplet state, of the proton channel
at 290 KeV bombarding energy is found to be a factor of 30 smaller than
the unpolarized cross section. This energy corresponds to a center of
mass energy of 145 KeV. It is of interest that this center of mass
energy is the energy at which the bulk of the d-d reactions should occur
if the temperature of the d-^{3}He plasma were 50 KeV, a figure often
mentioned as an appropriate temperature for a d-^{3}He reactor. This

suppression refers to the proton channel of the d-d reaction, but it may be presumed that the neutron channel is also suppressed by a similar factor since d(dn)^{3}He reaction is the mirror image of the d(dp)t reaction.

Although these results appear to give a direct experimental answer to the question of the spin dependence of the neutron production, some caution must be exercised in accepting them. The matrix elements that have been determined emperically are sums of matrix elements corresponding to different nuclear forces and it is possible that the smallness of the contribution from the quintuplet state is produced by a cancellation between either these component matrix elements or interference among different angular momentum states. In this case the cancellation may not persist at other energies so that an appropriate energy average over Maxwellian velocity distributions would yield much smaller suppressions.[8]

The next contribution to the determination of the spin dependence of the d-d reaction was due to Hale and Doolen who in 1983 calculated the relative cross section for various initial spin states from an R-matrix analysis of all the data bearing on this four nucleon system.[9,10] Their results are also presented in Table I and it is seen that they differ considerably from the conclusions drawn from Ad'yasevitch and Fomenko's matrix elements. In fact, from their results one sees that the proton reaction is only slightly suppressed in the quintuplet state at 300 KeV. The reason for the discrepancy is not clear. One possible explanation is that the R-matrix analysis makes use of some thick target data in which scattering in the target distorts the time angular distribution and artificially enhances the contribution from the quintuplet state. On the other hand, the assumption by Ad'yasevitch and Fomenko concerning the energy dependence of the matrix elements may be wrong and lead to incorrect matrix elements.

In order to resolve this question Hoffman and Fick in 1984[6] attempted to calculate the cross section from first principles employing a resonating group method. An essential ingredient in their analysis was the inclusion of the 4% D state on the ^{3}He nucleus. They found that the conversion of d d to n^{3}He can occur by the central force through this D state and further that the rate of this process was comparable to the rate of unpolarized d d reactions. The rough implication is that the rate of conversion of d d to this D state with all nuclei aligned is twenty five times stronger than the d-d reaction in other states such as the singlet state where no spin flip is needed. The reason for this surprisingly strong effect is not indicated in their paper although it must represent an important result. The conclusion of their paper is that their calculated results agree well with the R matrix analysis[10] and that there is no appreciable suppression of the d-d reaction by polarizing in the quintuplet state.

This work was criticized by Liu, Zhang, and Shuy in a 1985 Physical Review Letter[5] primarily on the grounds that the angular momentum of individuals nucleons in the ^5S d d channel will not match the angular momentum of the nucleons in the ^3He D state. They calculated the rates using a distorted wave Born approximation and found the contribution from this state was suppressed by one or two orders of magnitude, not enhanced. These results were reported last fall. The discrepancy between these two sets of workers is a factor of 10^3 to 10^4 between the contribution from the central force to the d d quintuplet channel. It is difficult to think that this large discrepancy is due merely to the different methods of calculations.

Zhang, Lui, and Shuy conclude that the principle contribution to the quintuplet d d channel is from the spin dependent forces. These are much smaller than the central force that is the main contributor to the unpolarized reaction. Hence, they feel that the d d reaction can be strongly suppressed (by a factor of at least ten) and that a neutron free D-^3He reactor may be indeed possible if the fuel spin is polarized.

We thus, have two votes for suppression, Zhang, Lui and Shuy[5] and Ad'yasevitch and Fomenko[7] and two votes for no suppression, Hale and Doolen[10] and Hofmann and Fick.[6] Each conclusion is based on a different method. Hence, it was impossible to determine from these papers whether a polarized neutron free reactor is possible.

We now present a table of the relative cross sections for the d(dp)t reaction calculated from Ad'yasevich and Fomenko's matrix elements. This table is taken from Ref. 11. We compare these results with those given by Hale and Doolen calculated from the R matrix analysis. In the table, σ_0 is the unpolarized cross section σ_{ij} is the cross section when one deuteron has a magnetic quantum number $m_s = i$ and the other has $m_s = j$. The ratio σ_{ij}/σ_0 is the enhancement of the d d reactivity. The values at center of mass energies different from 145 KeV are derived by assuming the energy dependence of the matrix elements is proportional to the penetration probability of the Coulomb barrier.

The Hale Doolen results are underlined. The Hale Doolen results have been averaged over angle. E_{cm} represents the center of mass energy. Only the σ_{11} and σ_{00} cross sections apply to a thermal situation. The cross sections σι ανφ σι/ι ψοθθεσρονφ το βεα{ φθιχεν σ οτε{σ~ θ is the angle between the relative motions of the deutrons and the magnetic field.

TABLE I

E_c(KeV))	σ_{11}/σ_0	σ_{00}/σ_0	σ_{10}/σ_0	σ_{1-1}/σ_0
25	0.067	2.27	$0.53-0.25\sin^2\theta$	$2.25+0.5\sin^2\theta$
	<u>1.15</u>	<u>1.17</u>	<u>0.914</u>	<u>0.933</u>
50	0.055	1.85	$0.85-0.4\sin^2\theta$	$1.43-0.66\sin^2\theta$
	<u>1.03</u>	<u>1.04</u>	<u>0.98</u>	<u>0.99</u>
150	0.035	1.20	$1.34-0.66\sin^2\theta$	$1.19+1.32\sin^2\theta$
	<u>0.933</u>	<u>0.93</u>	<u>1.04</u>	<u>1.03</u>

From this table one sees the large discrepancy in results for the quintuplet state cross section and the corresponding suppression of the D-D reactions. It is probably necessary to perform a direct experiment to resolve this discrepancy.

If instead of suppression one wishes to enhance the D-D cross section as in a D-D reactor or in a catalyzed D-D reactor one should polarize the deuterons in the m = 0 state. The corresponding cross sections are predicted to be enhanced by as much as a factor of two at lower energies, based on the extrapolation of the Ad'yasevitch results. Again the enhancement is not borne out by the R-matrix calculations. Further it should be noted that at the energies of interest for a D-D reactor E_c(KeV)) $\approx$ 150, even the first enhancement is reduced.

OTHER ADVANCED FUEL REACTIONS

Finally, it is possible to modify other advanced fuel nuclear reactivities. A possible reaction of interest to fusion is the ^{6}Li(p,^{3}He)^{4}He reaction. It proceeds through a 5/2$^-$ resonance state of ^{7}B$_e$. ^{6}Li has spin one so by polarizing the spins of p and ^{6}Li along the confining field one can show that the polarized cross section is fifty percent larger than the unpolarized cross section.

Similarly, the p ^{11}B reaction proceeds through a 2$^+$ resonant compound state of ^{12}C and its cross section can be enhanced by 14/9 ≈ 1.56 by polarizing the spins of p and ^{11}B parallel to the confining field.

CONCLUSION

Of all the possible benefits from spin-polarizing the fusion plasma the most attractive would be the suppression of the neutrons to make a nearly neutron free reactor. It is unfortunate that nuclear physics is not able at this time to tell us definitely whether this is possible. It is hoped that new experiments will lead to a resolution of this question.

ACKNOWLEDGMENT

This work supported by U.S. DoE Contract # DE-ACO2-76-CHO-3073.

REFERENCES

1. R. M. Kulsrud, E. J. Valeo, and S. C. Cowley, Physics of Spin-Polarized Plasmas, Nucl. Fus. 26:1443 (1986).

2. G. W. Shuy, Ali E. Dabiri, and H. Gurol, Conceptual Design of Deuterium - ^{3}He Fuell Tandem Mirror Reactor Satellite/Breeder System, Fus. Technol. 9:459 (1986).

3. E. J. Konopinski, and E. Teller, Theoretical Considerations Concerning the D+D Reactions, Phys. Rev. 73:823 (1948).

4. F. M. Beiduk, J. R. Pruett, and E. J. Konopinski, Theory of the D+D Reactions Part I Analysis of the Energy Dependence, Phys. Rev. 77:622 (1950).

5. J. S. Zhang, K. F. Liu, and G. W. Shuy, Fusion Reactions of Polarized Deuterons, Phys. Rev. Lett. 57:1410 (1986).

6. H. M. Hofmann and D. Fick, Fusion of Polarized Deuterons, Phys. Rev. Lett. 52:2038 (1984).

7. B. P. Ad'yasevitch and D. E. Fomenko, Analysis of the Results of the Investigation of the Reaction D(dp)T with Polarized Deuterons, Sov. J. of Nucl. Phys. 9:167 (1969).

8. G. M. Hale, private communication.

9. G. M. Hale and D. C. Dodder, A = 4 Level Structure from an R-Matrix Analysis of the Four-Nucleon System, in "Proc. 10th Intl. Conf. Few Body Problems in Physics," B. Zeitnitz, ed., Karlsruhe, W. Germany, (1983), p. 207.

10. G. M. Hale and G. D. Doolen, Cross Section and Maxwellian Reaction Rates for Polarized d + d Reactions, Los Alamos National Lab. Report LA-7971-MS (1984).

11. R. M. Kulsrud, H. P. Furth, E. J. Valeo, R. V. Budny, D. L. Jassby, B. J. Micklich, D. E. Post, M. Goldhaber, and W. Happer, Fusion Reactor Plasmas with Polarized Nuclei-II, in Plasma Phys. and Controlled Nucl. Fusion Research (Proc. 9th Intl. Conf. Baltimore 1982) Vol. 2, IAEA, Vienna (1983) p. 163.

APPLICATION OF POLARIZED NUCLEI TO FUSION

R.M. Kulsrud

Princeton Plasma Physics Laboratory
P O Box 451
Princeton, NJ 08544, USA

ABSTRACT

It is shown that the D-T fusion reaction can be modified by polarizing the nuclear spins. The ways in which this improves reactor performance are mentioned and the feasibility of the process of spin polarization of magnetic fusion is discussed.

I. THE D-T NUCLEAR REACTION

It is known that the cross section for a nuclear reaction depends on the relative orientation of the spins of the nuclei just before they react.[1] In general, in a fusing plasma the spins are randomly oriented with respect to each other (unpolarized) so that the appropriate cross section determining the rate of reactions is an average of the spin dependent cross section over all relative spin orientation. Because this averaged cross section is less than the cross section maximized over spin direction, it is in principle possible to increase the nuclear reaction rate by properly orienting the spins of the nuclei to attain this maximum.[2] A similar remark applies to the directionality of the reaction products. For an unpolarized plasma the reaction products are emitted isotropically, but different spin orientations lead to specific directional distributions for the reaction products. By polarizing one may thus control this directionality to some extent and this choice can also produce benefits for the performance of a nuclear reactor.

We illustrate these remarks by reference to one of the most important reactions for nuclear fusion, the deuterium-tritium

reaction. This reaction proceeds through a well defined quantum state
of ^{5}He which lies 107 KeV in energy above the energy of a free deuteron
and free tritium.[3] It has even parity and spin 3/2. One can analyze
the quantum states of the free d and t just prior to a nuclear collision
into angular momentum states in which the angular momentum is nearly all
supplied by addition of the d spin of one, and the t spin of one half.
The two angular momentum states have 1/2 and 3/2 units of angular
momentum. The latter state matches the angular momentum of the resonant
state in the compound nucleus produces a reaction, provided that the
coulomb barrier is penetrated. For the other state the chance of the
reaction is smaller by a factor of about one hundred and can be
neglected.[4] The spin dependence of the cross section is given
essentially by this simple consideration.

The spins of the d and t are defined relative to the confining
magnetic field. Let the probabilities for a deuteron to be in a state
with magnetic quantum number m be d_m (m= -1, 0, 1) and correspondingly
let t_m be the probability for a triton (m = - 1/2, 1/2). Then the cross
section for a d,t reaction can be written as[2,5]

$$\sigma = \sigma_{3/2} \ (a + 2/3b + 1/3c) \qquad\qquad (1)$$

where $\sigma_{3/2}$ is the maximum cross section and

$$a = d_1 t_{1/2} + d_{-1} t_{-1/2}$$

$$b = d_o \qquad\qquad\qquad\qquad\qquad (2)$$

$$c = d_1 t_{-1/2} + d_{-1} t_{1/2} \ .$$

a is the probability for parallel alignment, b for orthogonal alignment
and c for antiparallel alignment. Of course, a + b + c = 1. For random
orientation of spins a = b = c = 1/3, and σ_o = 2/3 $\sigma_{3/2}$. For total
polarization along B, $d_1 = t_{1/2} = 1$, a = 1, and $\sigma = \sigma_{3/2} = 3/2\sigma_o$. For
fractional alignment, say $d_1 = 1 - \varepsilon$, $d_o = \varepsilon$, $t_{1/2} = 1 - \varepsilon'$, $t_{-1/2} = \varepsilon'$,
$\sigma = (3/2 - \varepsilon' - \varepsilon/2 + \varepsilon\varepsilon') \sigma_o$. Further for orthogonal alignment $d_o = 1$,
$\sigma = \sigma_o$ while for antiparallel alignment $d_{-1} = t_{1/2} = 1$ say $\sigma = 1/2\sigma_o$.
The above cases illustrates the possibilities. The cross section ranges
between $1/2\sigma_o$ to $3/2\sigma_o$ and it is possible to decrease the cross section
by 2 or increase it by fifty percent.

A slightly more complicated analysis gives the directionality of the nuclear reaction products of the reactions, the alpha particle and the neutron.[5] These energies 3.5 MeV and 14.6 MeV are much larger than the kinetic energy of the compound nucleus, so for each reaction the alpha particle and the neutron leave with almost opposite directed velocities. Thus, we need only consider the angular distribution of one of them, say the neutron. The orbital state of the neutron-alpha particle system must be even, while the spin of the neutron is one half and that of the alpha particle is zero. Thus, they must be in a $D^{3/2}$ state, the neutron spin being oppositely directed to the orbital angular momentum. The spin state of the compound nucleus can be found from the initial spin states of the d and t. Then, resolving the 3/2 spin state of the compound nucleus, one finds the angular distribution of the n, alpha state. This gives the differential cross section as

$$\frac{d\sigma}{d\Omega} = \frac{\sigma_o}{2\pi} \left[3/4 \ a \ \sin^2\theta + \left(\frac{2}{3} b + \frac{1}{3} c\right)\left(1 + 3 \cos^2\theta\right) \right] \qquad (3)$$

where θ is the angle the neutron makes with the magnetic field. The resultant angular distribution of the neutron is symmetric under $\theta \to \pi - \theta$ so the alpha particle has the same distribution. Thus, for the case of parallel alignment the relative distribution of intensity is proportional to $\sin^2\theta$ per sterradian, and the reaction products are emitted roughly perpendicular to B while for the other two cases the relative distribution of intensities is proportional to $1 + 3 \cos^2\theta$ and the emission is primarily along B. If one is concerned about the spin of the neutron as well, one finds the distribution for spins parallel to B is different from that antiparallel to B and more complicated than Equation 3. However, it is worth remarking that for perfect alignment of the d and t the neutron is totally polarized for $\theta = \pi/2$.

II. DEPOLARIZATION

We now see that it is possible to increase the reactivity of a fusing plasma by fifty percent by polarizing the spins. However, it is important that the plasma once polarized remain polarized since there does not seem to be any way to polarize the plasma once it is ionized. Thus, it is necessary in case of a magnetically confined fusion reactor to polarize the plasma outside the reactor. If the depolarization were at all rapid in the plasma, then the mean polarization of the spins would be low and no change in reactivity would be attained.

Therefore, it is important to ascertain what physical processes can lead to depolarization of the plasma. There is a magnetic interaction of the spins with the confining field of the plasma but this is very small. The magnetic moments of the deuteron and tritium are

$$\mu_D = 0.86\mu_N$$

$$\mu_T = 5.94\mu_N$$

(4)

where $\mu_N = e\hbar/2m_p c = 6 \times 10^{-24}$ ergs/gauss. Thus, in a 10^5 gauss field the difference in the spin states is $\approx 2 \times 10^{-6}$ eV and is even smaller for the deuteron. These energies are far too small to produce any effect on the distribution of spins. Collisions should lead to randomization of spins. However, because the magnetic moments are very small the collisional rate of depolarization is very small. For example, if the tritons are all initially polarized in the 1/2 state, $t_{1/2} = 1$ then one can calculate the rate of depolarization by electrons[2] as

$$\frac{d\, t_{1/2}}{dt} = - n_e \langle \sigma v \rangle$$

(5)

where

$$\sigma = 200 \left(\frac{e^2}{m_p c^2} \right)^2 \approx 2 \times 10^{-29} cm^2 \ .$$

(6)

This is the fastest collisional rate. Summing over v one finds that the time scale for depolarization at $T_e = 10$ KeV, $n = 10^{15} cm^{-3}$, is 10^4 sec a very long time. A plasma confined for 10 sec would still have a polarization significantly greater than 99%.

The recognition of this remarkably slow depolarization due to collisions is probably responsible for the burst of interest in polarization as a useful process for fusion that occurred in 1982. The fact that collisions whose interaction energy (≈ 10 KeV) was eleven orders of magnitude larger than the interaction energy of the magnetic moment with the confining field, do not rapidly depolarize a plasma is at first appearance surprising and counter-intuitive. The strong feeling people had that polarization could not survive in a hot plasma led people to discount earlier suggestions[6] that polarization be used to improve the plasma fusion performance.

Other mechanisms that lead to depolarization are magnetic fluctuations that resonate with the precession frequency of the nuclear spins about the magnetic field inhomogeneities in the field, and the interaction of the polarized nuclei with surrounding walls.[5] The question of depolarization by waves will be discussed by Pegoraro.[7]

A possible effect of an inhomogeneous field on the depolarization of D has been suggested by Lodder.[8] The precession frequency of a deuteron is $\Omega_p = \mu B/\hbar = 0.86 \ eB/(2m_pc)$, that is 86% of its cyclotron frequency. Therefore, if the magnetic field has a shear with scale length L the deuteron sees a small oscillation in direction of the field at its cyclotron frequency Ω_{cD} as the deuteron moves in its gyrotron orbit of radius ρ. This near resonance with the precession frequency leads to a small oscillation in spin direction (or state). This oscillation is interrupted by random collisions that affect the gyration orbit and produces a random walk in spin direction, leading eventually to depolarization of the deuteron spin. The time of depolarization τ_L is[5]

$$\tau_L = \frac{2L^2}{\nu_o \pi^2} \frac{(\Omega_p - \Omega_{cD})^2}{\Omega_{cD}^2} \tag{7}$$

where ν_o is the rate of pitch angle scattering. In cgs units

$$\tau_L = \frac{6 \ L_2^2 \ B_4^2 \ T_4^{1/2}}{n_{14}} \ \text{sec.} \tag{8}$$

where L_2 is L in units of 10^2 cm, B_4 is B in 10^4 gauss, T_4 is T in 10^4 eV and n_{14} is n in 10^{14} cm^{-3}. Taking as typical values $B_4 = 5$, $T_4 = 1$, $n_{14} = 3$, and L_2 is 10 we get 5000 seconds. Thus, under standard conditions this process of depolarization is not important.

On the other hand wall interactions of the nuclei during plasma recycling are certainly important. The confinement time for a nucleus is much shorter than the time for it to make a nuclear reaction. A typical nucleus may leave the plasma penetrate the limiter or divertor plate, reside there for tens of milliseconds and then reenter the plasma. Depending on conditions at the limiter or on the pumping speed of the divertor this may happen ten or twenty times before the nucleus suffers a nuclear reaction, or it is finally removed from the confinement chamber and can be repolarized.

If the wall material in which the nucleus resides is metallic the nucleus will become quickly depolarized in only a few milliseconds and will reenter the confinement region as an unpolarized nucleus. Thus, during one "recycling" time the plasma will become depolarized.

Greenside, Budny and Post[9] have analyzed depolarization rates of hydrogen isotopes in various possible wall materials. They find that if a nonmagnetic material with no unpaired electrons is chosen, the depolarization time of a nucleus during its period in the material can be much longer. An attractive wall material is amorphous graphite for which the depolarization time could be as long as from one to five seconds. This is longer than the total residence time of a nucleus in the wall by a factor of fifty to a hundred. However, a graphite wall in a reactor cannot be treated as pure as it will probably contain many hydrogen isotope nuclei at any one time and these nuclei can interact with each other. Also the graphite will suffer damage from neutron bombardment and will gradually become impure. These impurities could also produce depolarization. Some simple experimental tests bolster the conclusions of Greenside, Budny and Post[7] that pure amorphous graphite (unloaded by hydrogen isotopes) does not rapidly depolarize hydrogen.[9] Unfortunately it is difficult to draw conclusions about the amount of depolarization that is produced by wall interactions with amorphous graphite under actual reactor conditions.

III. ADVANTAGES OF SPIN POLARIZED FUSION

What advantages in reactor performance may we expect if polarizing a large plasma proves feasible? Are these advantages worth the cost and effort of polarization of the plasma?

First one must distinguish between the cases of D-T fusing plasmas and other "advanced fuel" plasmas. The advantages to be gained by polarizing advanced fuel plasmas is more uncertain but they appear to be more substantial than those related to D-T fusion. I discuss them in my second lecture.

Second, it is recognized "that the gains from polarizing to enhance the cross section a factor of 1.5 are much larger in inertial confinement fusion[11] than is represented by this figure of 1.5. This is because the nuclear cross section enters to the third power in the required energy of the driver needed to achieve a specified gain. More detailed estimates show that the required energy of the driver may be reduced by as much as a factor of three. The reduction in cost is much

174

more. For this reason spin polarization of the D-T core of the pellet is of considerable interest to the ICF community. On the other hand the difficulties of achieving polarization of tritium at cryogenic temperatures and keeping it for a sufficient time are also considerable, (because of its radioactivity).

Third, the benefits of polarization to magnetic confinement fusion depend on conditions. If the situation is marginal, the gain in benefit from enhancing the nuclear reactivity by 1.5 can translate to a larger gain in marginal situations.[12] For example, one can write the power balance equation for ignition as

$$P_{nucl} = P_{loss} + P_{brem} \tag{9}$$

where P_{nucl} is the nuclear power, P_{brem} is the loss due to Bremsstrahlung and P_{loss} represents all other losses and is represented by the Lawson parameter $n\tau$. Let one be close to the ignition temperature for unpolarized plasma and let the required loss be P_{loss}. Then by Eq. (9) the required limit on the loss with polarization P_{loss}' is given by

$$\frac{P_{loss}'}{P_{loss}} = \frac{1.5 - b}{1 - b} \tag{10}$$

where $b = P_{brem}/P_{nucl}$ (unpolarized). If $b = 1/2$, then the loss can be twice as great as for an unpolarized plasma. Other marginal situations can be envisioned in which enhancement of the cross section by polarization could make the difference between ignition and nonignition.

Once a working reactor is achieved, polarization could best be employed to make the reactor cheaper. The critical question then becomes, is the reduction in cost less than the cost to polarize the plasma. For a lower limit to the reduction in cost one can assume the total magnetic field is reduced by a factor of the fourth root of 1.5 and the plasma is polarized so that, for fixed plasma beta, the nuclear reactivity is unchanged. This corresponds to a savings of about twenty percent in the cost of the magnetic system assuming the cost is proportional to the field squared.[13] Taking the magnetic system of reactor to cost about two hundred million, the savings is about fifty million dollars. On the other hand a proposed reactor system has been optimized using polarization and it has been found possible to reduce the size of the reactor by a third.[14] If the reactor core costs five hundred million, this would correspond to a saving of two hundred million dollars. Thus, a fair figure for the cost saving could be one hundred million dollars.

In order to have a basis to estimate the cost of a polarization, one must have a particular method of polarization in mind. Let us consider optical pumping by a tuned continuous power laser.[15] The number of particles that must be polarized per second depends on whether the wall interaction depolarizes the nuclei. If it does, then each nucleus must be polarized in one confinement time which we take as one second. A rough value for the number of particles in a gigawatt reactor is 10^{22} particles so that 10^{22} particles/sec must be polarized. On the other hand if the nuclei can be safely recycled twenty times without loss of polarization at the wall, then the rate is 5×10^{20}/sec. It is the nature of optical pumping that a photon at the resonance wave length of an alkali (say 8000°A for Rubidium) puts $\hbar$ units of angular momentum into a gas of deuterium or tritium atoms.[15] Since a fully polarized tritium atom has $\hbar$ units of angular momentum while a fully polarized deuterium atom has $3/2$ $\hbar$ units of angular momentum, roughly 3/2 eV must be supplied for each elementary polarization process. For the recycled plasma (5×10^{20}/sec) this means that polarizing lasers with powers of 120 watts are required while for uncycled plasma twenty times this power is required. Currently a one watt laser costs about $50,000 so assuming linear scaling in cost a one hundred watts would cost $5 million and twenty times this would cost $100 million. However, it is likely that these costs could be lowered by a factor of five if there is an incentive such as fusion.

These numbers are really very tentative and at best rough guesses. However, from them it is seems that wall interactions with polarizations are fairly critical to the polarization process being of practical use for magnetic confinement devices using D-T fusion. If recycling is permitted, it appears fairly certain that polarization can be employed to effect a considerable saving in the cost of fusion. Also it could play an important part of the program to achieve ignition.

IV. PRINCETON EXPERIMENTS

Efforts are being made at Princeton to develop these optical pumped polarized ion sources.[15,16]. Although sources have been described by Professor Grueebner[17] I should like to describe very recent results at Princeton which are quite exciting. The optical pumping method for polarizing hydrogen are quite old. However, recently, in connection with the possible application of polarization to fusion, an effort has been made to use lasers to polarize hydrogen in quantity. This is made possible by the tunable dye-laser as a source of intense light at the resonant frequency of the rubidium atom. The method is essentially to polarize a small sample of rubidium mixed with a large quantity of hydrogen. The rubidium then transfers its spin to the electron on the hydrogen atom which shares the spin with the hydrogen nucleus.

These experiments were originally carried out by Knize and Happer.[15,16] They uncovered two difficulties. The first was rapid depolarization on walls; to overcome this they worked at pressures of ten to a hundred torr to reduce diffusion to the walls. The experiment was carried out in a small vessel (100 cm^3) at a temperature sufficient to vaporize rubidium. The second difficulty was that it was difficult to dissociate molecular hydrogen at high pressure so that the densities of atomic hydrogen were only 10^{12} cm^{-3}.

In the late fall of 1985 they decided to overcome these difficulties by disassociating the hydrogen in a separate tube from the polarization tube. The walls of the latter were covered with paraffin which produced much less depolarization. Unfortunately, while the new apparatus was being constructed funding was terminated and the experiment was halted at the end of 1985. No further work was done in 1986 but at the beginning of this year new funding was found and the experiment continued. The apparatus was completed and the hydrogen density was raised by a factor of ten in the first two months.[18] In March the apparatus was improved and hydrogen densities of $10^{15}/cm^3$ are now expected.[18] It is next planned to polarize the hydrogen with their laser and to attempt to polarize deuterium. It appears that an exciting new source of atomic polarized hydrogen is becoming available.

V. CONCLUSION

Progress in the development of the theory and experiments on polarized fusion has continued. At present the prospects for the employment of spin polarized nuclei to aid the development and use of D-T fusion seem considerably brighter than they at first appeared.

ACKNOWLEDGMENT

This work supported by U.S. DoE Contract # DE-AC02-76-CHO-3073.

REFERENCES

1. M. Goldhaber, On the Probability of Artificial Nuclear Transformations and Its Connection with the Vector Model of the Nucleus, Proc. Cambridge Philos. Soc. 30:561 (1934).

2. R. M. Kulsrud, H. P. Furth, E. J. Valeo, M. Goldhaber, Fusion Reactor Plasmas with Polarized Nuclei, Phys. Rev. Lett. 49:1248 (1982).

3. J. P. Connor, T. W. Bonner, and J. R. Smith, A Study of the $H^3(dn)$ He^4 Reaction, Phys. Rev. Lett. 88:468 (1952).

4. R. E. Brown, N. Jarmie, and N. Hardekopf, Nuclear Reactions among the Hydrogen Isotopes at Low Energies, IEEE Trans. Nucl. Sci. 30:1164 (1983).

5. R. M. Kulsrud, E. J. Valeo, and S. C. Cowley, Physics of Spin Polarized Plasma, Nucl. Fusion 26:1443 (1968).

6. E. Medi, At a Meeting of European Fusion, Groupe de Liaison, (Feb. 26, 1963).

7. F. Pegoraro, Depolarization Mechanics, in "Proceedings of the Course/Workshop Muon Catalyzed Fusion on Fusion with Polarized Nuclei," Plenum Press, Erice, Italy (April 3-9, 1987).

8. J. J. Lodder, On the Possibility of Nucler Spin Polarization In Fusion Reactor Plasmas, Phys. Lett. 98A:179 (1983).

9. H. S. Greenside, R. V. Budny, and D. E. Post, Depolarization of D-T Plasmas by Recycling in Material Walls, J. Vac. Sci. Technol. 2:619 (1984).

10. D. J. Leopold, J. B. Boyce, P. A. Fedders, and R. E. Norberg, Deuteron and Proton Magnetic Resonance in a-Si:(D,H), Phys. Rev. B 26:6053, 1982.

11. R. M. More, Nuclear Spin-Polarized Fuel in Inertial Fusion, Phys. Rev. Lett. 51:396 (1983).

12. B. J. Micklich and D. L. Jassby, Implications of Polarizeu Deuterium-Tritium Plasmas for Toroidal Fusion Reactors, Nucl. Tech./Fusion 5:162 (1984).

13. D. L. Jassby, private communication.

14. J. L. Cecchi, private communication.

15. R. J. Knize and J. L. Cecchi, Time Resolved Optical Pumping Study of Hydrogen Discharges, Phys. Rev. A 33:3595, (1986).

16. R. J. Knize and J. L. Cecchi, Initial Results Utilizing Optical Pumping with a Laser to Produce Spin-Polarized Hydrogen, Phys Lett. A 113:255 (1985).

17. W. Grueebner, Sources of Polarized Nuclei, in "Proceedings of the Course/Workshop Muon Catalyzed Fusion and on Fusion with Polarized Nuclei" Plenum Press, Erice, Italy (April 3-9, 1987).

18. R. J. Knize, private communication.

DEPOLARIZATION OF SPIN POLARIZED PLASMAS BY COLLECTIVE MODES

F. Pegoraro

Scuola Normale Superiore
56100 Pisa, Italy

ABSTRACT

A magnetically confined, fusing plasma with coherently polarized spin nuclei can be subject to instabilities due to the anisotropy in velocity space of the distribution function of the charged products of the fusion reactions. This anisotropy results from the polarization of the fusing nuclei. The characteristics of these instabilities depend strongly on the spatial inhomogeneity of the plasma.

In this presentation a D-T plasma with parallel polarized spins is explicitly considered and it is shown that a significant rate of triton spin depolarization can be produced, if adequate amplitudes of the fluctuating magnetic fields are achieved. The extension of these considerations to a D-^{3}He spin polarized plasma is also outlined.

1. INTRODUCTION

As already pointed out in Dr. Kulsrud's lecture, if the nuclear spins in a fusing plasma are polarized in an appropriate manner, the nuclear reactions are modified in such a way as to enhance the performance of the reactor. This possibility was heralded in a letter [1] published in 1982, which was followed by a number of papers reporting on different aspects of the physical processes relevant to polarized plasmas. An extensive list can be found in the recent "Nuclear Fusion" article [2] by Kulsrud and collaborators.

The significance of spin polarized plasmas depends on our understanding and solving a variety of problems which belong to nuclear, atomic and plasma physics. The nuclear physics part aims at determining the spin dependence of the cross section at low energies of the three most relevant fusion reactions (see Table I). Since the dependence of the D–D reaction on the spins of the fusing particles is still not clear, we will restrict ourselves to D–T and to D–^{3}He spin polarized plasmas in this presentation. The spin dependence of the D–T and D–^{3}He reactions is rather similar as T and ^{3}He are mirror nuclei. An exhaustive discussion of these features is given in [2]. We recall that the D–^{3}He reaction is of great practical importance for the achievement of an almost neutron free fusion reactor (the few remaining neutrons being produced by the fusion reactions between the deuterons), but requires higher ignition temperatures and longer plasma energy confinement times than the D–T reaction ("advanced fuel" plasmas are investigated in Ref. [3]).

Table I: Nuclear reactions in a fusing plasma

$$D + T \rightarrow \quad ^4He \ (3.5 \ Mev) + n \ (14.1 \ Mev)$$

$$D + D \rightarrow \begin{cases} ^3He \ (0.82 \ Mev) + n \ (2.45 \ Mev) \\ \\ T \ (1.01 \ Mev) \ + \ p \ (3.02 \ Mev) \end{cases}$$

$$D + {}^3He \rightarrow \quad ^4He \ (3.6 \ Mev) + p \ (14.7 \ Mev)$$

The atomic physics part is aimed at establishing efficient methods for producing coherently polarized fuel nuclei and for bringing them inside the plasma, where collisions and wall effects must be sufficiently weak so as to ensure that the depolarization time is significantly longer than the particle fusion time. These effects have been studied in detail for a magnetically confined plasma (see e.g. [2]). The conclusion was reached that, if appropriate materials can be used, sufficiently long depolarization times may be hoped for. At the same time the necessity was stressed of a direct experimental test under fusion conditions.

Besides assessing the advantages in reaching ignition that can result from the use of polarized fuels, plasma physics is intrinsically relevant to the spin polarization problem since collective plasma phenomena determine the magnitude of the fluctuating magnetic fields. The polarization state of the nuclei is extremely susceptible to the changes in the magnetic field direction which are resonant with its spin precession frequency. Magnetic fluctuations, due to collective modes with frequencies in the range of the nuclei cyclotron frequencies and of their first harmonics, can match the local value of the spin precession frequency (which varies inside the plasma if the magnitude of the equilibrium magnetic field $\underline{B}_0$ is not constant) and lead to spin depolarization on short time scales. The relationship between the values of the relevant ion cyclotron and spin precession frequencies is shown in Table II. The key feature of this depolarization mechanism is that, while magnetic fluctuations at a thermal level can be shown [2] to be insufficient to cause fast depolarization, magnetically confined plasmas are systems far from thermodynamic equilibrium and can support fluctuations well above the thermal level. The dominant rôle of plasma fluctuations, i.e. of the fluctuations of the density, temperature and of the self-consistent electromagnetic fields, is well known to the magnetic fusion community as these fluctuations generate anomalous particle and energy transport phenomena that far exceed those due to elementary interparticle collisions.

Table II: Cyclotron and spin-precession frequencies

Species	Cyclotron frequency	Spin precession frequency
D	Ω_D	$\Omega_D^P \simeq 0.86\,\Omega_D$
T	$\Omega_T = \frac{2}{3}\Omega_D$	$\Omega_T^P \simeq 5.96\,\Omega_D$
^3He	$\Omega_{^3\text{He}} = \frac{4}{3}\Omega_D$	$\Omega_{^3\text{He}}^P \simeq -4.25\,\Omega_D$
^4He (α)	$\Omega_\alpha = \Omega_D$	(not relevant)
H (p)	$\Omega_p = 2\,\Omega_D$	(not relevant)

The role of enhanced magnetic fluctuations was stressed in a letter [4] published in 1983 where a possible self-limiting process of the polarization of a D-T plasma was indicated. This process is associated with the anisotropic distribution in velocity space of the energetic α-particles produced by the fusion reactions. In [4] only the case in which the spins are parallel polarized so as to increase the nuclear cross section was explicitly analyzed. The anisotropy is intrinsic to the polarization process itself and follows from simple considerations of parity and angular momentum conservation. The anisotropic distribution can interact resonantly with magnetosonic waves with frequencies in the ion cyclotron range. This resonant interaction leads to energy being transferred to the waves and thus, ultimately, to enhanced magnetic fluctuations.

In two following papers [5], [6], a more exhaustive analysis of the instability condition was performed taking into account the all-important effects of the plasma inhomogeneity on the occurrence and properties of unstable normal modes. It was found that the actual excitation of unstable normal modes, in the frequency range of the triton precession frequency, depended on a balance between the destabilizing contribution of the energetic, non-thermal, fusion produced α-particles and the stabilizing influence of the resonances at the harmonics of the cyclotron frequency of the less energetic, thermal ions that form the bulk of the plasma. The outcome of the competition between these two effects was found to depend on geometrical factors characterizing the plasma column and, in particular, on its aspect ratio. On the other hand, waves around the deuteron precession frequency were found to be stable in plasmas with finite values of the ratio between the electron pressure and the magnetic field pressure. By estimating the resulting depolarization rate of the triton spins, it was concluded that the magnetic normal modes identified in [5] and [6] if unstable, and possibly even if weakly damped, would pose a serious threat to the viability of polarized nuclear fuels in magnetic confinement configurations.

This presentation will review the analysis and the results obtained in [4, 5, 6] and outline their extension to the case of a D-^{3}He plasma. We will show that the main differences that must be accounted for when extending the results obtained in a D-T plasma are:

i) in the D-^{3}He reaction both fusion products are charged and must be confined inside the plasma. The fusion produced protons are lighter and have energies approximately four times greater than those of the α-particles.

ii) the cyclotron frequency of ^{3}He is approximately twice as great as that of T while its spin precession frequency is a factor 0.71 smaller than that of T and has the opposite sign.

iii) the required ignition temperatures, and thus the temperature of the bulk ions, are higher by a factor close to three.

On the basis of the more efficient ion cyclotron damping that results from the above changes of the relevant parameters, we will argue that magnetic fluctuations in a D-^{3}He plasma are less likely to cause a fast nuclear spin depolarization.

2. HEURISTIC MODEL: HOMOGENEOUS PLASMAS

In a magnetically confined plasma a high energy ion population may drive high frequency collective modes unstable in the ion-cyclotron range of frequency. The presence of non thermal features in the energetic ion distribution function, such as non-monotonicity and/or anisotropy in velocity space, can be the source of the excitation energy that is necessary to sustain the growth of the mode amplitude. A net nuclear spin polarization of a plasma is in itself a non-thermal feature which is magnified by the effects of the occurring fusion reactions. In the D-T (and in the D-^{3}He) reaction the maximum enhancement of the fusion rate is obtained when both the spin of the deuteron and the spin of the triton, (of the ^{3}He nucleus) are aligned either parallel or antiparallel to the equilibrium magnetic field in a state of total spin 3/2. Then the conservation of parity and of angular momentum requires the α-particle and the neutron (proton) produced by the decay of the unstable ^{5}He (^{5}Li) isotope to be in a d state of orbital angular momentum. This leads to an anisotropic distribution of the α-particles and of the neutrons (protons) in velocity space. After summing over the possible orientations of the neutron (proton) spin, the distribution is found to be proportional to $\sin^2\psi$ where ψ is the angle between the particle velocity and the magnetic field direction. The anisotropy is preserved during the first phase of the slowing down of the α-particles and, for the D-^{3}He reaction, of the protons. The anisotropy of the neutrons in the D-T reaction is irrelevant as far as the excitation of unstable magnetic fluctuation is concerned, but plays a rôle in determining the damage caused to the plasma chamber wall by the neutron flux, as discussed in dr. Pedretti's lecture.

In order to gain a simple understanding of the relevant modes and of their damping and destabilizing mechanisms, we first refer to a spatially homogeneous, magnetized plasma with two bulk ion species $i = 1,2$ in which a small population of high energy ions, denoted by the subscript h, is also present. The modes of interest must be electromagnetic modes that can interact resonantly with the high-energy ions and that are not damped, or at most only weakly damped, by the bulk plasma. Thus we consider modes with frequency close to one of the harmonics of the hot ion cyclotron frequency Ω_h. Further, in order to avoid significant parallel electron Landau damping and transit time damping, we restrict ourselves to modes that have a vanishing electric field parallel to the equilibrium magnetic field and that propagate almost perpendicularly to $\underline{B}_0$. Then the modes of interest turn out to be magnetosonic (compressional, fast Alfvén) waves. In a low β plasma with a single ion species (β is the ratio between the kinetic and the magnetic pressure) they obey the dispersion relation

$$\omega^2 = k^2 c_A^2 \quad , \tag{1}$$

where c_A is the Alfvén velocity. If the plasma has two ion components, the ion cyclotron motion leads to the more structured dispersion relation

$$\frac{\omega^2}{k^2 \bar{c}_A^2} \times \frac{\omega^2 - \bar{\bar{\Omega}}^2}{\omega^2 - \Omega_{Hy}^2} = 1 \tag{2}$$

which is derived (see e.g. [4,6]) using the cold plasma conductivity tensor. Here $\bar{\bar{\Omega}} = \alpha_1 Z_1 \Omega_2 + \alpha_2 Z_2 \Omega_1$ is the cutoff frequency, $\Omega_{Hy} = (\Omega_1 \Omega_2 \bar{\bar{\Omega}}/\bar{\Omega})^{1/2}$ is the hybrid resonance frequency with $\Omega_{1,2}$ the cyclotron frequency of the two bulk ion species, $\bar{\Omega} = \alpha_1 Z_1 \Omega_1 + \alpha_2 Z_2 \Omega_2$, $\alpha_{1,2}$ is the ratio between the particle densities of the two ion species and the electron density ($\alpha_1 Z_1 + \alpha_2 Z_2 = 1$) and $Z_{1,2}$ are the corresponding atomic numbers. The relevant Alfvén velocity is given by $\bar{c}_A = (\Omega_e \bar{\Omega})^{1/2} d_e$ with $d_e = c/\omega_{pe}$ the electron inertial skin depth and Ω_e the electron cyclotron frequency. Taking $\Omega_2^2 \geq \Omega_1^2$, we have $\Omega_1^2 \leq \Omega_{Hy}^2 \leq \bar{\bar{\Omega}}^2 \leq \Omega_2^2$. For either $\omega^2 \gg \Omega_2^2$, high frequency range, or $\omega^2 \ll \Omega_1^2$, low frequency range, Eq. (2) reduces to the form given by (1). In these frequency ranges the waves are elliptically polarized. In the intermediate frequency range $\Omega_2^2 > \omega^2 \geq \Omega_1^2$, $\omega^2 > \Omega_1^2$, their polarization changes from linear ($\omega^2 = \Omega_{Hy}^2$, $\omega^2 = \alpha_1 Z_1 \Omega_2^2 + \alpha_2 Z_2 \Omega_1^2$) to right circular ($\omega^2 = \Omega_1^2$, Ω_2^2) to left circular ($\omega^2 = \bar{\Omega}^2$). In the frequency interval $\Omega_{Hy}^2 < \omega^2 < \bar{\bar{\Omega}}^2$ the waves are evanescent: for $\omega = \bar{\bar{\Omega}}$, $k^2 = 0$ (cut off frequency), for $\omega = \Omega_{Hy}$, $k^2 = \infty$ (hybrid resonance frequency).

The resonant interaction between the compressional Alfvén waves and the high energy ion population and the damping (not included in the cold plasma dielectric tensor) caused by the bulk electrons and ions add an imaginary part γ to the frequency. Without entering into the details of the explicit evaluation of γ, which is strongly affected by the spatial inhomogeneity of the plasma, we write $\gamma = \gamma_e + \gamma_i + \gamma_h$ and examine the magnitude and properties of these three contributions. The electron damping arises from the interaction between the electron (orbital) magnetic moment and the parallel gradient of the mode magnetic field, and turns out to be proportional [6] to the (small) ratio $\beta_e = 4\pi n_e T_e / B^2$,

$$\gamma_e / \omega \propto \beta_e \zeta_e \exp(-\zeta_e^2) \tag{3}$$

where $\zeta_e = \omega / |k_\parallel| v_{the}$, $k_\parallel$ is the parallel wave number and v_{the} is the thermal electron velocity. Waves that propagate almost perpendicularly to the equilibrium magnetic field have been chosen in such a way that ζ_e can be made greater than one. In this case the electron damping rapidly becomes negligible. The damping due to the bulk ions originates from the resonances $\omega = p \Omega_i + \delta\omega_i$ between the waves and the ions at the pth harmonic of the ion cyclotron frequency Ω_i (here $\delta\omega_i$ is the Doppler shift in the particle frame). The modes have perpendicular wave lengths much longer than the mean gyroradius ρ_i of the bulk ions so that the damping due to high order harmonics is small. In fact the damping γ_i^p due to the pth-harmonic is proportional to $[p/(p-1)!]\ (k_\perp^2 \rho_i^2)^{p-1}$ where $k_\perp^2 \rho_i^2 \simeq (\omega/\Omega_i)^2 (v_{thi}/\bar{c}_A)^2 = (\omega/\Omega_i)^2 \beta_i$, $k_\perp$ is the perpendicular wave number and $\beta_i = 4\pi n T_i / B^2 \ll 1$ is the ratio between the bulk ion pressure and the magnetic pressure. Modes in the high-frequency range resonate with the higher harmonics and are only weakly damped by the bulk ions. Both the electron and the ion damping are increasing functions of the bulk particle β-values.

The contribution γ_h of the hot ions to the imaginary part of the mode frequency originates from those particles that satisfy the resonant condition $\omega = p \Omega_h + \delta\omega_h$ where, again, p is the harmonic index and $\delta\omega_h$ is the Doppler shift of the wave seen in the particle frame (in a homogeneous plasma $\delta\omega_h = k_\parallel v_\parallel$, where $v_\parallel = v \cdot \underline{B}_0/B_0$). Since the density ratio n_h/n_e is small, only the resonant contribution of the hot ions needs to be considered. Solving the Vlasov equation [4,6] shows that γ_h/ω is of the form

$$\frac{\gamma_h}{\omega} \propto \frac{n_h}{n_e} \Gamma(\frac{p \Omega_h}{k_\perp v_h}, \langle \frac{\partial f_{ho}}{\partial v_\perp^2} \rangle) \tag{4}$$

Here Γ is a function that depends on the perpendicular mode number $k_\perp$

times the mean gyroradius v_h/Ω_h of the hot ions and on a weighted average $< \partial f_{ho}/ \partial v_\perp^2>$ over the resonant ions of the derivative in the plane perpendicular to $\underline{B}_0$ of the equilibrium distribution function f_{ho} in velocity space. We take f_{ho} to be given by $f_{ho} = \hat{f}_{ho}(\epsilon)F_h(\epsilon,\mu)$, where ϵ is the particle energy and $\mu = m_h v_\perp^2/2B_0$ the (orbital) magnetic moment. Here $\hat{f}_{ho}$ is isotropic and has the characteristic slowing down dependence, [e.g. for the α-particles produced in a D-T plasma $\hat{f}_{ho}(\epsilon) \propto (\epsilon^{3/2} + \epsilon_c^{3/2})^{-1}$ for 0.67 Mev $\underset{\sim}{\backsim} \epsilon_c \underset{\sim}{<} \epsilon \underset{\sim}{<} \epsilon_\alpha = 3.5$ Mev]. The anisotropic part F_h depends on the polarization state of the reacting nuclei. For coherently parallel polarized spins $F_h = \mu B_0/\epsilon$. The derivative of f_{ho} is negative (i.e. it contributes a stabilizing term to γ_h) while the derivative of F_h is positive. In [4] and [6] it is shown that the contribution of the anisotropic part can prevail so that $< \partial f_{ho}/ \partial v_\perp^2>$ is positive, with a growth of order $\gamma_h/|\omega| \sim 10^{-1} n_h/n_e$.

3. MAGNETOSONIC NORMAL MODES IN AN INHOMOGENEOUS PLASMA

The ratio γ_h/ω derived in the previous section scales as n_h/n_e which is, by hypothesis, a small quantity. The rate at which energy is transferred to the modes from the high-energy ion population is therefore slow and the mode energy balance can be significantly altered either by energy convection, which is often an important factor in inhomogeneous configurations, or by changes, caused by the plasma inhomogeneity, both in the bulk plasma and in the resonant interaction between the mode and the hot ions.

In order to ascertain the rôle of energy convection, it must be determined whether magnetosonic waves in the ion cyclotron frequency range exist as localized, regular normal modes in the plasma configuration of interest. Obviously even in the absence of localized regular normal modes, regular perturbations do occur in a inhomogeneous plasma. However, in general these perturbations are less easily destabilized than localized normal modes. In fact, transient perturbations of the wave-packet type eventually leave the region where the plasma contributes a positive growth rate since in an inhomogeneous configuration the magnitude and the sign of γ_h, and therefore the wave packet amplification, varies within the plasma. On the other hand, stationary perturbations depend on an energy flux to the plasma as a boundary condition in order to have regular amplitudes.

The existence and the properties of the normal modes are rather sensitive to the geometry of the magnetic equilibrium configuration. We refer to a toroidal, axisymmetric configuration with a large, toroidal, magnetic field and a (smaller) poloidal field generated by the current flowing in the plasma. The magnitude of the magnetic field is not constant inside the plasma and varies approximately as $1/R$ where R is the distance from the symmetry axis of the configuration. In order to study the effects introduced by the inhomogeneity of the configuration, it will suffice to assume simply that the magnetic surfaces have circular, concentric, poloidal cross sections. Then we adopt a set of coordinates where ϕ and θ are the toroidal and the poloidal angles respectively and r is the radial coordinate, with r_0 the minor radius of the torus and $R = R_0 + r \cos \theta$, R_0 being the major radius of the torus.

A detailed account of the analysis of the mode structure in such a configuration lies outside the scope of this presentation and can be found in Ref. [6] where this rather involved problem is solved in two stages, starting from the second order differential dispersion equation for the component along $\underline{B}_0$ of the perturbed magnetic field. First a periodic, cylindrical model configuration, that is viewed as referring to a torus of large aspect ratio ($R_0/r_0 \gg 1$) and vanishing poloidal field, is analyzed. On the basis of the information obtained from this model, the full dispersion equation in a toroidal configuration is subsequently solved, by a suitable approximation procedure, in the high-frequency range ($\omega^2 \gg \bar{\Omega}^2$). A similar calculation applies to the low frequency range ($\omega^2 \ll \Omega_{Hy}^2$). In these frequency ranges, localized normal modes, with a phase fast oscillating in θ, are found.

More specifically, the main mode properties are as follows: in the cylindrical model, where θ is an ignorable coordinate, the magnetic field is constant and the only surviving inhomogeneity arises from the radial dependence of the plasma density $n = n(r)$, the normal modes have large poloidal mode numbers m and occupy a circular annulus in the poloidal plane centred around $r = \bar{r} \equiv -2(d \ln [n(r)/n_0]/dr)^{-1}_{r=\bar{r}}$ where $n_0 = n(r=0)$. The width Δ of the annulus is given by $\Delta^{-1} = [(2m)^{1/2}/\bar{r}]$ $[3/4 - 1/2(d \ln [n'(r)/n_0]/d \ln [n(r)/n_0])^{1/4}_{r=\bar{r}}$, where $n'(r) = dn(r)/dr$, and is of order $(\Delta/r_0) \sim (d_i/r_0)^{1/2} \sim m^{-1/2}$, where d_i is the ion inertial skin depth. The real part of the mode frequency is given approximately by $\omega_0^2 = \bar{c}_{A0}^2 [n_0/n(\bar{r})] (m/\bar{r})^2$. In the toroidal configuration the magnitude of the magnetic field is proportional to $1/R$ and thus depends explicitly on θ. The resulting θ dependence of the dispersion equation is generally weak for the high and low frequency ranges. The circular annulus that is obtained in the cylindrical model is shifted and distorted, but remains well confined inside the plasma provided the density profile is not too close to a member of a family of "secular" profiles that satisfy the "resonance" condition $\Delta^{-2} = \pm (m\,s)/\bar{r}^2$, with s an integer. When this condition is approximately satisfied, the mode drifts outwards. In addition, for all profiles, a correction $\Delta k_\parallel$ is introduced to the parallel wave number of order $k_\parallel (r/R_0)^2 \cos\theta$ and must be included when computing the electron transit time damping.

This shows that in the high and low frequency ranges, localized normal modes do occur in a toroidal configuration with the exception of narrow regions in the space of density profiles. After estimating the time it takes these modes to drift outwards, and comparing it with the expected mode growth rate as given e.g. in the previous section, we do not consider these regions to be physically significant.

In the intermediate frequency range the dependence of B_0 on the distance from the axis of symmetry of the torus introduces two closely spaced surfaces where $\omega = \Omega_{Hy}$ and $\omega = \bar{\bar{\Omega}}$ respectively. As shown by the "local" dispersion relation (2) the properties of the waves change drastically around these frequencies and thus the perturbation procedure adopted in the high and in low frequency ranges fails. In particular the modes are reflected at the cutoff surface and undergo mode conversion at the hybrid resonance surface. In [6] the conclusion was drawn that in the intermediate frequency range modes localized within a portion of an annulus (in analogy to the high and to the low frequency ranges) are only possible if the cutoff surface is approximately at the middle of the plasma column. However modes that are reflected by the cutoff surface have necessarily large values of the parallel wave number and are thus subject to significant electron transit time damping. On this basis it was concluded that fluctuations with frequencies in the intermediate range cannot attain sufficiently large amplitudes to cause a fast spin depolarization rate.

4. MODE FREQUENCY AND STABILITY IN TOROIDAL D-T (D-^{3}He) SPIN POLARIZED PLASMAS

The frequency ω of the modes that can induce spin flip transitions must be such that the resonance condition

$$\omega = \Omega_{1,2}^P + \ell\,\Omega_{1,2} + \delta\omega_{1,2} \tag{5}$$

is satisfied within the annulus where the modes are localized. Here $\Omega_{1,2}^P$ is the spin precession frequency of the two (fusing) bulk ion species, ℓ is the harmonic index and $\delta\omega_{1,2}$ is the Doppler shift of the waves as seen in the frame of the bulk ions. The combination $\Omega_{1,2}^P + \ell\,\Omega_{1,2}$ is a function of the distance from the symmetry axis since $\Omega_{1,2}^P, \Omega_{1,2} \propto B_0 \propto 1/R$. For the relevant plasma conditions, $\delta\omega_{1,2} \ll \Omega_{1,2}^P$. Moreover the weight of the resonances at the cyclotron harmonics with $\ell \neq 0$ is small, as shown in Sec. 2 in connection with the bulk ion cyclotron damping. Thus the resonance condition effectively reduces to $\omega \sim \Omega_{1,2}^P$. For a D-^{3}He plasma the above conclusion is perhaps too restrictive (higher ignition temperatures are required leading to higher values of β_i) and e.g. the $\ell = \pm 1$ resonance could be important in the spin depolarization process.

For a D-T plasma the resonance condition $\omega \sim \Omega_D^P$ implies that the polarization of the deuterons will not be significantly affected by collective magnetic fluctuations since $\Omega_T < \Omega_D^P < \Omega_D$, that is, since the relevant modes fall into the intermediate frequency range where the electron transit time damping is likely to have a strong stabilizing effect. In a D-T plasma we are thus led to consider modes that can interact with the triton spins. The resonance condition $\omega \sim 5.94\,\Omega_D$ does not fix the frequency uniquely, since Ω_D is a function of R. In order to be able to induce spin-flip transitions of both circulating and trapped[*] tritium nuclei we choose $\omega \sim 6\,\Omega_D(R_0 + \bar{r})$.

In a toroidal plasma the balance between the destabilizing effect of the resonant interaction with the hot ions and of the stabilizing one with the bulk ions, is substantially modified with respect to the simple picture given in Sec. 2 for a homogeneous model. First the Doppler shift $\delta\omega_{1,2}$ acquires a new contribution due to the particle magnetic curvature drift. This results in a change of the resonant surfaces in velocity space and thus, ultimately, in a change in the weighted average $< \partial f_{ho}/ \partial v_\perp^2>$. Secondly, a new source of energy is available to the modes. It originates from the density gradients of the hot ions (fusion reactions are faster in the central part of the column where the plasma is hotter and denser) and contributes a positive term to γ_h. Thirdly, the resonant interaction with both the hot and the bulk ions can occur at different locations

---

[*] We recall that in a high temperature (collisionless) toroidal plasma a finite fraction of the particles does not move freely along field lines but is trapped in a neighborhood of the minimum value of B_0 along a field line. Their motion is thus restricted to the low field side of the plasma column.

through different harmonics. In a D-T plasma if the p = 6 resonance with the α-particles occurs at the low field portion of the annulus ($R \sim R_0 + \bar{r}$), the p = 5 resonance may occur e.g. around the midplane of the column ($R \sim R_0$) and the p = 4 at the high field side of the annulus ($R \simeq R_0 - \bar{r}$). The resonances with the deuterons occur at the same locations, but have a smaller width since the deuterons are less energetic. The resonances with the tritons involve higher harmonics ($\Omega_T = 2/3 \Omega_D$) and are thus less important. This global nature of the resonant interaction has two main consequences. Firstly, cancellations between stabilizing and destabilizing contributions from different resonances with the α-particles can occur. Secondly, even for high-frequency modes, resonances with lower harmonics of the bulk ion cyclotron frequency are involved. This makes the ion cyclotron damping more effective.

The actual location and occurrence of these resonances depend in a rather detailed way on the physical and geometrical characteristics of the plasma column, in particular on its aspect ratio as it determines the value of $\bar{r}/R_0$, once the shape of the density profile is prescribed. The result of the competition between destabilizing and stabilizing terms is thus difficult to assess in general.

In [6] a detailed numerical analysis was performed for a D-T plasma in a reference device with R_0 = 520 cm, r_0 = 130 cm, n_{eo} = 4 $\times$ 10^{14} cm^{-3}, $B_0(R_0)$ = 50 kG, and a quasi-parabolic density profile. The α-particle density was given by $n_{\alpha o}/n_{eo} \simeq 2 \times 10^{-3}$ and had a profile substantially more peaked than that of the electrons. A normal mode with frequency $\omega = 5 \, \Omega_\alpha(R_0)$ was considered and it was found to be unstable with a growth rate $\gamma_\alpha/\omega \sim 3 \times 10^{-5}$ due to both the non-thermal features of the α-particle distribution, i.e. to their anisotropy in velocity space and to the spatial inhomogeneity of their density. The ion cyclotron damping was found to be smaller than γ_α for the considered device, but was shown to become of the same order for $r_0/R_0 \gtrsim \sqrt{2}/5$.

No comparable detailed analysis has been performed for a D-^{3}He burning plasma configuration so as to include consistently the changes in the single particle properties (cyclotron and spin precession frequencies), those in the plasma properties (bulk plasma temperature, beta values) and those in the geometry of the envisaged confinement configuration (aspect ratio e.g.). The following qualitative features can however be identified. Modes that can depolarize the spin of trapped and circulating ^{3}He nuclei, must have a frequency $\omega \sim 4.25 \, \Omega_D \, (R_0 + \bar{r})$, if $\ell = 0$ is chosen in Eq. (5)). For such modes, moving along the annulus towards higher values of B_0, at a distance R from the magnetic axis such that $\omega = 4\Omega_D(R)$, the $p = 4$ resonance with the α-particles (potentially destabilizing) and with the deuterons (stabilizing) is encountered. It coincides with the $p = 3$ resonance with the ^{3}He nuclei (stabilizing), and with the $p = 2$ resonance with the hot protons (potentially destabilizing). At smaller values of R, the $p = 3$ resonance with the α-particles and the deuterons is encountered followed by the $p = 2$ resonance with the ^{3}He nuclei. At even smaller values of R the $p = 2$ resonance with the α-particles and with the deuterons may occur, which coincides with the $p = 1$ resonance with the hot protons. By comparing these resonance possibilities with those occurring in a D-T plasma we observe that: i) two species (α-particles and protons) can give a destabilizing contribution. However, in view of their very different characteristic velocities (the protons are approximately four times faster) it is not clear whether modes can be found for which the contributions of the two energetic species can add up without strong cancellations; ii) the damping resonances with the bulk ions (in particular ^{3}He ions) occur through lower harmonics and are thus more effective. In addition the value of β_1 is expected to be larger. On the other hand, if we take $\ell = 1$ in Eq. (5), the relevant mode frequency is higher, $\omega = 5.58 \, \Omega_D$, and the bulk ion damping consequently smaller. In particular, choosing $\omega = 6\,\Omega_D(R_0 + \bar{r})$, the $p = 3$ resonance with the hot protons can contribute to the mode growth rate. However as stated before, these latter modes are less effective in inducing spin flip transitions.

Since $\Omega_D < \Omega_{3_{He}}$, in a D-^{3}He plasma, modes that can depolarize the deuteron spin $(\ell = 0.86 \, \Omega_D)$ fall in the low frequency range and not in the intermediate frequency range as is the case in a D-T plasma. Thus, localized normal modes with $\omega = \Omega_D(R + \bar{r})$ could be considered. These modes have a p = 1 resonance with the α-particles and with the deuterons and the latter will have a strong stabilizing effect.

On the basis of these qualitative considerations we may argue that in a D-^{3}He plasma magnetic fluctuations are less likely to achieve amplitudes sufficiently large to cause a fast nuclear spin depolarization than they are in a D-T plasma. If confirmed, this result could be important for the development of advanced fusion reactors, as their ignition requirements are considerably eased if the nuclear spins are polarized [7].

5. SPIN DEPOLARIZATION RATE

A simple estimate of the depolarization rate of the triton spins induced by the fluctuating magnetic field of the unstable modes in an ignited D-T plasma with the spins aligned parallel to the equilibrium field, can be obtained by referring to the homogeneous model of Sec. 2. The transition probability w per precession period between the triton spin states is

$$w = (\frac{2\pi}{\hbar})^2 | < -\frac{1}{2} |\mu_T \cdot \underset{\sim}{B}| \frac{1}{2} > |^2 \frac{\delta(\Omega_T^P - \omega)}{\Omega_T^P} \quad , \tag{6}$$

where μ_T is the nuclear magnetic moment of the triton. Then, within the homogeneous model, the depolarization rate $\nu_{dep} = \Omega_T^P w/2\pi$ of the triton spins can be parametrized as [6]

$$\nu_{dep} \underset{\sim}{=} \frac{\pi}{2} \Omega_T^P (\frac{\tilde{B}_\perp}{B_0})^2 (\frac{\Omega_T^P}{\Delta\omega}) \; s^{-1} \tag{7}$$

Here $\tilde{B}_\perp$ is the left circularly polarized perpendicular component of the fluctuating magnetic field[(*)] and $\Delta\omega$ is an effective frequency width, that depends on the expected features of the unstable spectrum.

It is also possible to define $\Delta\omega$ so that it accounts for the effects of the inhomogeneous toroidal configuration analyzed in Secs (3) and (4), where the triton spins interact with the fluctuating magnetic field only when the tritons cross the region of the annulus where $\Omega_T^P \underset{\sim}{=} \omega$. Assuming that this region occurs at the low field side of the annulus and that it has an angular width $\bar{\theta}$, we recover [6] Eq. (6) with $\Delta\omega \underset{\sim}{=} \Omega_T^P \bar{r} \sin \bar{\theta}/(2R_0)$ after averaging ν_{dep} over the poloidal motion of the tritons along field lines. The tritons cross the annulus where the mode is localized as a result of their radial diffusion through the plasma. It is thus convenient to define a spatially averaged depolarization rate $< \nu_{dep}>$ which accounts for the radial width of the mode and is given by $< \nu_{dep}> \sim \nu_{dep} (\Delta/r_0)$.

The depolarization rate $<\nu_{dep}>$ can now be computed in terms of the saturation level $\tilde{B}_\perp$ of the fluctuating fields. This value is not generally known and must be estimated e.g. by relating it to the efficiency of the energy transfer from the hot ions to the unstable waves or in general by invoking some non-linear saturation mechanism. Under reasonable assumptions, as shown in [6], we find that

$$<\nu_{dep}> \tau_{diff} \underset{\sim}{\geq} 1 \tag{8}$$

where τ_{diff} is the radial particle diffusion time. This implies that

───────────

[(*)] The right circular polarized component should be considered in the case of ^3He.

the triton spin polarization is destroyed in a time at least as short as the time it takes the particles to diffuse out of the plasma. Since these particles re-enter the plasma (recycling), and since in practice many recyclings must be considered, the triton depolarization produced by the unstable fluctuations is significant.

6. CONCLUSIONS

In this presentation we have shown that plasma collective effects play a special and very important rôle in determining the depolarization rate of the nuclear spins in a magnetically confined, polarized plasma. These collective processes are intrinsic to a polarized plasma as the enhanced magnetic fluctuations are produced by non thermal features that originate from the spin polarization itself.

We have also shown that, while in a uniform plasma the existence of unstable modes can be assessed simply, in a realistic configuration, the competition between stabilizing and destabilizing contributions seems to depend on a number of finely tuned, geometrical and physical properties of the plasma confinement configuration. This makes it difficult to present a simple and generally applicable assessment of the mode stability. However the complexity of the resulting picture should not be over-stressed. In fact the existence of localized normal modes in a toroidal configuration has been proved. Possibly, this result is even more important than the actual evaluation of the mode growth rate as it shows that the systems possess proper modes of oscillation capable of destroying the polarization of the spins of the nuclei. These proper modes, if not heavily damped, can be readily excited by different mechanisms, such as e.g. non linear processes, not included in this presentation.

Finally we have shown that smaller fluctuation amplitudes can be expected in an ignited D-^{3}He plasma than in a D-T plasma as the damping mechanisms are likely to be more effective. This is a positive result for advanced fuel reactors, where the use of spin polarized nuclei may substantially reduce the value of the temperature and of the confinement time required for ignition.

Acknowledgements

This presentation is based on results obtained in collaboration with B. Coppi, S. Cowley, P. Detragiache and R. Kulsrud.

193

References

[1] R.M. Kulsrud, H.P. Furth, E. J. Valeo and M. Goldhaber, Phys. Rev. Lett. $\underline{49}$, 1248 (1982)

[2] R.M. Kulsrud, E. J. Valeo and S.C. Cowley, Nucl. Fusion $\underline{26}$, 1443 (1986)

[3] B. Coppi, Phys. Scripta, $\underline{T2/2}$ 590 (1982)

[4] B. Coppi, F. Pegoraro and J.J. Ramos, Phys. Rev. Lett. $\underline{51}$, 892 (1983)

[5] B. Coppi, S. Cowley, P. Detragiache, R.S. Kulsrud and F. Pegoraro, Comments Plasma Phys. Contr. Fus. $\underline{9}$, 49 (1985)

[6] B. Coppi, S. Cowley, R. Kulsrud, P. Detragiache and F. Pegoraro, Phys. Fluids $\underline{29}$, 4060 (1986).

[7] B. Coppi and G. Vlad, Massachusetts Institute of Technology Report PTP 82/16 (1982). Unpublished.

UNCOLLIDED NEUTRON FLUENCES IN A MIRROR-LIKE MACHINE
(MFTF-B) WITH POLARIZED DT PLASMAS

C. Di Nicola* and E. Pedretti+

*Fusion Neutronics Consultant
Via A. Ascari 208
00142 Roma, Italy

and

+ENEA, Dip. TIB/FICS
CRE Casaccia
S.P. Anguillarese km 301
C.P. 2400
00100 Roma, Italy

1. INTRODUCTION

According to Kulsrud et al.[1], polarized DT plasmas could be interesting
for a fusion reactor because polarization of deuteron and triton spins would
influence both the fusion rate and the angular distribution of the fusion
products (14 MeV neutrons and 3.5 MeV alpha particles). In particular, with
parallel polarization (in which all the deuteron and triton spins are parallel to the confining magnetic field $\vec{B}$) the fusion cross section would increase by a factor 1.5 and the angular distribution would be described by
$(1/2)\sin^2\alpha$ (where α is the angle comprised between $\vec{B}$ and the line of flight
of either fusion product), which is equivalent to saying that alpha parti-
cles and neutrons would be emitted preferentially in a direction perpendicular to $\vec{B}$. On the contrary, with transverse polarization (in which all the
deuteron spins are normal to $\vec{B}$) the fusion cross section would remain un-
altered and the angular distribution would vary as $(1 + 3 \cos^2\alpha)/2$, which
means that in this case the fusion products would be emitted preferentially
along the direction of $\vec{B}$.

Changes caused by plasma polarization in the confinement of 3.5 MeV alpha particles in tokamaks have been studied by Bittoni et al.[2], who have
found that, for IA $\geqslant$ 5 MAmp (where I is the plasma current and A is the
aspect ratio), the fraction of confined alphas increases (decreases) with
transverse (parallel) polarization by about 12% with respect to the case of
unpolarized plasma.

Variations due to plasma polarization in the distribution of uncollided
14 MeV neutrons on the first wall of toroidal machines have been investigated by Micklich and Jassby[3], who have considered a line-plasma in a rectangular vacuum vessel; by Fubini et al.[4], who have made calculations for the

extended circular plasma in the circular vacuum vessel of the FTU (Frascati Tokamak Upgrade) machine; by Pedretti et al.[5], who have treated the case of JET with circular plasma; and by Di Nicola and Pedretti[6], who have dealt with the more complex case of JET with a D-shaped plasma characterized by parabolic radial profile of ion density and by gaussian radial profile of ion temperature.

Neutronic calculations required by the evaluation of a commercial tokamak reactor with polarized fuels have been performed by Finn et al.[7], who have considered a D-shaped plasma with uniform spatial source distribution.

Concerning the mirror configuration of fusion devices, neutronic calculations aimed at analyzing the effect of plasma polarization on the performance of a mirror reactor blanket have been carried out by Takahashi and Lazareth[8], but, to our knowledge, no calculations of uncollided neutron distributions in mirror machines with polarized fuel have been reported.

For this reason we have thought that it might be of interest to establish how plasma polarization would influence the neutron distribution in a mirror machine, like the MFTF-B facility[9], in which the superconducting magnets, to be used for generating the magnetic field $\vec{B}$ in the central cell, are installed within the vacuum vessel[10], so that some concern about possible damage due to enhanced neutron fluence on the inner side of the magnets might be justified, at least in principle.

An additional reason of interest for this kind of investigation is due to the fact that in mirror reactor concepts the breeding blanket is placed inside the magnets, pretty close to the plasma, where effects of fuel polarization can be expected to be higher than on the far-away magnets.

Thus, instead of calculating the uncollided neutron fluence on the vacuum vessel of the machine (as previously done for toroidal devices), we have determined the neutron fluence on ideal cylindrical surfaces coaxial with the machine, both along a generatrix and along a radius (at the end of the reference surfaces).

After defining the problem in Sect.2 and specifying the calculational formalism in Sect.3, we report the results in Sect.4 and discuss them in Sect.5, where some comments and conclusions are also given.

2. DEFINITION OF THE PROBLEM

With reference to the actual geometry of the MFTF-B machine (see for instance Fig.2 of ref.10) we have schematized the problem as indicated in Fig.1, which is characterized by symmetry with respect to the z axis and by symmetry with respect to the vertical midplane.

In the terminal region, the plasma radius has been assumed to be given by the formula:

$$r_p(z) = \left\{ (1+g)+(1-g)\cos\left[\pi(z+l_v)/l_E\right] \right\} a/2 \qquad (1)$$

where

$$l_v = \begin{cases} -l_C & \text{for} \quad l_C \leq z \leq (l_C+l_E) \\ \\ l_C & \text{for} \quad -l_C \leq z \leq -(l_C+l_E) \end{cases} \qquad (2)$$

196

and g is a numerical factor, such that $1 \geq g \geq 0$.

For the ion density we have assumed the following radial and axial parabolic profiles:

$$n_i(r,z) = n_{io} \, q_n(r,z) \tag{3}$$

where n_{io} is the peak value and

$$q_n(r,z) = \begin{cases} \left[1-(r/a)^2 \right] & \text{for } -l_C \leq z \leq l_C \\[2em] \left[1-(r/r_p(z))^2 \right]\left[1-((z+l_V)/l_E)^2 \right] \end{cases} \tag{4}$$

is the shape factor in which l_V is defined by eq.(2).

For the ion temperature we have adopted the following gaussian profiles:

$$T_i(r,z) = \begin{cases} T_{io} \, \exp\left[-2(r/a)^2 \right] & \text{for } -l_C \leq z \leq l_C \\[2em] T_{io} \exp\left[-2(r/r_p(z))^2 -2((z+l_V)/l_E)^2 \right] \end{cases} \tag{5}$$

where T_{io} is the peak value and, again, l_V is defined by eq.(2).

The assumptions embodied in eqs.(1) to (5) are not based on experimental evidence, even because MFTF-B has not been put into operation yet; these equations have been assumed, for calculational purposes, only as first approximation of what can reasonably be expected in the real operation of a linear machine as peculiar as MFTF-B.

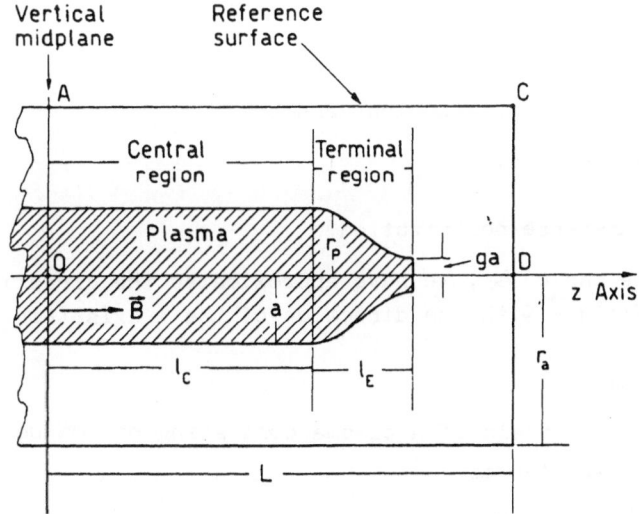

Fig. 1. Geometrical schematizazion of the mirror-like machine.

We thought that it might be interesting to establish how plasma polarization would influence the distribution of uncollided neutrons along the
generatrix AC and along the radius CD of the reference surface.

Concerning the plasma, we have considered two types, i.e. the extended
plasma shown in Fig.1, and a line-plasma (a=0) having the same total length
(l_C+l_E) and the same total neutron yield of the extended plasma.

The reason for considering these two types of plasma is twofold:
i) frequently, the results of neutron measurements are interpreted in terms
   of line-plasma, especially when the neutron emission from the plasma is
   highly peaked on axis;
ii) the simpler case of the line-plasma can be used as a reference, to check
   the results obtained in the more complex case of extended plasma.

It is evident that the line-plasma model is very useful for quickly
obtaining a general idea of the results looked for (notice that, for some
purposes, the line-plasma results are more than acceptable); but, it is
also evident that the extended plasma model, especially when based on appropriate assumptions on the plasma properties, leads to more accurate results.

3. FORMALISM

In general, the uncollided neutron fluence at a point W of given coordinates, due to a plasma of volume V, can be expressed as

$$\psi(W) = (1/4\pi) \int_V (Y_S/X^2) \; f(\alpha) \; dV \qquad (6)$$

where Y_S is the specific neutron yield, X is the distance of the observation
point W from the plasma element dV (located at a point P of coordinates variable within V) from which neutrons are emitted with angular distribution
$f(\alpha)$.

For an unpolarized plasma, the angular distribution of the neutrons is
described by the numerical factor

$$f(\alpha)_{unp} = 1. \qquad (7)$$

For a completely polarized plasma the angular factor is given by[1]

$$f(\alpha)_{par} = (3/2) \; \sin^2 \alpha \qquad (8)$$

in the case of parallel polarization, and by

$$f(\alpha)_{tra} = (1 + 3 \; \cos^2 \alpha)/2 \qquad (9)$$

in the case of transverse polarization.

To define the angle α , we have approximated the direction of the confining magnetic field $\vec{B}$ with the direction of the z axis (Fig.1).

3.1. Extended plasma

The total neutron yield of a plasma consisting of 50% deuterium and
50% tritium can be written as

$$Y = (T_S/4) \int_V n_i^2 \; S_{DT} \; dV \qquad (10)$$

198

where T_s is the duration of the neutron production (assumed to be constant during the shot) and S_{DT} is the DT fusion reactivity that we have determined by the Hively's formula[11]:

$$S_{DT} = \exp(a_1/T_i^c + a_2 + a_3 T_i + a_4 T_i^2 + a_5 T_i^3 + a_6 T_i^4) \qquad (11)$$

in which a_1, a_2, ..., a_6 and c are constants.

For an extended plasma like the one shown in Fig.1, the volume integral appearing in eq.(10) can be calculated as follows. By using the cylindrical coordinates (r, ϑ, z) shown in Fig.2, one has $dV = r\, d\vartheta\, dr\, dz$ and hence, by recalling eqs.(3), (4) and (5)

$$Y = T_s n_{io}^2 (1_C I_C + I_T) \qquad (12)$$

where

$$I = \int_0^a F_C(r)\, dr \qquad (13)$$

and

$$I = \int_{1_C}^{1_C+1_E} dz \int_0^{r_p(z)} F_T(r,z)\, dr \qquad (14)$$

in which the integrands are given by

$$F_C(r) = \left[1-(r/a)^2\right]^2 r\, S_{DT}(r) \qquad (15)$$

for the central region, and by

$$F_T(r,z) = \left[1-(r/r_p(z))^2\right]^2 \left[1-((z+1_v)/1_E)^2\right]^2 r\, S_{DT}(r,z) \qquad (16)$$

for the terminal region.

As shown by eq.(6), the ingredients required for calculating the uncollided neutron fluence $\psi(W)$ are the distance X, the angle α and the specific neutron yield which, in the case of an extended plasma, is simply given by

$$Y_s = T_s n_i^2 S_{DT} / 4. \qquad (17)$$

For an observation point W_G, located on the generatrix AC at a distance d from the midplane xy (Fig.2), one has:

$$X_G^2(r, \vartheta, z) = r^2 + r_a^2 - 2 r_a r \sin\vartheta + (d-z)^2 \qquad (18)$$

and

$$\alpha_G(r, \vartheta, z) = \arccos\left[(d-z)/X_G\right]. \qquad (19)$$

Similarly, for an observation point W_R, located at a distance h from the z axis along the radius CD (Fig.2), one has:

$$X_R^2(r, \vartheta, z) = r^2 + h^2 - 2 h r \sin\vartheta + (L-z)^2 \qquad (20)$$

and

$$\alpha_R(r, \vartheta, z) = \arccos\left[(L-z)/X_R\right]. \qquad (21)$$

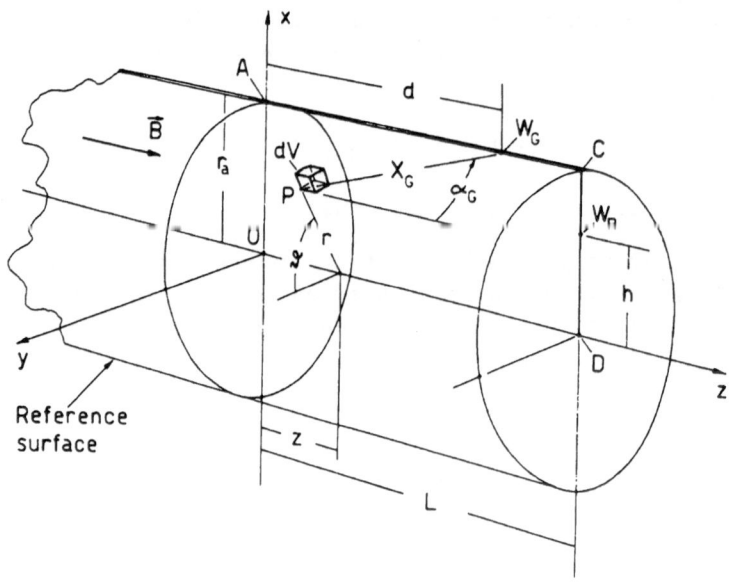

Fig. 2. Geometry used for the extended plasma.

By inserting eqs.(17) and (3) into eq.(6) one obtains, for the uncollided neutron fluence at W_G :

$$\psi(d) = T_s \, n_{io}^2 \, I(d) \, / \, (8\pi)$$
(22)

with

$$I(d) = I_1(d) + I_2(d) + I_3(d)$$
(23)

where

$$I_1(d) = \int_{-l_C}^{l_C} dz \int_{-\pi/2}^{\pi/2} d\vartheta \int_0^a F(r,\vartheta,z) \, dr \, ,$$
(24)

$$I_2(d) = \int_{l_C}^{l_C+l_E} dz \int_{-\pi/2}^{\pi/2} d\vartheta \int_0^{r_p(z)} F(r,\vartheta,z) \, dr \, ,$$
(25)

and

$$I_3(d) = \int_{-(l_C+l_E)}^{-l_C} dz \int_{-\pi/2}^{\pi/2} d\vartheta \int_0^{r_p(z)} F(r,\vartheta,z) \, dr \, ,$$
(26)

in which the integrand is given by

$$F(r,\vartheta,z) = q_n^2 \, S_{DT} \, f(\alpha_G) \, r \, / \, x_G^2 \, .$$
(27)

200

The uncollided neutron fluence $\psi(h)$ at the point W_R is obtained by replacing X_G and α_G with X_R and α_R in the integrand, eq.(27).

3.2. Line-plasma

Under the assumption that the line-plasma, of length $2(l_C+l_E)$, be characterized by uniform neutron emission, one can write:

$$Y_S = Y / \left[2(l_C+l_E) \right] \tag{28}$$

and hence, for the uncollided neutron fluence at the observation point W_G

$$\psi_L(d) = \frac{Y}{8\pi(l_C+l_E)} \int_{-(l_C+l_E)}^{l_C+l_E} \left[f(\alpha_{GL})/ X_{GL}^2 \right] dz \tag{29}$$

where the index L denotes line-plasma,

$$X_{GL}^2 = (d-z)^2 + r_a^2 \tag{30}$$

is the distance of the observation point W_G from the plasma element dz (located on the z axis), and

$$\alpha_{GL} = \arcsin(r_a/X_{GL}). \tag{31}$$

The uncollided neutron fluence $\psi_L(h)$ at the point W_R is obtained by replacing, in eq.(29), X_{GL} and α_{GL} with the corresponding quantities

$$X_{RL}^2 = (L-z)^2 + h^2 \tag{32}$$

$$\alpha_{RL} = \arcsin(h/X_{RL}). \tag{33}$$

Notice that, for an unpolarized plasma, i.e. for $f(\alpha) = 1$, the integral appearing in eq.(29) can be determined analytically. In this case eq. (29) becomes

$$\psi_{L,unp}(d) = \frac{Y}{8\pi(l_C+l_E)r_a} \left\{ \text{arctg}\left[(1-d)/r_a\right] + \text{arctg}\left[(1+d)/r_a\right] \right\}. \tag{34}$$

For an unpolarized plasma, a similar equation can be derived for the uncollided neutron fluence at the observation point W_R.

4. RESULTS OF THE CALCULATIONS

For the geometry of the plasma (Fig.1) we have used the following values, derived from ref.9 :

$l_C = 6.3$ m, $l_E = 0.7$ m, $a = 0.3$ m and $g = 1/6$.

As for the reference surface of length $L = 7.15$ m, we have considered two radii:

i) $r_a = 2.5$ m, corresponding to the inner side of the superconducting magnets;

ii) $r_a = 0.4$ m, which can be thought as representative of the first wall of a breeding blanket, like the one considered for MARS (Mirror Advanced Reactor Study)[12].

As regards the neutron generation capability of the DT plasma, we have assumed the following peak values of ion density and temperature[9]:
$$n_{io} = 5 \times 10^{19} \text{ions/m}^3 \quad \text{and} \quad T_{io} = 15 \text{ KeV.}$$

With $T_s = 1$ sec/shot, the neutron yield, determined by eq.(12), comes out to be $Y = 8.77 \times 10^{16}$ neut/shot.

In performing calculations for both unpolarized and polarized plasmas, we have utilized the property

$$f_{tra}(\alpha) = 2 f_{unp}(\alpha) - f_{par}(\alpha). \tag{35}$$

All the volume integrals, not amenable to analytic calculation, have been determined by the Simpson method. The results of the calculations are shown in Figs.3 to 6, in which we present the uncollided neutron fluences as functions of the independent variable x defined by

$$x = \begin{cases} d & \text{along the generatrix AC} \\ L + (r_a - h) & \text{along the radius CD} \end{cases} . \tag{36}$$

The curves reported in Figs.3 and 4 (extended plasma) were obtained by the expression of $\psi(d)$, eq.(22), in the range $0 \leqslant d \leqslant L$, and by the expression of $\psi(h)$, eq.(22) with X_G and α_G replaced by X_R and α_R, in the range $r_a \geqslant h \geqslant 0$.

Similarly, the curves shown in Figs.5 and 6 (line-plasma) were obtained by the expressions of $\psi_L(d)$ and $\psi_L(h)$, i.e. by eq.(29) with appropriate expressions of X and α.

5. DISCUSSION AND CONCLUSIONS

In the case of extended plasma and of reference surface with large radius (Fig.3), we observe that the neutron fluence due to the unpolarized plasma exhibits the typical behavior that can be expected on the basis of the $1/X^2$ law.

Then, we notice that along the generatrix AC there are two "null" points (N and M, at $x \simeq 3.6$ and 5.2 m, respectively), i.e. points where plasma polarization produces no effect on the uncollided neutron fluence. Between the null points we have

$$\psi_{par} > \psi_{unp} > \psi_{tra} \tag{37}$$

but the differences are negigible from a practical point of view.

Outside the null points interval, the inequalities (37) are reversed and the highest neutron fluence occurs for transverse polarization at the beginning of the generatrix AC, where we have $\psi(0) \simeq 5.4 \times 10^{14}$ neut/m^2/shot, which is about 4% higher than the corresponding neutron fluence due to unpolarized plasma.

These results seem to indicate that, under the assumptions made for the plasma properties, there are no reasons to be concerned about possible damages to the superconducting coils caused by plasma polarization.

In the case of extended plasma and reference surface with small radius (Fig.4), we observe that, in agreement with the $1/X^2$ law, the neutron fluen-

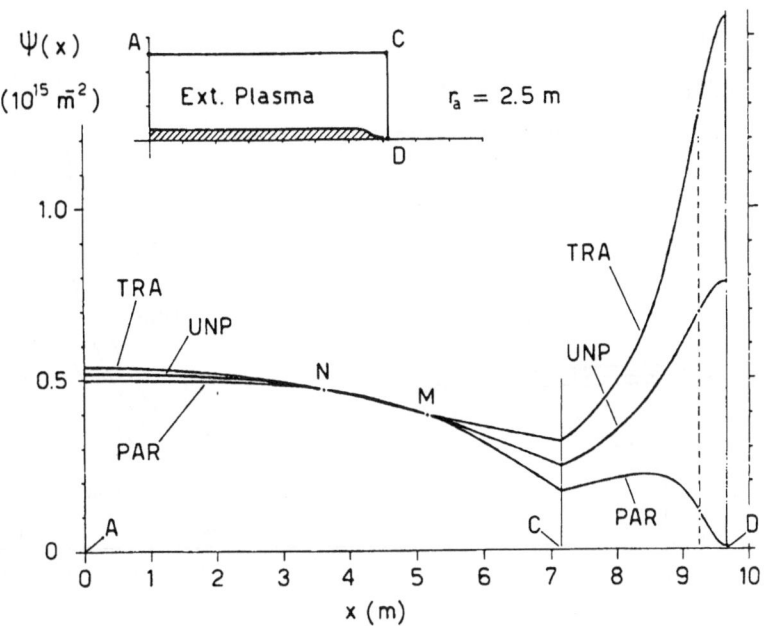

Fig. 3. $\psi(x)$ vs. x on reference surface with large radius for unpolarized and polarized DT extended plasmas (schematic insert in scale).

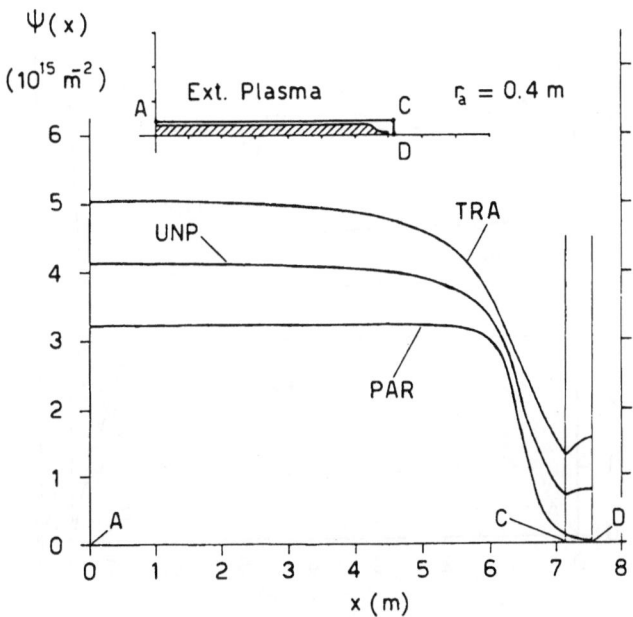

Fig. 4. $\psi(x)$ vs. x on reference surface with small radius for unpolarized and polarized DT extended plasmas (schematic insert in scale).

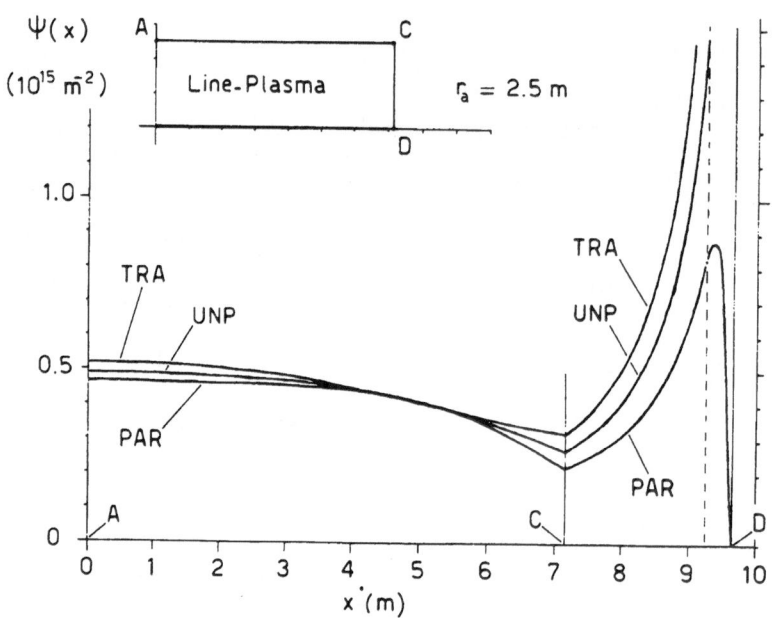

Fig. 5. $\psi(x)$ vs. x on reference surface with large radius for unpolarized and polarized DT line-plasmas (schematic insert in scale).

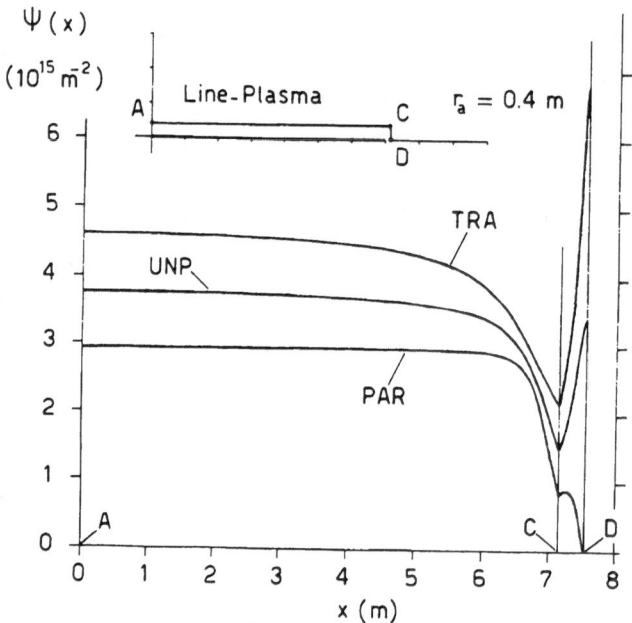

Fig. 6. $\psi(x)$ vs. x on reference surface with small radius for unpolarized and polarized DT line-plasmas (schematic insert in scale).

ces along the generatrix AC are appreciably higher than in the first case. Obviously, the neutron fluences along the radius CD (whose length is 0.4 m) are now the same which were obtained in the last 0.4 m of the previous case.

Along the generatrix AC there are no "null" points and the differences in neutron fluence caused by plasma polarization are always relatively large. These results indicate that the beneficial effect of transverse polarization on the breeding in a blanket placed at $r_a = 0.4$ m might be appreciable.

The results obtained with the line-plasma (Figs.5 and 6) are qualitatively very similar to those obtained with the extended plasma; quantitative differences can be explained in terms of the $1/X^2$ law and by keeping in mind that while neutron production is uniform along the line-plasma, in the terminal region of the extended plasma the neutron production rapidly drops off because of the decreasing axial profiles of ion density and of ion temperature.

For tokamak machines, Finn et al.[7] conclude that "polarized fuels are not likely to be used in commercial reactors".

Although no exhaustive investigation exist for mirrors, it is probable that the same conclusion hold for mirror reactors too.

However, in research devices such as CIT[13], Ignitex[14] and MFTF-B itself, plasma polarization might allow the achievement of conditions (higher temperatures, lower magnetic fields, lower betas, etc.) which could be essential in determing the success of the experiment, otherwise not obtainable.

REFERENCES

1. R.M.Kulsrud,H.P.Furth and E.J.Valeo,"Fusion reactor plasmas with polarized nuclei", Phys.Rev.Lett., 49:1248 (1982).
2. E.Bittoni,A.Fubini,M.Haegi,"Enhancement of alpha-particle confinement in tokamaks in the case of polazided DT nuclei", Nucl.Fusion, 23:830 (1983).
3. B.J.Micklich & D.L.Jassby,"Implications of polarized DT plasmas for toroidal fusion reactors",Nucl.Technology/Fusion, 5:162 (1984).
4. A.Fubini,E.Pedretti,C.Di Nicola,"Calculation of uncollided neutron fluences on the first wall of toroidal machines with polarized DT plasmas, Proc.Workshop on Fusion Blanket Technology, Erice, June 1983.
5. E.Pedretti,A.Fubini & C.Di Nicola,"Calculation of the uncollided neutron fluence on the first wall of the JET machine with polarized DT plasmas", Proc.Symp. on Polar.Phenomena in Nucl.Physics, Osaka, Aug.1985.
6. C.Di Nicola & E.Pedretti,"Effect of deuteron and triton spin polarization on JET with a D-shaped DT fusion plasma", Proc.4th Int.Conf. on Emerging Nuclear Energy Systems, Madrid, June 1986.
7. P.A.Finn,J.N.Brooks,D.A.Ehst,Y.Gohar,R.F.Mattas & C.C.Baker,"An evaluation of polarized fuels in a commercial DT tokamak reactor", Fusion Technology, 10:902 (1986).
8. H.Takahashi & O.W.Lazareth,"Effect of polarized fusion on the neutronics of a mirror fusion reactor",Trans.Am.Nucl.Soc., 44:149 (1983).
9. C.D.Henning,Edit.,"A proposal for a national mirror fusion program plan", UCAR-10042-86,Lawrence Livermore Nat.Lab., Jan.1986.
10. J.W.Gerich,"The vacuum vessel for the tandem mirror fusion test facility", UCRL-5301, Livermore, March 1986.

11. L.M.Hively,"Convenient computational forms for maxwellian reactivities", Nucl.Fusion, 17:873 (1977).

12. J.D.Gordon,"Mirror advanced reactor study engineering overview", Nucl. Eng.& Design/Fusion, 3:119 (1986).

13. J.Sheffield et al.,"Physics guidelines for the Compact Ignition Tokamak", Fusion Technology, 10:481 (1986).

14. R.Carrera,E.Montalvo & M.N.Rosenbluth,"Ohmic ignition in compact tokamak experiments", ·Bull.Am.Phys.Soc., 31:1565 (1986).

THE EFFECTS OF POLARIZED PLASMA NUCLEI IN THE TOKAMAK CONFIGURATION

M. Haegi

Associazione EURATOM-ENEA sulla Fusione
C R E Frascati
CP. 65
00044 Frascati, Roma, Italy

and

E. Bittoni

ENEA C R E "E. Clementel"
Via Mazzini, 2
40138 Bologna, Italy

ABSTRACT

The effects of polarization of deuterium and tritium nuclei in tokamaks are studied. Two cases are analysed: either both nuclei have spins aligned along the magnetic field, or deuterium only is polarized perpendicular to B. The number of the confined alpha particles is calculated and found to be nearly the same in both cases i.e. about 20% higher than the non polarized case, showing thus a moderate improvement. However, for perpendicular spins, the total wall loading due to the fast-alpha losses decreases by a factor of about two, and the inner-wall and blanket loading due to the fast neutrons is also reduced.

The finding [1,2] that the depolarization time of the deuterium and tritium nuclei in a thermonuclear plasma may exceed the (D,T) fusion reaction time by a comfortable margin, together with the availability of new techniques [2,3,4] which are expected to produce amperes of highly polarized deuterium and tritium nuclei, makes it worth while to proceed to quantitative calculations of the confinement of the resulting alpha particles.

Let us here consider the (D,T) reaction only [1]. Two cases are of particular relevance:

a) If the deuterium and tritium nuclei are polarized parallel to the magnetic field **B**, the angular distribution of the emerging neutrons and alphas is $\sin^2\theta$, where θ is the pitch angle relative to **B**. In this case, the (D,T) cross-section is enhanced by a factor of 1.5, but the (D,D) reaction is strongly suppressed.

b) If the deuterium nuclei are polarized perpendicular to the magnetic field, the angular distribution of the emerging neutrons and alphas is $1 + 3\cos 2\theta$. The (D,T) cross-section is unchanged. This tendency of the fusion products to emerge preferentially along the field lines has, in a tokamak device, two desirable consequences: first, the alphas, by following more closely the field lines, improve their confinement and the magnetohydrodynamic stability of the plasma; secondly, the fast neutrons are emitted more tangentially to the curved magnetic field lines, thus decreasing the flux to the inner part of the torus. This is useful because it is desirable to minimize the neutron flux to this "crowded" part which is hard to attain and to cool.

One of the conclusions of a previous paper, taking an isotropic nuclear spin distribution [5] , is that, for relatively peaked profiles and a low aspect ratio, the contained fraction F of the alphas depends mainly on the product IA of plasma current and aspect ratio, and to a lesser extent on I and A independently, and on q and T_0, where q is the safety factor at the edge of the plasma and T_0 the central ion temperature. In this paper, we have repeated the calculations for case a), in which the nuclear spin of deterium and tritium is parallel to B, and for case b), in which the deuterium spin is perpendicular to B. The depolarization time has been assumed much larger than the fusion reaction time.

To calculate the contained fraction F of the alphas which are produced as a function of the product IA, we have assumed, as in Ref. [6], a stationary plasma with Spitzer-like resistivity, radial parabolic distributions, axisymmetric toroidal magnetic field, and no alpha slow-down. The behaviour of F as a function of IA in the case of the unconfined alpha-orbit losses (first-orbit losses) for parallel and perpendicular nuclear spins is shown in Figs 1a and b. The same function is shown in Figs 2a and b for parallel and perpendicular spins, taking into account the sum of first-orbit and all banana-orbit losses. Banana-orbit losses refer to alpha particles born somewhere on this orbit, initially confined. These alphas may then subsequently be lost by pitch-angle scattering, ripple diffusion, MHD plasma instabilities, etc. Loss by pitch-angle scattering onto banana orbits at lower energies is ignored.

In these figures, we see that F is essentially a function of IA in the range of $2.5 \leq A \leq 4$. We have drawn identical conclusions for $2 \leq T_0 \leq 15$ keV and $2.5 \leq q \leq 4$. Thus, for our present purpose, we shall further consider that, within the above range of parameters, as for the isotropic case, the contained fraction F depends mainly on the product IA.

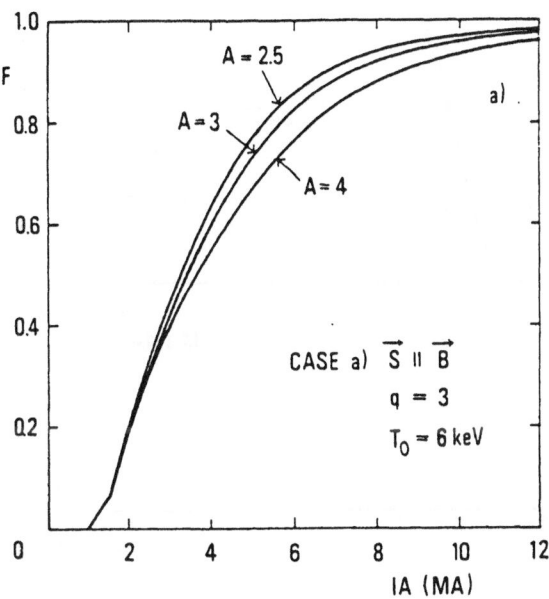

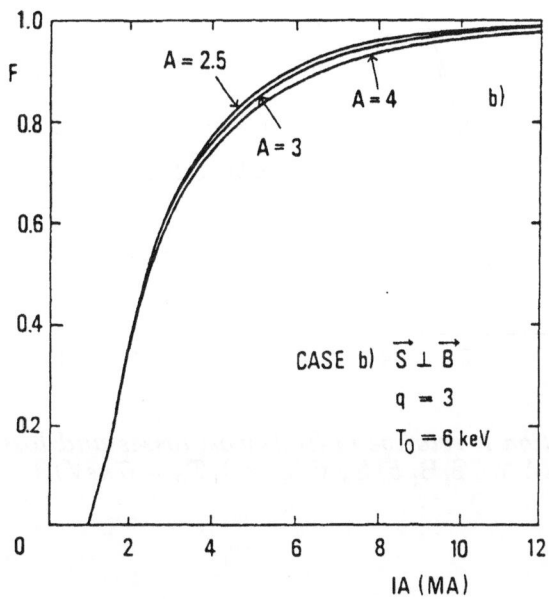

Fig. 1 *Contained fraction F relative to first-orbit losses as a function of IA.*
a) $S_\parallel B$*; b)* $S_\perp B$ *(S = Nuclear spin, q = 3, T_0 = 6 keV)*

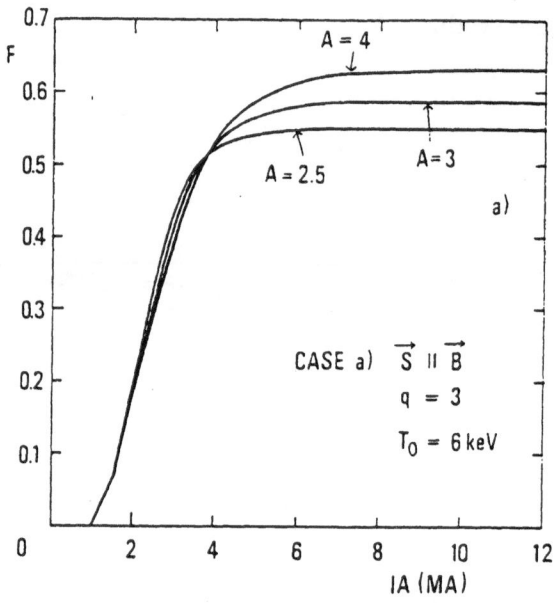

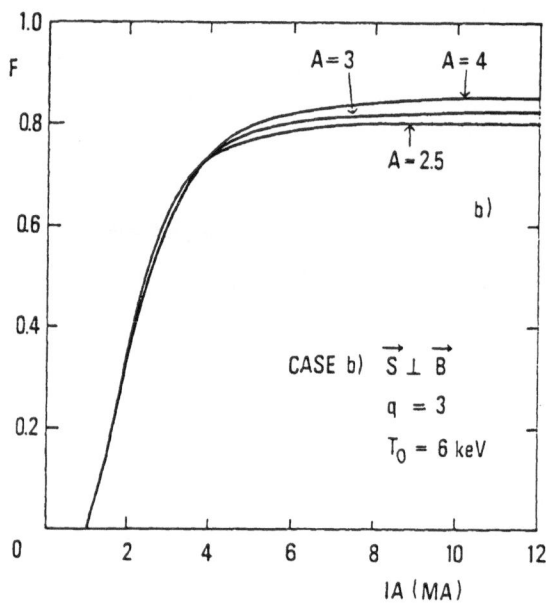

Fig. 2 *Contained fraction F relative to first-orbit losses and losses of all banana orbits as a function of IA. a)* $S_{\parallel} B$, *b)* $S_{\perp} B$ *(q = 3, T_o = 6 keV)*

210

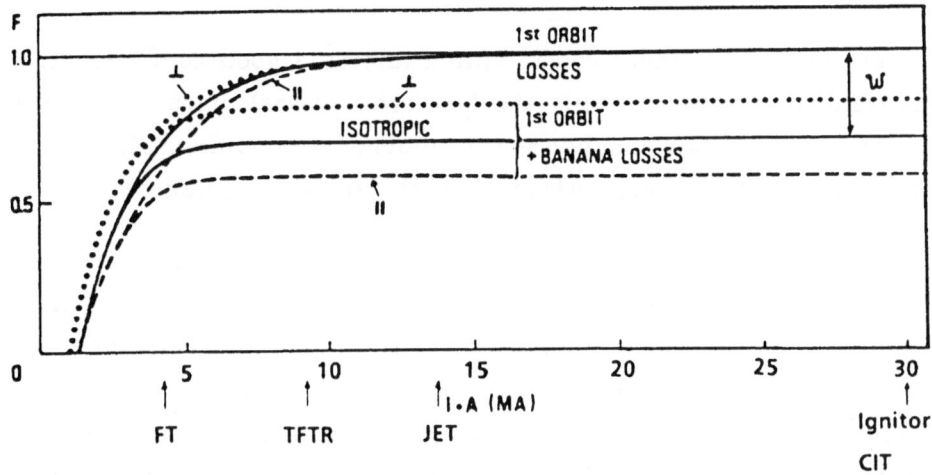

Fig. 3 *Contained fraction F of the alphas for istropic, parallel (‖) deuterium and tritium, and perpendicular (⊥) deuterium nuclear spin (q = 3, A = 3, T_0 = 6 keV). W is the wall loading due to fast alpha*

The results are summarized in Fig. 3 for A = 3, q = 3, T_0 = 6 keV. On the one hand, considering only the first-orbit losses (which is the most optimistic assumption), we find that:

for the parallel and perpendicular-spin case, the losses are very similar to those of the isotropic case.

On the other hand, considering the sum of the first-orbit losses and the losses of all banana orbits, a rather pessimistic assumption is drawn [7,8], and in the present case we have found that:

for the parallel-spin case, the losses increase by about 20% with respect to the isotropic case; for the perpendicular-spin case, at a high value of IA, these losses *decrease by a factor of two*, and at a low value of IA, there is also an improvement: F = 70% is reached for IA = 3.7 MA, instead of ~ 6 MA, as is the case for the isotropic distribution.

CONCLUSIONS

Where the device under consideration affords a good confinement of the alphas (taking into account only the first-orbit losses), parallel spins would enhance the reaction rate and, hence, the number of the confined alphas, by about 1.5. Instead, perpendicular spins lead to no substantial advantage.

In the alternative case of poor alpha confinement (considering both first-orbit and banana losses), parallel spins still enhance the reaction rate by a factor of 1.5, but slightly reduce the alpha confinement; on the other hand, perpendicular spins do not change the cross-section, but increase the confinement by a factor of about 1.2. Hence, the total number of confined alphas is nearly the same for both spin orientations, being about 1.2 times higher than in the isotropic case. For perpendicular spins, however, the total wall loading W due to the fast-alpha losses decreases by a factor of about two, and the inner-wall and blanket loading due to the fast neutrons is also reduced.

REFERENCES

[1]   Kulsrud, R.M., Furth, H.P., Valeo, E.J. Budny, R.V., Jassby, D.L. et al., in *Plasma Physics and Controlled Nuclear Fusion Research* (Proc. 9th Int. Conf. Baltimore, 1982) Vol. 2, IAEA, Vienna (1983) 163;

Kulsrud, R.M., Furth, H.P., Valeo, E.J., Goldhaber, M., *Phys Rev Lett* **49** (1982) 2048

[2]   Kulsrud, R.M., Vaelo, E.J., Cowley, S.C., *Nucl Fusion* **26**, 11 (1986) 1443

[3]   Bhaskar, N.D., Happer, W., McCleland, T., *Phys Rev Lett* **49** (1982) 25

[4]   Cline, R.W., Greytak, T.J., Kleppner, D., *Phys Rev Lett* **47** (1981) 1195

[5]   Bittoni, E., Haegi, M., Santini, F., Segre, S.E., *Nucl Fusion* **22** (1982) 1675

[6]   Bittoni, E., Fubini, A., Haegi, M. *Nucl Fusion* **23**, 6 (1983) 830

[7]   Goldston, R.J., White, R.B., Boozer, A.H., *Phys Rev Lett* **47** (1981) 647

[8]   Tani, K., Takizuka, T., Azumi, M., Kishimoto, H., *INTOR Phase IIA*, 4th Workshop, IAEA, Vienna (March 1982), 136. EUR-FU-BRU-XII 132/82/EDVII (internal document; final report to appear 1984).

SOURCES OF POLARIZED NUCLEI

W. Grüebler

Institute for Intermediate Energy Physics
Eidgenössische Technische Hochschule, ETH-Hönggerberg
8093 Zürich, Switzerland

The basic principles, status, and new methods for the production of neutral nuclear spin polarized hydrogen atoms and of positive and negative polarized hydrogen ions are reviewed. The new developments are described and the performance of different types of sources is discussed. In particular, the improvement of the polarized atomic hydrogen beam density by cooling the atoms to low velocity, and their use for polarized targets and ion sources are discussed. The different ionization processes are presented, and suggested. The presently known techniques suggest H^+ and D^+ beam intensities in the mA range whereas H^- and D^- currents of a few 100 μA seem attainable.

INTRODUCTION

The production of polarized hydrogen ions for the use in accelerators has a long history. The status of the polarized ion sources has periodically been reviewed over the past twentyfive years during Symposia on Polarization Phenomena [1-6] and much can be learned from these review papers about the basic principles and the technical problems. Intense beams of polarized negative hydrogen ions have become particularly important for the use in high energy physics because of the ease with which these beams can be inserted or extracted in cyclotrons and synchrotrons by stripping the electrons in thin foils [7]. The problems and possible developments have further been discussed in the workshop on High Intensity Polarized Proton Sources in Ann Arbor [8], Vancouver [9] and Montana [10] as well as during the Conferences on High Energy Spin Physics in Brookhaven [11], Marseille [12] and Protvino [13]. On the other hand, the suggestion that great benefits can be expected by the injection of polarized hydrogen in fusion reactors [14] shows the general high interest in very intense sources of polarized hydrogen atoms and ions.

PRINCIPLES FOR THE PRODUCTION OF POLARIZED HYDROGEN ATOMS AND IONS

At present, an increasing variety of methods for the production of polarized atoms and ions has been used and more are proposed. However, all the methods have in common that at first an ensemble of polarized neutral atoms is prepared, which will either be used for targets or will be ionized selectively to positive or negative ions.

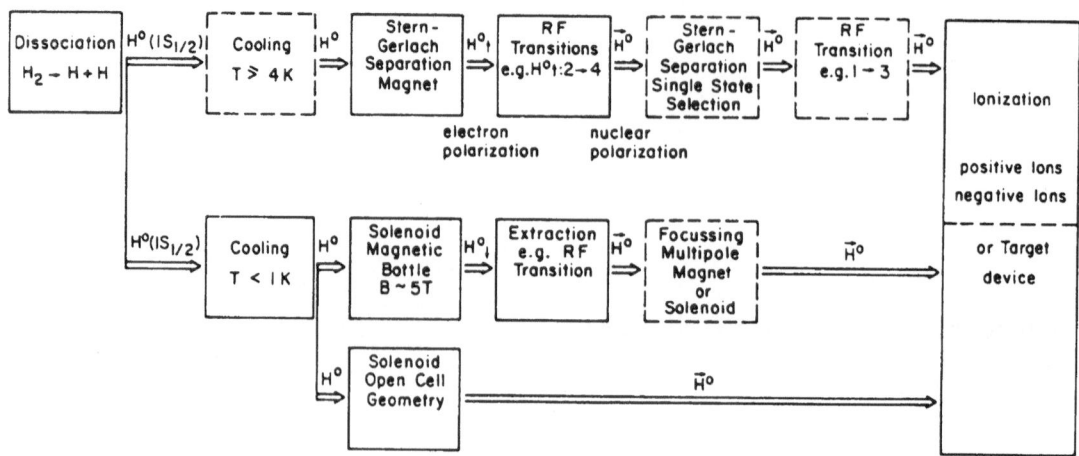

Figure 1: Schematic diagram for the production of polarized hydrogen atoms and ions by the conventional atomic beam method (upper part) and the ultracold method using stable atomic hydrogen (lower part). The boxes with dashed lines are optional.

Ground State Atomic Beam Sources

A schematic diagram of the conventional atomic beam source is shown in the upper part of Fig. 1. Hydrogen atoms are generated by dissociation of molecules in a rf or microwave discharge. An atomic beam of thermal velocity is formed by a nozzle and a skimmer. This beam is polarized in passsing through an inhomogenous magnetic field (usually a sextupole field). This Stern-Gerlach separation of magnetic substates of the hydrogen atoms delivers a fully electron polarized atomic beam. The application of rf transitions between hyperfine states of the atoms selects different nuclear polarization and provides rapid spin reversal. This nuclear polarized neutral beam can be used as a nearly 100 % polarized hydrogen jet target or can be compressed in a vessel. For polarized ions the atoms are converted to positive or negative ions by different processes which are discussed in a later chapter. This simple conventional scheme, which is shown in the upper part of Fig. 1 by the solid line boxes is used in many operational polarized ion sources producing an atomic beam-intensity of $\sim 10^{16}$ atoms s^{-1} and over 100 μA positive ions or about 5 μA negative ions. In the last 5 years substantial effort has been made to cool the atomic beam from room temperature to T > 4 K in order to obtain higher density and intensity of the atomic beam. Further improvement involves the single state selection by a second Stern-Gerlach separation and an additional rf transition to reverse the polarization direction. These new improvements, which are shown in Fig. 1 by the dashed line boxes, are now operational. This progress will be discussed in a later chapter.

In the ultracold method shown in the lower part of Fig. 1, the atoms are first cooled on helium-lined walls to about 0.5 K. The thermal energy of these atoms is much smaller than the potential energy of the atoms in a strong magnetic field. Atoms in state 1 and 2 are completely repelled at the entrance of a superconducting solenoid with B $\gtrsim$ 5 T, whereas the lowest hyperfine states 3 and 4 are pulled into the solenoid. These two states are trapped in the magnetic potential well. The inhomogeneous magnetic fringe field at the entrance selects the spin states with 100 % efficiency. In addition, the strong magnetic field confines the atoms and inhibits spin depolarization due to collisions. The confined atoms, however, have first only a electron polarization. Since the state 4 is a mixed state the atoms in this state recombine first, leaving a residual gas of atoms in state 3 with a high nuclear polarization. In a helium-lined vessel the life-time of these atoms is of the order of hours. The recombination of atoms in state 4 is slow therefore the elimination of these atoms takes a long time (some ten minutes) after filling the magnetic bottle and hence also nuclear polarization is obtained only after this time.

In laboratory devices this stable atomic hydrogen has been observed with a density of 10^{17} atoms cm^{-3} [15]. Densities as high as $4.5 \cdot 10^{18}$ atoms cm^{-3} have been observed after compression of the gas to small volumes [16,17]. Kleppner et al. [18] have proposed several applications of the stable atomic hydrogen gas for polarized targets and ion sources. They suggest to use an open cell geometry and pass a fast beam of Cs atoms through the dense gas in order to produce negative ions. Unfortunately the high density provided in a open cell cannot be used to work effectively for this charge exchange since a density-length product higher than about $2 \cdot 10^{15}$ atoms cm^{-2} causes a secondary charge exchange (e.g. neutralization) and the H^- ions are lost before they can be extracted from the cell. However, even if operated at reduced density, this scheme may represent a significant improvement compared with the conventional colliding beam technique.

Another approach is to pass a fast unpolarized proton beam through the cell. In this case a proton picks up a polarized electron by charge exchange. The polarization is then transferred to the nucleus by passage through a Sona magnet system in which the Zeeman states, but not the hyperfine states, are inverted [19]. Finally, the neutral atoms are converted to negative ions by a second charge exchange collision. This scheme is similar to the so called optical pumped sources (cf. this paragraph).

If the presence of the high magnetic field can be tolerated the open cell can also be directly used as internal target located within an accelerator or storage ring, so that the high energy particle pass through the polarized trapped hydrogen gas repeatedly.

Some applications require extracting the atoms from the cell. Soon after the experimental discovery of the stable gaseous state of atomic hydrogen [20] it was proposed that the atoms stored in the magnetic bottle could be extracted by flipping their electron spin to state 1 or 2 by resonant microwaves [20]. An atom in such a state is expelled from the magnetic bottle. It was suggested that this magnetic acceleration and subsequent focusing either by multipole magnet or solenoid fringe field would result in a polarized, high density and nearly monochromatic atomic beam with high nuclear polarization. Such a beam could be used for polarized ion sources, polarized atomic beam jet targets and atomic beam storage rings [21]. The transition rate caused by electron spin resonance must be matched to the desired escape rate, too high a rate can destabilize the gas. Niinikoski et al. have shown [22] that working at a cell density below 10^{16} atoms cm^{-3}, it seems possible to form pulses of a few milliseconds length with a 10 % duty cycle and to obtain densities of more than 10^{13} atoms cm^{-3} at the focusing point. It is expected that the beam has a mean velocity of about 200 m s^{-1}, a velocity spread of $\Delta v/v_o = 0.05$ and a nuclear polarization of nearly 100 %. A target based on this method is presently under construction at the University of Michigan [23].

The use of stable atomic hydrogen for an ion source would provide the capability of storing the atoms in between the output pulses. The pulsed extraction has the potential of injecting polarized hydrogen ions into synchrotrons with intensity equal to that of unpolarized beams. In d.c. beam sources the ultracold hydrogen might offer little improvement compared with the best conventional polarized atomic beam sources, because the losses in the cooling and stabilization might offset the advantages gained by the better focusing and lower velocity of the atoms. A feasibility study for an atomic beam source based on stable atomic hydrogen has been carried out at CERN [24]. Finally, it has to be pointed out that none of these proposed devices are operational yet and it is clear that a fair number of challanging problems of principal and technical nature has to be solved before an operational stage will be reached.

Polarized Ion Sources Based on Metastable Atoms (Lamb-Shift Sources)

The basic physics of these sources and the technical details of their operation have been extensively reviewed [25,26]. The general Lamb-shift scheme is shown in Fig. 2.

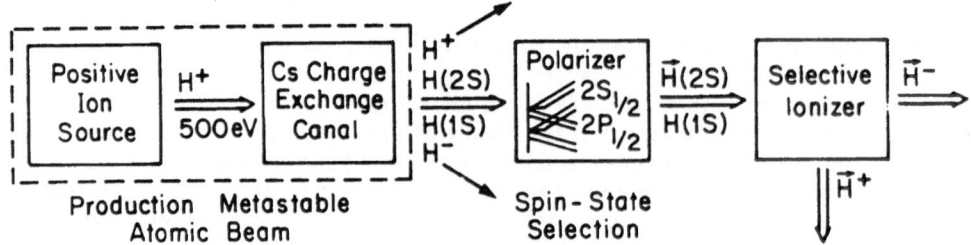

Figure 2: Schematic diagram of the polarized ion source using metastable atoms (Lamb-shift source).

Protons obtained from either a rf ion source, a duoplasmatron or ECR ionizer pass through a cesium charge exchange canal where a substantial fraction is converted into metastable $2S_{1/2}$ atoms. The charged components emerging from the cell are deflected out of the neutral beam by electrostatic deflector plates. The metastable atoms in the beam are polarized by eliminating the undesired spin states by transitions from a $2S_{1/2}$ state to a $2P_{1/2}$ state induced either by static fields or a combination of static and rf fields. The $2P_{1/2}$ state decays spontaneously to the $1S_{1/2}$ ground state. After this spin-state selection the metastable beam is selectively ionized by passing the beam through argon gas for negative ions and iodine vapour for positive ions.

The Lamb-shift sources have been well investigated in many laboratories and many sources have been operational at tandem accelerators and cyclotrons for over a decade. During this time the design has been optimized in nearly all components, but the progress in the last five years has been slow. In the workshops devoted to polarized ion sources [8-10] the reasons for the limitations have been discussed extensively. The collective worldwide experience gained from Lamb-shift source laboratories indicates strongly that there is a fundamental current limit which applies to all such sources. Besides losses from space-charge quenching and collisional quenching in the production of the metastable beam, the beam divergence from the unpolarized positive ion source is found to be another limiting factor.

This brightness problem of the 500 eV H^+ (or 1.0 keV D^+) source as well as the quenching problems limit the present polarized source output to a maximum current of about 3 μA [27,28]. This is also confirmed by emittance measurements performed at Giessen [29]. From the knowledge of quenching effects and brightness of low energy positive ion sources it seems unlikely that the $\vec{H}^-$ currents of Lamb-shift sources can be improved by more than a factor 2 to 3 of the present maximum. For the production of polarized positive ions photoionization has been proposed by de Jong [30], which could increase the current output by a large amount.

Production of Polarized Hydrogen Atoms by Spin Exchange Collisions (e.g. Optically Pumped Sources)

One of the most elegant methods to polarize electrons and nuclei of atoms is the use of optical pumping by laser beams. Unfortunately this powerful method cannot be applied directly to hydrogen atoms since there exists no appropriate laser. On the other hand electron polarized atoms can act as donor of polarized electrons which are attached to a proton beam in order to form a beam of neutral electron polarized hydrogen atoms. This idea was first suggested by Zavoiskii [31], who in 1957 proposed as a donor thin ferromagnetic foils, magnetized to saturation. Haeberli [32] proposed in 1965 to replace the foil by a vessel containing polarized hydrogen atoms or an optically pumped alkali vapour. With the advent of powerful lasers polarized alkali vapours can be produced [33-35] in large quantities for the spin exchange target. Today also dense targets of stable atomic hydrogen are available for the spin exchange collision [18,20].

216

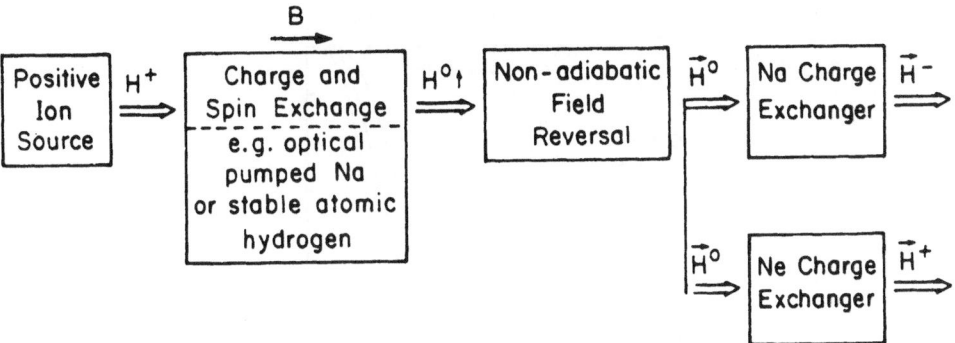

Figure 3: Schematic diagram of the production of polarized hydrogen atoms and ions by spin exchange collisions (e.g. optically pumped source).

The spin exchange scheme for a polarized ion source is shown in Fig. 3. Unpolarized positive ions, preferably produced in a ECR ion source pass through a target cell filled with a vapour or gas of electron spin polarized atoms. The protons pick up a polarized electron in a charge exchange process. The spin exchange has to be carried out in a strong magnetic field $B \gtrsim 1$ T in order to prevent depolarization from atoms formed in a excited state. The neutral electron polarized hydrogen beam is nuclear polarized by inducing a transition between Zeeman states in a nonadiabatic field passage (Sona transition [29]). Positive ions are produced by charge changing the atomic hydrogen beam in a neon cell. For negative ion production, an alkali or alkaline earth vapour (e.g. sodium) is used as an additional electron donor.

A polarized negative ion source based on an optically pumped sodium target using a single frequency dye laser has been operational for several years at the KEK synchroton [36,37]. Mori et al. [36] report a polarization of $\sim 80\%$ for a 10 μA H$^-$ beam and about 40 % polarization for a beam intensity of 25 μA. Similar project are under investigation in Vancouver [32] and Los Alamos [39].

At the Institute for Nuclear Research in Moscow, a duoplasmatron and neutralizing gas replace the positive ion source in Fig. 3. The neutral hydrogen atoms with an energy of 5 keV are then brought into the magnetic field surrounding the polarized sodium target where they are first reionized in helium gas prior to the attachment of polarized electrons from the sodium. The pulsed solenoid can be operated at fields up to 1.6 T and a pulse width of 30 μs with a repetition rate of 1 Hz. A peak intensity of the pulse of 1 mA has been reported [61] for protons with a polarization of 65 %. Polarized H$^-$ pulses with a peak intensity of 60 μA have been observed [62].

While this kind of sources is using thin targets with density of about 10^{13} atoms cm^{-3} and high magnetic field in the collision region a new interesting suggestion is made by Anderson et al. [40]. They propose to use a thick polarized target (electron spin polarization Na, Cs, or hydrogen target, thickness $\sim 10^{16}$ to 10^{17} atoms cm^{-2}) and low magnetic field in the collision region. In this case the electron and nuclear spin is coupled together by the hyperfine interaction to form a total angular momentum F. This requires a field B much less than the critical field B_c. A fast unpolarized hydrogen ion captures a first polarized electron either in a pure or a mixed state (e.g. state 1 and 2 or state 3 and 4). In the mixed state the hyperfine interaction mixes spins, converting some of the electron spin polarization into nuclear spin polarization. The H$^\circ$ atoms in the pure state cannot capture a second electron from the polarized target to form an H$^-$ ion since the only bound state occurs with electron spin antiparallel. However, the H$^\circ$ atoms in the mixed states can form H$^-$ ions by electron attachment. Since either electron may be detached in a subsequent collision, the repeated attachment of a polarized electron followed by detachment result in the entire beam being collisionally pumped into the pure state. For this aim, there is the necessary condition that the collision frequency be much less than the hyperfine frequency (0.3 -

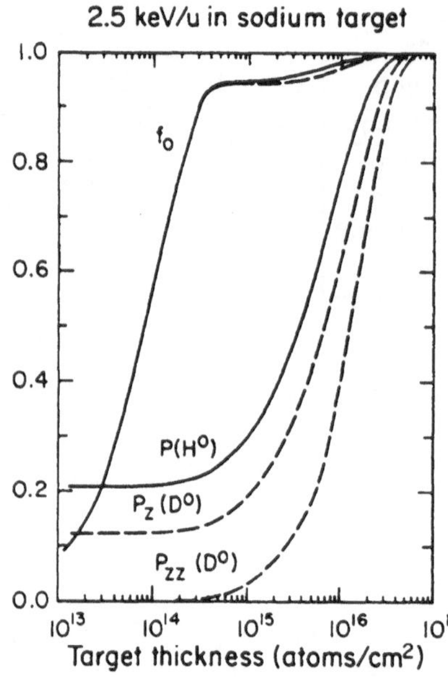

2.5 keV/u in sodium target

Figure 4: Polarization and neutral fraction f_o for 2.5 keV/u H$^+$ (solid lines) and D$^+$ (dashed lines) as a function of polarized sodium-vapour target thickness. The quantities p_z and p_{zz} are the vector and tensor polarization of the deuteron.

1.5 GHz). The problems of producing high intensity polarized hydrogen beams by collisional pumping are presently being studied at the Lawrence Berkeley Laboratory [41]. The expected nuclear polarization for H^o and D^o atoms as well as the neutral fraction of atoms as a function of a polarized sodium target thickness is shown in Fig. 4.

IMPROVEMENTS OF THE ATOMIC BEAM APPARATUS

Separation of the $m_j = +1/2$ and $m_j = -1/2$ States

This is a reminder of the method to produce an electron polarized hydrogen beam. Figure 5 shows the Rabi diagram where the relative energy $W/\Delta W$ is plotted versus the magnetic field B in units of the critical field of hydrogen atoms $B_c = 507$ G. In an inhomogeneous magnetic field the atoms with $m_j = +1/2$ are separated from $m_j = -1/2$, since a force

$$F = \mu_{eff} \cdot grad\ B \tag{1}$$

is acting on the hydrogen atoms. The quantity μ_{eff} is the effective magnetic moment of the particle. In atomic beam apparatus mostly magnetic sextupole fields are used for this Stern-Gerlach separation.

Atoms in the states 1 and 2, having $m_j = +1/2$ and $\mu_{eff} = -\mu_B$, are focused along the axis z of the sextupole magnet. Such trajectories calculated for T = 295 K, $B_o = 1.0$ T, entrance radii $r_a = 0, 1, 2$ mm, and angles $\alpha = 0°, \pm1°, \pm2°, \pm3°$ are shown in Fig. 6. Atoms in the states 3 and 4 with $m_j = -1/2$ and $\mu_{eff} = +\mu_B$ entering the magnet are deflected out of the atomic beam and are separated completely from the atoms in the $m_j = +1/2$ states. Most critical are atoms entering with small angles with respect to the magnetic axis. A few such trajectories are shown in Fig. 7 for T = 295 K, $B_o = 1.0$ T, $r_a = 0, 1, 2$ mm and $\alpha = 0°, \pm1°$.

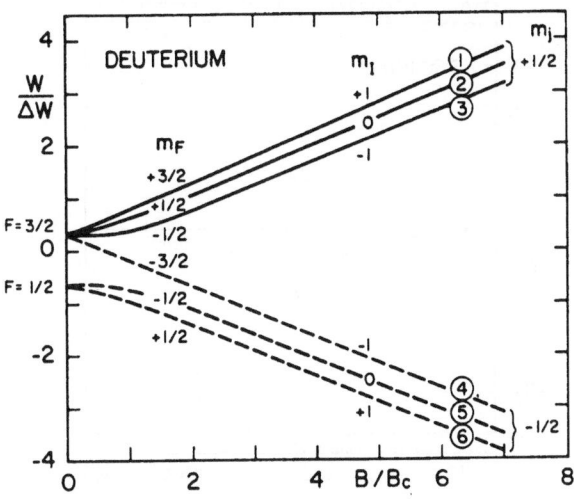

Figure 5: Energy-level diagram of the hydrogen atom in a magnetic field (Rabi diagram).

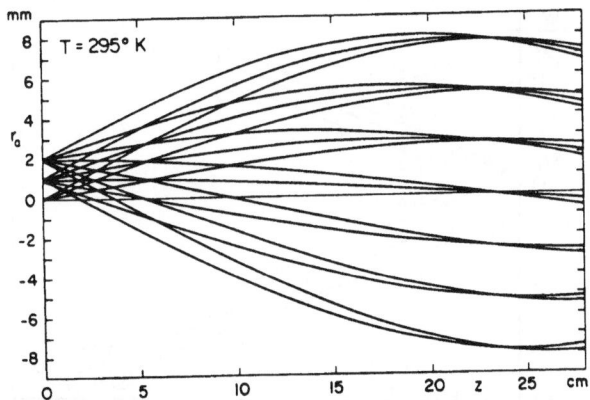

Figure 6: Calculated trajectories for $m_j = +1/2$ states in a sextupole magnet for $T = 295°$ K, $B_o = 1.0$ T, $r_o = 8$ mm and $r_a = 0, 1, 2$ mm and $\alpha = 0°, 1°, 2°, 3°$.

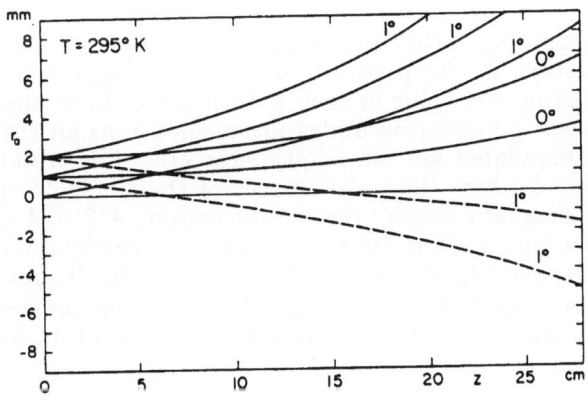

Figure 7: Calculated trajectories for the $m_j = -1/2$ states in a sextupole magnet for $T_o = 295$ K, $B_o = 1.0$ T, $r_o = 8$ mm and $r_a = 0, 1, 2$ mm and $\alpha = 0°$ and $1°$.

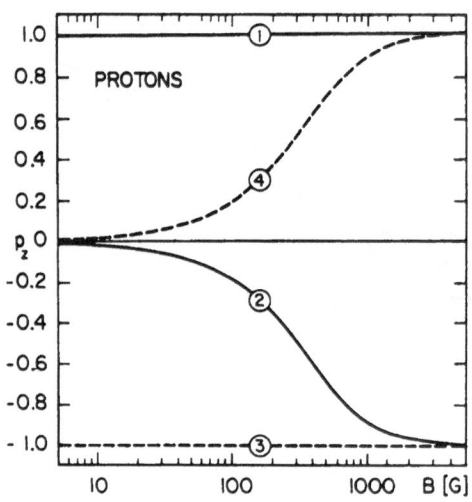

Figure 8: Polarization p_z of protons in the hydrogen atoms as a function of external field for the 4 single substates.

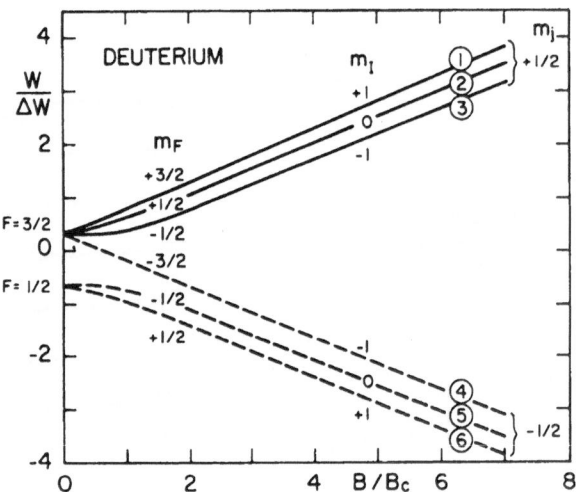

Figure 9: Energy-level diagram of the deuterium atoms in a magnetic field.

Nuclear Polarization for Hydrogen

It is obvious from Fig. 5 that the separation of atoms in the states 1 and 2 from those in 3 and 4 results, in the presence of a strong magnetic field, in zero nuclear polarization since the proton spins in each pair of states have opposite directions. In order to obtain nuclear polarization under these conditions an rf transition has to be induced between a populated and an empty substate. The situation is most clearly demonstrated in Fig. 8 where the polarizations of the single components are plotted as a function of the magnetic field of the ionizer region. A 2 to 4 transition, induces a complete proton polarization. The same aim can be reached by the multilevel transition 1 to 3 in a weak magnetic field, (weak field transition WF), but the polarization is inverted. While in both cases a strong magnetic field in the ionizer region is required, only a reduced polarization can be achieved without transition by the use of a weak B-field in the ionizer.

Production of Polarized Atomic Deuterium Beams

The production of polarized deuterons by polarized atomic deuterium beams is

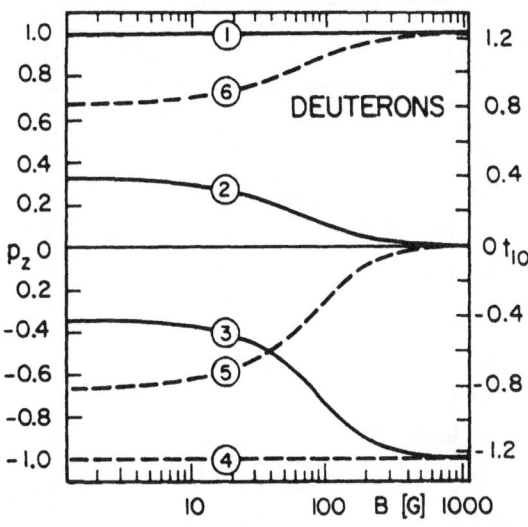

Figure 10: Vector polarization p_z or t_{10} of deuterons in deuterium atoms as a function of external magnetic field for the 6 single substates.

particularly interesting since also in this case high vector and/or tensor polarization can be obtained. Polarized atomic deuteron beams can reach tensor polarization p_{zz} values near -2.0. Further all different vector and tensor polarization components of spin -1 particles be produced.

Figure 9 shows the Breit-Rabi diagram for atomic deuterium. The critical field B_c is here 117 G. The separation of the electronic substates 1, 2 and 3 with $m_j = +1/2$ from the substates 4, 5 and 6 with $m_j = -1/2$ can be performed again with sextupole magnets. The production of the nuclear polarization occurs with the aid of rf transitions induced between different populated and empty substates. The vector polarization p_z or t_{10} of single substates as a function of the magnetic field is shown in Fig. 10. The tensor polarization p_{zz} or t_{20} as a function of the target magnetic field is presented in Fig. 11. Application of an rf transition in the weak field region, where F is a good quantum number and where m_F levels are nearly equally spaced, results in interchanging populations of states m_F and states $-m_F$, i.e. the interchanges of substates 1↔4, 2 ↔ 3 and 5 ↔ 6 takes place. Technically these weak-field transitions (WF) are simple to be accomplished by the adiabatic-passage method. Transitions involving only two levels are carried out in medium or strong magnetic field. They are performed in a field strong enough that F is no longer a good quantum number. The required combination of transitions depends on the application.

Cooling of the Atomic Beam

The density of ρ of the atomic beam is inversely proportional to the most probable velocity and hence proportional to $T^{-1/2}$. The solid angle Ω accepted by a sextupole magnet is given for a Maxwell velocity distribution by the relation

$$\Omega = 2.1 \frac{\mu_B \cdot B_o}{kT} \tag{2}$$

with μ_B the Bohr magneton and B_o the magnetic field on the pole tip.

Therefore the density of an atomic beam after the separation magnets is strongly dependent on the temperature:

$$\rho \propto T^{-3/2} \tag{3}$$

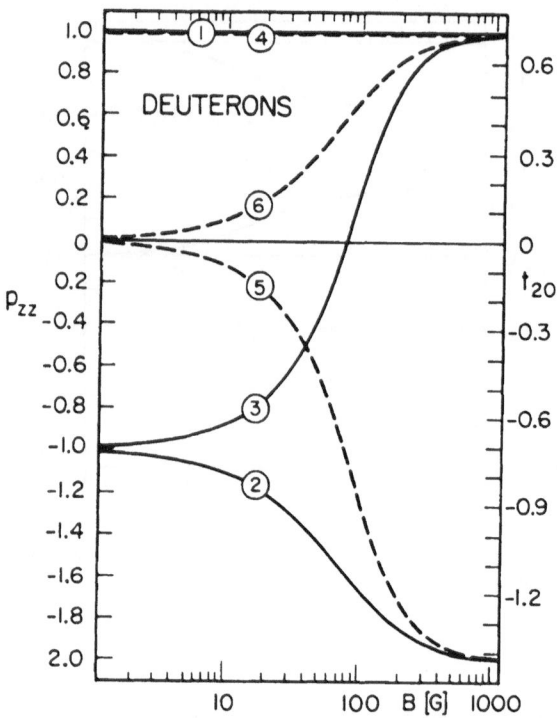

Figure 11: Tensor polarization p_{zz} or t_{20} of deuterons in deuterium atoms as a function of external magnetic field for the 6 single substates.

For liquid nitrogen temperature a density gain factor of 7.5, for liquid hydrogen temperature a factor of 57 and for L^4He a factor of 590 results compared with room temperature. These density gains can only be realized if there is no intensity loss from the dissociator due to the cooling and if the geometry of the beam forming apertures and the separation magnets are adjusted adequately. A further limitation arises from the geometrical constraints of the device in which the atomic beam is intended to be used.

Several attempts have been made to cool the atoms in the beam forming stage by cooling the nozzle. At Bonn a 14 mm long copper canal is cooled by liquid nitrogen (77 K) [42]. The velocity v_o of the beam measured corresponds to a temperature of 145 K. At Argonne (ANL) a copper block was clamped around the pyrex nozzle and cooled to approximately 30 K by a closed cycle ^4He refrigerator system [43]. The measured velocity v_o is equivalent to 110 K. These examples clearly show the inefficiency of this simple cooling procedure. Apparently the number of collisions on the wall is too small and/or the surface does not reach the measured temperature.

Better results should be obtained by cooling the atoms in a separate cooled chamber. The nozzle is then mounted on this accomodator. Care has to be taken to cause an optimum number of collisions on the wall in order to cool the atoms but prevent an unnecessarily large recombination rate.

For the investigation of the problems arising in the production of cooled high density atomic hydrogen beams we have built at ETH a test bench for an atomic beam apparatus. The experimental set-up is shown in Fig. 12. The hydrogen molecules are dissociated at room temperature in a Pyrex tube by a 27 MHz rf discharge and the produced atoms are transferred through a short Teflon tubing to a copper accommodator. The accommodator is cooled by a closed cycle ^4He-refrigerator. A nozzle, a skimmer and a collimator at the entrance in the magnet chamber form the particle beams. Two sextupole magnets allow to study the effect of the desired atomic states. A velocity se-

222

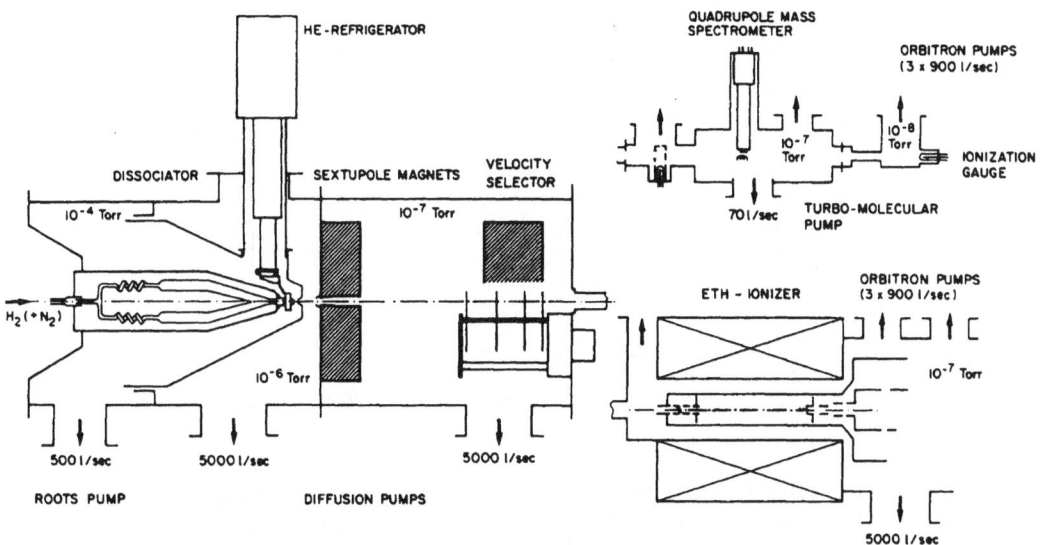

Figure 12: The ETH experimental test bench arrangement for cooling a high density atomic beam. On the right hand side, the diagnostic elements (upper part) and the strong field ionizer of the polarized ion source are shown (lower part).

lector is used to measure the velocity distribution of the produced atomic or molecular beams. These investigations are carried out for judging the efficiency of the cooling, the properties of the beam forming elements and the degree of dissociation obtained in the atomic beam. The measurement of the density of the beam is accomplished by a quadrupole mass spectrometer with a cross beam ion source. The intensity of the beams is measured in a compression tube containing an ionization gauge. The velocity measurements have been tested with molecular hydrogen and deuterium beams as well as with helium beams.

Numerous geometries of the accomodator have been investigated, the goal being to simultaneously optimize thermal accommodation, recombination and beam formation. Good results are obtained with a conical channel 20 mm long having an exit aperture of 3 mm diameter.

The nature of the accommodator surface is a crucial point to avoid recombination of the atoms. The material of the accommodator, however, is not of primary importance. Oxydized or unoxydized Cu, Al, Teflon coating on these materials show the same behaviour at low temperatures, leading us to the conclusion that under practical conditions of gas and vacuum cleanness the role of frozen or adsorbed species dominates. We found that doping the gas with a small amount of N_2 allows to create and maintain a good recombination inhibiting surface at about 34 K.

A typical example for a cooled atomic beam is shown in Fig. 13. The experimentally determined width (FWHM) is about half of the width of the Maxwell distribution. The increase of the most probable velocity compared to the accommodator temperature and the decrease of the width arises from the gasdynamical character of the beam forming and expansion of the atomic gas from the accommodator into the vacuum. A more detailed description of this investigation is given in ref. 44. Similar investigations have been carried out at the BNL [45] cooling the accommodator in the region between 4 and 10 K.

The Separation Magnets

At ETH we have developed two types of sextupole magnets 10 and 15 cm long, which are designed in a modular technique such that the pole pieces can be changed

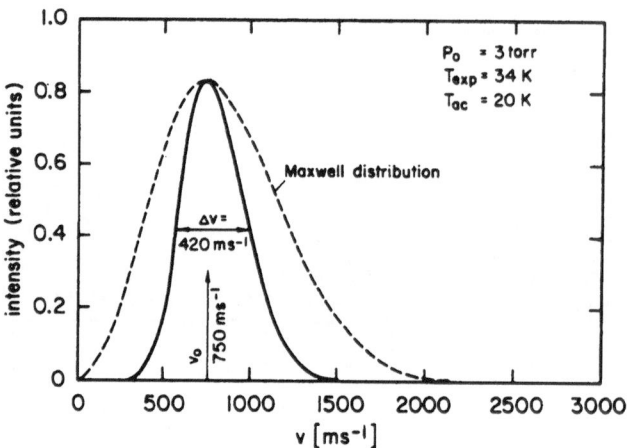

Figure 13: The velocity distribution of an atomic hydrogen beam measured at the ETH atomic beam apparatus. Temperature of the accommodator 20 K. Pressure in the dissociator $p_o = 3$ torr.

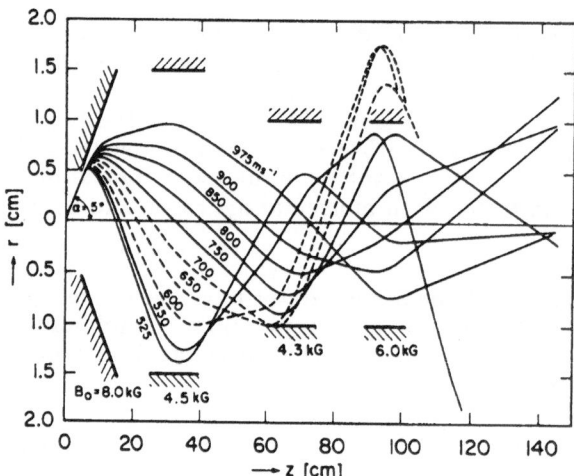

Figure 14: Calculated trajectories between 525 and 975 ms^{-1} and an angle of 5° from the axis. The plot gives the radial distance of the atoms from the symmetry axis of the magnets.

in a simple way. Computer programs have been developed to analyze a variety of configurations by the acceptance diagram technique and to calculate single trajectories of electron polarized atoms for lens systems of up to four sextupole magnets. In practive, we have in an atomic beam a velocity distribution similar to Fig. 13. The behaviour of the trajectories of atoms starting on the axis with an angle of 5° and velocities between 525 ms^{-1} and 975 ms^{-1} (which corresponds about to Δv in Fig. 13 is shown for a four sextupole magnet system in Fig. 14. Most of the atoms are 40 cm after leaving the magnet system still within a reasonable diameter, however, a part of atoms from the velocity spectrum is lost due to overfocusing of these particles.

Intensive computer simulations for optimizing the geometry of the sextupole lens system, using the velocity distribution found from cooled beams and for realistic

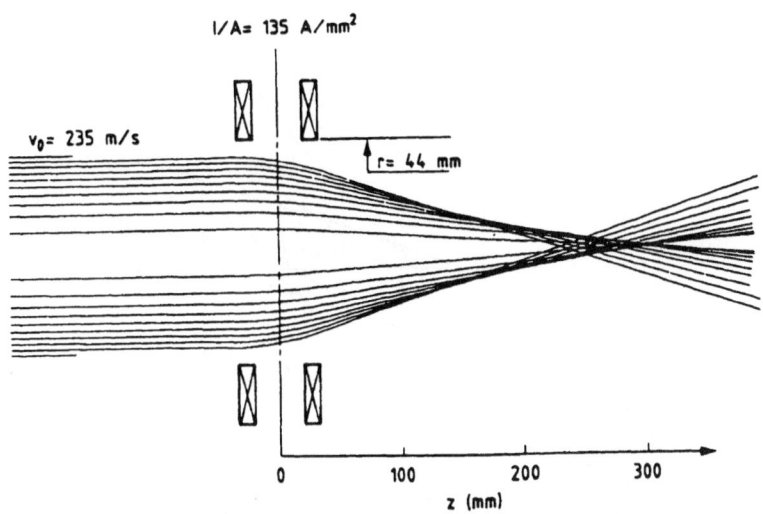

Figure 15: Tracks of hydrogen atoms entering into a gradient coil pair parallel to the axis.

starting conditions, have been performed in order to obtain a maximum acceptance and transmission of the separation magnets for high density beams for targets.

The CERN group [47] has investigated the possibility to use solenoid fields to focus neutral atomic hydrogen beams. It has been shown that the fringe field of the storage solenoid of ultracold atoms can be used directly. A considerable gain in the acceptance was achieved by introducing a smaller solenoid with a opposite current at the opening of the stabilization solenoid. This extra coil was dimensioned to trop the superposed field to zero at its own centre point. In this way the radial component of the field gradient was greatly increased giving an improved focusing effect.

The same group [47] studied also the focusing properties of a gradient coil pair, composed of two similar short solenoids with opposite currents. Figure 15 depicts the results of a track tracing calculation where atoms with a velocity of $v_o = 235$ ms^{-1} were considered. This lens type could also be most useful as large acceptance lenses in cooled conventional atomic beams.

The usual method to design a magnet system is the calculation of trajectories as shown in Fig. 14 and 15. This method makes it difficult to extract global and systematic information on the behaviour of either the whole system, or of a single element, if parameters of the system are changed. We have found that the application of the acceptance diagram technique to atomic beam systems has great advantages in order to match the optical properties of the magnet system with other elements of a source or target [46]. A typical example of such an acceptance diagram for a two component magnet system is shown in Fig. 16. Here the acceptance diagram of all components of the system has been transfered to the collimator position at the entrance of the first magnet, and the area of the overlap region (i.e. the acceptance area) of the whole system is shown. At present the most limiting element of the ETH polarized ion source is the ionizing region.

Results

At ETH, the improvements obtained so far by an atomic beam cooled to a temperature of approximately 34 K and the new design of the separation and focusing sextupole magnet system have resulted in a atomic beam intensity of 10^{17} atoms s^{-1} with a peak density of $2 \cdot 10^{12}$ atoms^{-3} at the ionizer position. Calculations and observed beam properties suggest that a density of 10^{13} atoms cm^{-3} can be reached in a suitable geometry.

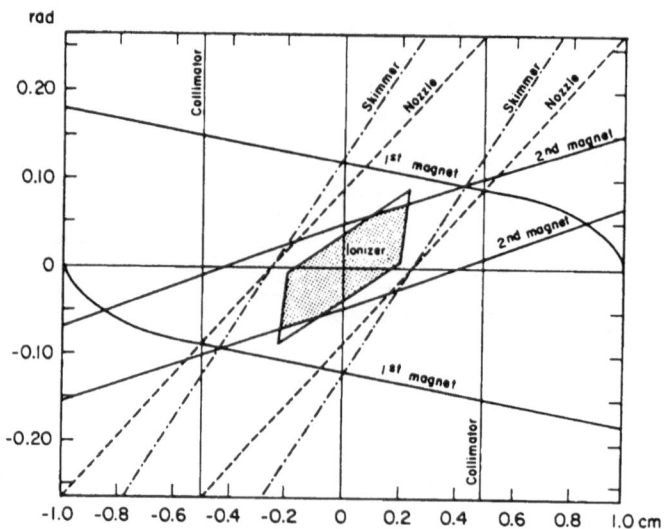

Figure 16: Acceptance diagram for a two magnet system. For details see ref. 46.

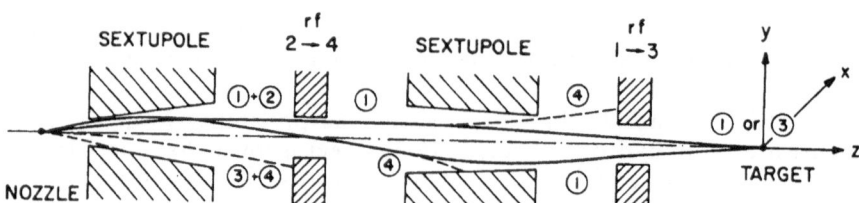

Figure 17: Production scheme for single substates.

The cooled atomic beam has been installed at the ETH polarized ion source. First experiences show that the atomic beam optic has to be slightly adjusted to get a better overlap with the electron beam in the ionizer. So far, an increase in ion beam intensity by a factor of 4 has been observed [60]. Further work is needed to reach the design aim of more than 1 mA positive or 50μA negative polarized hydrogen ions.

NUCLEAR POLARIZATION AND SINGLE STATE SELECTION

The use of multimagnet systems for the separation and focusing of electronic states of an atomic hydrogen beam allows a high flexibility in the nuclear polarization and polarization reversal by selecting single states.

A single Zeeman state can be separated by the use of two Stern-Gerlach sextupole magnets with an intermediate rf transition. This scheme is shown in Fig. 17. An additional rf transition at the end allows to change the sign of the polarization. Although the intensity drops in this scheme by a factor 2 it has the interesting feature that the polarization is preserved in any magnetic field. This makes it particularly simple to direct the polarization vector of the particles in x-, y- or z-direction by three sets of Helmholtz coils mounted and activated correspondingly.

The different combinations of necessary B-fields, rf transitions, Stern-Gerlach separation magnets (S-G) as well as the relative intensity I, the maximum polarization P_{max} and the figure of merit $I \cdot p^2$ are summarized in Table 1.

For deuterium beams a variety of different vector and tensor polarization combinations can be produced. The required combination of transitions depends on the

Table 1. Production of Polarized Atomic Hydrogen Beams

Mode	Ionizer or target B-field	Stern-Gerlach separation (S-G) and rf-transitions	Substates	P_{max}	I	Ip^2
a	weak	S-G	1 + 2	+1/2	1	1/4
b	weak	S-G, $1 \rightarrow 3$ (WF)	2 + 3	-1/2	1	1/4
c	strong	S-G, $2 \rightarrow 4$	1 + 4	+1.0	1	1.0
d	strong	S-G, $1 \rightarrow 3$ (WF)	2 + 3	-1.0	1	1.0
e	any	S-G, $2 \rightarrow 4$, S-G	1	+1.0	1/2	1/2
f	any	S-G, $2 \rightarrow 4$, S-G, $1 \rightarrow 3$ (WF)	3	-1.0	1/2	1/2

strong B-field: $>> 507$ Gauss

application. From a large variety of schemes a number of interesting combinations of Stern-Gerlach separations and rf transitions are shown in Table 2. Mode (b) and (c) give a pure vector polarization but with different signs. This can be used to switch the sign of the polarization periodically. The modes (d) to (g) produce a combined vector and tensor polarization with $|p_z| = 1/3$ and $|p_{zz}| = 1$. Two of these modes can be combined to a measuring scheme such that either the sign of the vector polarization or the tensor polarization or both signs are switched. The mode (h) with a pure tensor polarization $p_{zz} = -2$ can be combined either with the unpolarized mode (a) or with the pure tensor polarization mode (i) with $p_{zz} = +1$. In spite of the reduced intensities of mode (i) and (h) totally the highest figure of merit results from this combination, namely $I \cdot (\Delta p_{zz}) = 2/3 \cdot 9 = 6$. The modes (k) and ($\ell$) give $p_{zz} = +1$ and $|p_z| = 1$ with inverted sign of p_z. The pure states 1 and 4 allowing ionization in any field can be produced in the modes (m) to o.

Since the required magnetic field in the ionizer is in any case relatively low, the sign of the vector polarization can also be easily reversed by changing the field direction in the ionizer or target region. In this case the sign of the tensor polarization is unchanged.

IONIZATION OF ATOMIC BEAMS

The different used and proposed ionization processes are collected in Table 3. The corresponding ionization cross sections are indicated in the second column. For comparison the relations for the expected polarized ion intensity per unit of the ionizing beam intensity are given in the third column. For these relations a conventional atomic beam of $2 \cdot 10^{16}$ $\vec{H}^o$ s^{-1} with $v_o = 3 \cdot 10^5$ cm s^{-1} collimated to a cross section of 1 cm^2 is assumed. This results in a beam density n of about 10^{11} atoms cm^{-3}. In the next column observed ionizing particle beam intensities are shown, and in the fifth column observed polarized positive or negative dc ion beam intensities are indicated for the conditions of the preceding columns. In a further column ion beam intensities observed with cooled atomic beams with higher density and/or pulsed source mode are shown for specific sources. In the last column a comparison is made for the corresponding dc beam currents.

The first process in Table 3 produce positive ions. The process 2 requires a space charge compensation which can be obtained by a deuterium plasma beam. This has been accomplished by the Moscow group in a pulsed arc source [48]. Although a strong magnetic field in the ionization region is not absolutely required (e.g. by using a atomic beam in a single state) a strong magnetic field in the ionizer often is desired for the confinement of the electrons.

Table 2. Production of Polarized Atomic Deuterium Beams

Mode	Ionizer or target B-field	Stern-Gerlach separation (S-G) and rf-transitions	Substates	Vector pol. p_{zz}	Tensor pol. p_{zz}	Relative intensity I	Figure of merit $I \cdot p_z^2$	Figure of merit $I \cdot p_{zz}^2$
a	strong	S-G	1 + 2 + 3	0	0	1	0	0
b	strong	S-G 1→4 (WF)	2 + 3 + 4	-2/3	0	1	4/9	0
c	strong	S-G 2→6, 3→5	1 + 5 + 6	+2/3	0	1	4/9	0
d	strong	S-G 3→5	1 + 2 + 5	+1/3	-1	1	1/9	1
e	strong	S-G 3→5 2→3 (WF) 1→4 5→6	3 + 4 + 6	-1/3	+1	1	1/9	1
f	strong	S-G 2→6	1 + 3 + 6	+1/3	+1	1	1/9	1
g	strong	S-G 2→6 3→2 (WF) 1→4 6→5	2 + 4 + 5	-1/3	-1	1	1/9	1
h	strong	S-G 1→4 (WF) S-G 3→5	2 + 5	0	-2	2/3	0	8/3
l	strong	S-G 1→4 (WF) S-G 2→6	3 + 6	0	+1	2/3	0	2/3
k	strong	S-G 3→5 S-G 2→6	1 + 6	+1	+1	2/3	2/3	2/3
ℓ	strong	S-G 3→5 1→4 (WF) S-G 2→3	3 + 4	-1	+1	2/3	2/3	2/3
m	any	S-G 2→6 3→4	1 + 4	0	+1	2/3	0	2/3
n	any	S-G 3→5 2→6 S-G	1	+1	+1	1/3	1/3	1/3
o	any	S-G 3→5 2→6 1→4 (WF)	4	-1	+1	1/3	1/3	1/3

strong B-field: >>117 Gauss

Table 3. Ionization Processes

Process	$\sigma[\text{cm}^2]$	Intensity relation for a conventional atomic beam $I(H^0)=2\cdot10^{16}\ \text{s}^{-1}\ \text{cm}^{-2}$ $\left\{ n\approx10^{11}\ \text{cm}^{-3},\ v_0=3\cdot10^5\ \text{cm s}^{-1} \right\}$	Ionizing beam intensity $[\text{mA/cm}^2]$	$I(H_{ion})$ $[\mu A]$	$I(H_{ion})$ special conditions (cooled atomic beam, pulsed etc.) $[\mu A]$	$I(H_{ion})$ corresponding DC-beam $[\mu A]$
1) $H^0 + e^- \xrightarrow{1\ keV} H^+ + 2e^-$	$0.7\cdot10^{-16}$	$I(H^+) \approx 0.05\ \mu A$ per $\text{mA/cm}^2\ e^-$	$2.5\cdot10^3$	125	500 a)	500
2) $H^0 + D^+(+e^-) \xrightarrow{2\ keV} H^+ + D^0(+e^-)$ space charge compensation	$20\cdot10^{-16}$	$I(H^+) \approx 5\ \mu A$ per $\text{mA/cm}^2\ H^+$	100		10000 b)	0.5
3) $H^0 + Cs^0 \xrightarrow{50\ keV} H^- + Cs^+$	$6\cdot10^{-16}$	$I(H^-) \approx 1.5\ \mu A$ per part. $\text{mA/cm}^2\ Cs^0$	2	3	30 c)	0.015
4) $H^0 + D^-(+Cs^+) \xrightarrow{2\ keV} H^- + D^0(+Cs^+)$ space charge compensation	$20\cdot10^{-16}$	$I(H^-) \approx 5\ \mu A$ per $\text{mA/cm}^2\ D^-$	20			
5) Process 1) $+H^+ + Na^0 \xrightarrow{5\ keV} H^- + Na^+$ $\eta=5\%$	$3\cdot10^{-16}$	$I(H^-) \approx 0.003\ \mu A$ per $\text{mA/cm}^2\ e^-$	$2.5\cdot10^3$	6	25 a)	25

a) dc beam, cooled atomic beam (accommodator T=35 K) ETHZ [60]
b) pulsed peak current, average current during 100 μs = 5 mA, repetition rate 1 Hz, cooled atomic beam ; nozzle cooling with LN$_2$ (Moscow) [48]
c) pulsed peak current during 600 μs, repetition rate ~1 Hz, cooled atomic beam; nozzle cooling (Brookhaven) [54]

The process 1 in Table 3 is the most common method to produce polarized positive ions. The cross section for this process is not particularly high, however, a large number of electrons can be confined in a well shaped, strong magnetic field. While older versions of this type of ionizer had typically electron current densities of 0.2 A/cm^2 with an ionization efficiency of 0.3 %, a modern electron impact ionizer yields electron current densities of several A/cm^2 with efficiencies of 3 to 5 % (e.g. the ETH ionizer [49] which produces with a cooled beam 500 μA H$^+$).

A most efficient method to ionize light and heavy ion is the use of electron cyclotron resonance (ECR) ionizer. In this type of ionizer a plasma of cold ions with hot electrons is produced in the ionization region. The applied electron cyclotron resonance conditions produce a high electron density, which can be 10^3 times higher than in a normal electron impact ionizer. Electron energy of several keV can be obtained. Gas efficiency $n_g = I^+/(I^+ + I_g)$ of approximately 70 % has been observed for H$_2$ molecules, which should give an efficiency of 35 % for H-atoms. Recently the use and possible depolarization effect in a ECR ionizer habe been investigated [50]. Although this study resulted in a positive conclusion, experimental tests are necessary for a final conclusion.

The colliding beam method for the production of H$^-$ refers to the processes 3 and 4 of Table 3. A single electron is exchanged from a neutral atom or negative charged ion to a polarized hydrogen atom with thermal velocity. The colliding beam method using neutral Cs atoms for the direct conversion of H^o to H$^-$ has been developed at Wisconsin [51-53]. At the AGS in Brookhaven a similar source with a moderate cooled atomic beam produces in a pulsed mode peak H$^-$ currents of up to 30 μA [54].

At BNL a ionizer is under development for a polarized H$^-$ source based on the resonant charge exchange process [55]. Here the neutral Cs beam is replaced by a space charge compensated D$^-$ beam (cf. Table 3, process 4).

The intense beams of polarized positive hydrogen ions produced by electron impact ionizers can be used for the production of H$^-$ ions by a near-resonant double charge exchange process (cf. Table 3, process 5). Since this conversion from H$^+$ to H$^-$ is mainly a two step process, it has to be performed in a strong magnetic field in order to avoid depolarization via the hyperfine interaction during the neutral state of the atoms. The equilibrium fraction yields F$^\infty$ for different alkali and lakline earth vapours as a function of energy of the incident hydrogen ion beam has been measured by Schlachter et al. [56,57]. For more than a decade we have used in our laboratory sodium vapour at an operational energy of 5 keV. At this energy F$^\infty$ = 0.07 for H$^+$ and 0.10 for D$^+$. A beam current of 6 μA of H$^-$ has been observed from the conventional ETH polarized ion source [58]. Using the cooled beam the output is increased to 25 μA as obtained by scaling the measurements actually performed on a simplified, 30 % less efficient system than in ref. [58]. Larger equilibrium yields F$^\infty$ are observed at lower energies, in particular for charge exchange in the alkaline earth vapours Sr and Ca (40 to 50 %) and by the alkali vapours Cs and Rb (20 to 30 %) [57]. These new results are shown in Fig. 18 together with the corresponding alkali metal results. The design of an appropriate deceleration system, which brings the ions to the required low energy between a few hundred eV and 1 keV in a high transmission oven, could improve the double charge exchange yield by a substantial factor.

An interesting approach in this direction is the design of a neutral injector line based on negative deuteron, for a Tokamak fusion reactor [59]. Positive deuterons from a ECR source are converted to negative ions by passing through a supersonic cesium jet at an energy of 1 keV. The whole production line is immersed in a more or less uniform magnetic field which improves the beam transport between the source and the double charge exchanger. A schematic diagram of this Grenoble design is shown in Fig. 19. A charge exchange efficiency of 20 to 25 % is measured in this device. Details

230

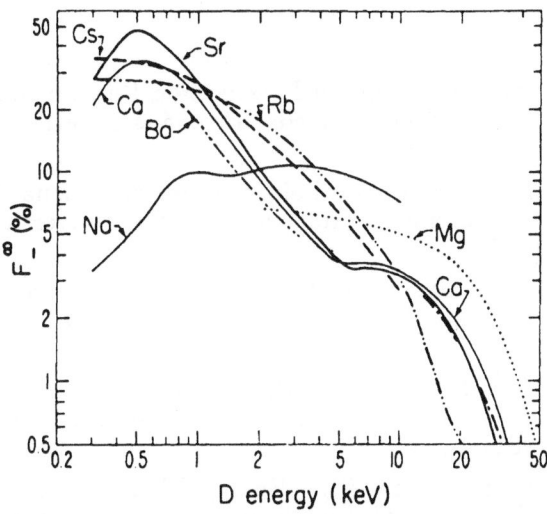

Figure 18: A collection of equilibrium double charge exchange yields F^∞ for different alkali and alkaline earth vapours as a function of deuteron energy (from ref. 57).

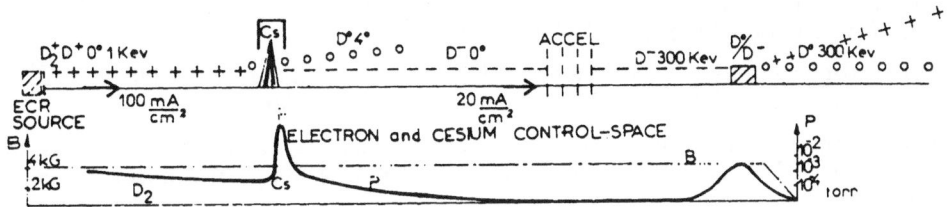

Figure 19: The magnetized cesium D^- line with supersonic cesium cell and ECR ion source (from ref. 59).

are given in ref. 59. For situation where the use of alkali vapours is too dangerous, xenon can be used with reasonable conversion yield. Between an energy of 2 and 4 keV a H^- yield from 4 to 6 % is experimentally observed [59].

CONCLUSION

In conclusion, we have at present three different types of operational polarized ion sources. An abundance of new ideas have been presented in special workshops and in the literature. This variety makes it sometimes difficult to decide in which direction one should proceed. One should also keep in mind that the realization of technical progress is tedious and sometimes frustrating compared to the satisfaction that the creation of new ideas can generate.

I am very grateful to all my colleagues all over the world who have provided me with the newest information about developments in the field. I am also very much indebted to Drs. P.A. Schmelzbach and V. König for many fruitful discussions and help in preparing this review.

REFERENCES

[1] R. Fleischmann, 2nd Int. Symp. on Polarization Phenomena of Nucleons, Karlsruhe, P. Huber and H. Schopper, eds. Birkhäuser Verlag Basel and Stuttgart (1966), p.21.

[2] H.F. Glavish, Third Int. Symp. on Polarization Phenomena in Nuclear Reactions, Madison, H.H. Barschall and W. Haeberli eds. The University of Wisconsin Press Madison (1971), p.267.

[3] T.B. Clegg, Fourth Int. Symp. on Polarization Phenomena in Nuclear Reactions, Zürich, W. Grüebler and V. König eds. Birkhäuser Verlag Basel and Stuttgart (1976), p.111.

[4] W. Grüebler and P.A. Schmelzbach, Proc. Fifth Int. Symp. on Polarization Phenomena in Nuclear Physics, Santa Fe, G.G. Ohlsen et al., eds. AIP Proc. No. 69, AIP New York (1981), p. 848.

[5] W. Haeberli, Proc. Int. Symp. High Energy Physics with Polarized Beams and Polarized Targets, Lausanne, C. Joseph and J. Soffer, eds. Birkhäuser Basel (1981), p.199.

[6] W. Grüebler, Proc. Sixth Int. Symposium on Polarization Phenomena in Nuclear Physics, Osaka, M. Kondo et al., eds. Suppl. Journ. Phys. Soc. Japan 55 (1986), p.435.

[7] C. Hojvat et al., IEEE Trans. Nucl. Sci. 26:3149 (1979).

[8] Proc. Polarized Proton Ion Sources, Ann Arbor, A.D. Krisch and A.T.M. Lin eds., AIP Conf. Proc. No. 80, AIP New York (1982).

[9] Proc. High Intensity Polarized Proton Source, Vancouver 1983, G. Roy and P. Schmor eds., AIP Conf. Proc. No 117, AIP New York (1984).

[10] Proc. Intern. Workshop on Polarized Sources and Targets, Montana 1986, S. Jaccard and S. Mango eds., Helv. Phys. Acta 59:513-806, Birkhäuser Verlag Basel (1986).

[11] Proc. High Energy Spin Physics, Brookhaven, G.M. Bunce ed., AIP Conf. Proc. No 95, AIP New York (1983).

[12] Proc. Int. Symposium on High Energy Spin Physics, Marseille, J. Soffer ed., Journ. Phys. 46:C2 (1985).

[13] Proc. 7th Int. Symp. on High Energy Spin Physics, Profvino (USSR) 1986, in press.

[14] R.M. Kulsrud, H.P. Furth, E.J. Valeo and M. Goldhaber, Phys. Rev. Lett. 49:1248 (1982).

[15] D. Kleppner, ref. 8, 111.

[16] H.F. Hess, D.A. Bell, G.P. Kochanski, D. Kleppner and T.J. Greytak, Phys. Rev. Lett. 51:483 (1983) and Phys. Rev. Lett. 52:1520 (1984).

[17] R. Sprik, J.T.M. Walraven and I.F. Silvera, Phys. Rev. Lett. 51:479 and 942 (1983).

[18] D. Kleppner and T.J. Greytak, ref. 11, 546.

[19] P.G. Sona, Energ. Nucl. 14:295 (1967).

[20] I.F. Silvera and J.T.M. Walraven, Phys. Rev. Lett. 44:164 (1980).

[21] T.O. Niinikoski, Proc. Int. Symp. on High Energy Physics with Polarized Beams and Targets, Lausanne, C. Joseph and J. Soffer eds., Birkhäuser Verlag Basel (1981), p.191.

[22] T.O. Niinikoski, S. Penttilä, J.-M. Rieubland and A. Rijllart, Proc. 4th Int. Workshop on Polarized Target Material and Techniques, W. Meyer ed., Physikalische Institut, Universität Bonn (1985), p.183.

[23] R.S. Raymond, P.R. Cameron and D.G. Crabb, ref. 15, 202.

[24] T.O. Niinikoski, S. Penttilä, J.-M. Rieubland and A. Rijllart, ref. 3, 139.

[25] T.B. Clegg, ref. 8, 21 and refs. therein.

[26] T.B. Clegg, ref. 9, 63.

[27] P. Schiemenz, F.J. Eckle, G. Eckle, G. Graw, H. Kader and F. Merz, ref. 6, 1056.

[28] A. Isoya, T. Nakashima, K. Sagara and H. Nakamura, ref. 6, 1054.

[29] A. Hofmann, H. Baumgart, J. Günzi, E. Huttel, N. Kniest, E. Pfaff, G. Reiter, S. Tarraketta and G. Clausnitzer, ref. 6, 1058 and private communication G. Clausnitzer.

[30] T.B. Clegg, ref. 9, 103.

[31] E.K. Zavoiskii, Soviet Physics JETP 5:338 (1957).

[32] W. Haeberli, Proc. 2nd Int. Symp. on Polarization Phenomena in Nuclear Reactions P. Huber and H. Schopper eds., Birkhäuser Verlag Basel (1966), p.64.

[33] L.W. Anderson, Nucl. Inst. Meth. 167:363 (1979).

[34] L.W. Anderson and J.T. Cusma, ref. 9, 133.

[35] W. Happer, E. Miron, R. Knize and J. Cecchi, ref. 9, 114.

[36] Y. Mori, K. Ikegami, Z. Igarashi, A. Takagi and S. Fukumoto, ref. 3, 123.

[37] Y. Mori, A. Takagi, K. Ikegami and S. Fukumoto, ref. 6, p.453.

[38] C.D.P. Levy, M. McDonald, P.W. Schmor and S.Z. Yao, Proc. Intersection between Particle and Nuclear Physics, Steamboat Springs, R.E. Mischke ed., AIP Conf. Proc. No 123, AIP New York (1984).

[39] R.R. Stevens, Jr., private communication.

[40] L.W. Anderson, S.N. Kaplan, R.V. Pyle, L. Ruby, A.S. Schlachter and J.W. Stearns, Phys. Rev. Lett. 52:609 (1984) and J. Phys. B At.Mol.Phys. 17:229 (1984).

[41] A.S. Schlachter, private communication.

[42] H.G. Mathews, A. Kruger, S. Penselin and A. Weinig, Nucl. Inst. Meth. 213:155 (1983).

[43] P.F. Schultz, E.F. Parker and J.J. Madsen, ref. 4, 909.

[44] W. Grüebler, Proc. Workshop on Polarized Targets in Storage Rings, Argonne, Report ANL-84-50, 223 (1984) and ref. 9, 1.

[45] A. Hershcovitch, A. Kponou and T.O. Niinikoski, ref. 10, 526.

[46] W.Z. Zhang, P.A. Schmelzbach, D. Singy and W. Grüebler, Nucl. Instr. Meth. A240:229 (1985).

[47] M. Ellilä, T.O. Niinikoski and S. Penttilä, Report CERN-EP/85-78 and submitted to Nucl. Inst. Meth.

[48] A.S. Belov, S.K. Esin, S.A. Kubalov, V.E. Kuzik, A.A. Stepanov and V.P. Yakushev Pis'ma Zh. Eksp. Teor. Fiz. 42:319 (1985) and ref. 13.

[49] P.A. Schmelzbach, W. Grüebler, V. König and B. Jenny, Nucl. Inst. Meth. 186:655 (1981) and W. Grüebler, ref. 9, 1.

[50] T.B. Clegg, V. König, P.A. Schmelzbach and W. Grüebler, Nucl. Instr. Meth. A238:195 (1985).

[51] D. Hennies, R.S. Raymond, L.W. Anderson, W. Haeberli and H.F. Glavish, Phys. Rev. Lett. 40:1234 (1978).

[52] W. Haeberli, M.D. Barker, C.A. Gossett, D.G. Mavis, P.A. Quin, J. Sowinski, T. Wise, Nucl. Inst. Meth. 196:319 (1982).

[53] T. Wise and W. Haeberli, ref. 11, 615.

[54] A. Kponou, J.G. Alessi and Th. Sluyters, Contr. to Particle Accelerator Conf. Vancouver, May (1985).

[55] J.G. Alessi, Contr. to Particle Accelerator Conf. Vancouver, May (1985).

[56] A.S. Schlachter, K.R. Stadler and J.W. Stearns, Phys. Rev. A22:2494 (1980).

[57] R.H. McFarland, A.S. Schlachter, J.W. Stearns, B. Lin and R.E. Olson, Phys. Rev. A26:772 (1982).

[58] W. Grüebler and P.A. Schmelzbach, Nucl. Inst. Meth. 212:1 (1983).

[59] R. Geller, B. Jacquot, C. Jacquot and Sermet, Nucl. Instr. Meth. 175:261 (1980).

[60] P.A. Schmelzbach, D. Singy, W. Grüebler and W.Z. Zhang, Nucl. Instr. Meth. A251:407 (1986).

[61] A.N. Zelenskii, A.S. Kokhanovskii, V.M. Lobashev and V.G. Polushkin, ref. 6, p.1064.

[62] P.W. Schmor, ref. 10, p. 643.

PROSPECTS FOR POLARIZING SOLID DT

W. Heeringa

Kernforschungszentrum Karlsruhe
Institut für Kernphysik I
P O Box 3640
7500 Karlsruhe
Federal Republic of Germany

INTRODUCTION

The fusion cross section of light particles with energies in the keV·
region is determined by resonances: particle-unstable states in the compound nucleus. If one resonance dominates, as in the D-T case, a clear increase in fusion cross section can be achieved by the employment of properly
aligned polarized particles. The employment of nuclear polarized fuel for
nuclear fusion reactors has become a topic of interest since the paper of
Kulsrud et al.,[1] in which it was shown that in magnetically confined plasmas the polarization decay rate is probably smaller than the fusion rate.
Several papers about the behaviour of polarized plasmas appeared since then;
these proceedings contain a contribution by Pegoraro on this subject. Also
in inertial confinement fusion it is estimated that the nuclear polarization is largely retained until fusion takes place.[2,3]

The low-energy D-T cross section is almost completely determined by
the $I^\pi = 3/2^+$ resonance at 107 keV, where I and π denote the nuclear spin
and parity. At low energies the orbital angular momentum $\ell = 0$. Since the
deuteron spin is 1 and the tritium spin is 1/2, the reaction only occurs
for particles with parallel spins. For unpolarized particles the statistical weight of this situation is 4, that of the anti-parallel situation is
2. Hence in the ideal case a 50% increase of the cross section can be achieved with polarized particles. For partially polarized tritons and deuterons the maximal increase of 50% appears to be reduced proportionally to
the size of both polarizations. For example, if both deuterons and tritons
are polarized to 50%, the cross section increase amounts to 0.5x0.5x50% =
12.5%.

A second important aspect of polarized fuel is the anisotropy which
occurs in the angular distribution of the reaction particles. The anisotropy in the neutron flux distribution may appear to be advantageous for reactor design. It is possible to suppress the emission in the direction of
the quantization axis (by polarizations parallel to it) or to enhance the
emission in this direction (by perpendicularly polarized deuterons), see
Ref.1. In the latter case, however, the anisotropy is only small and there
is no gain in cross section. This topic is addressed in a contribution by
Pedretti to these proceedings.

In magnetic confinement reactors the polarized fuel may be injected in two ways: as a neutral particle beam or in the form of solid beads. Calculations and experiments with unpolarized solid beads show the advantages of this injection method: higher plasma densities and longer confinement times can be achieved.[4,5] The inertial confinement scheme employs pellets with DT-fuel inside an enclosure of some other material. Recent designs, especially in connection with ion beams, conceive multilayered pellets with an inner layer of solid DT-mixture and a relatively large empty space in the centre.

Hence in both confinement schemes the employment of solid DT-fuel is foreseen. Therefore it is interesting to investigate the possibilities to polarize solid DT-fuel. In nuclear and particle physics well-established techniques are available to produce polarized solid samples. Two difficulties will be encountered in trying to apply these methods to DT-mixtures. One is given by the fact that hydrogen occurs as para- or as ortho-hydrogen. The other results from the ß-decay of tritium. The heat deposited by the ß-particles in the sample makes it difficult to cool it to the temperature region required for the polarization process.

NUCLEAR POLARIZATION IN SOLIDS

Thermal Equilibrium Polarization

Magnetic moments can be oriented in space by putting them in a magnetic field. This also holds for nuclei with a magnetic moment, i.e. nuclei with spin I ≠ 0. The polarization P of a sample of nuclei in thermal equilibrium with their surroundings at the temperature T is given by:

$$P = \frac{2I+1}{2I}\coth\left(\frac{2I+1}{2I}\frac{\mu B}{kT}\right) - \frac{1}{2I}\coth\left(\frac{\mu B}{2IkT}\right),$$ (1)

where μ is the magnetic moment of the nuclei and B the magnetic field. In order to obtain sizeable polarizations the thermal energy kT should be lowered until it becomes of the magnitude of the magnetic energy μB. This is shown in Fig.1. for the three hydrogen isotopes at a magnetic field of 15 T. This field is about the limit of what is obtainable nowadays for small to medium-size superconducting solenoids. Below 25 mK the proton and tritium polarizations rise above 50%. At this temperature the deuterium polarization is still very low, however. One has to cool below 5 mK in order to obtain a deuterium polarization above 50%. The spins and magnetic moments of the hydrogen isotopes are listed in Table 1.

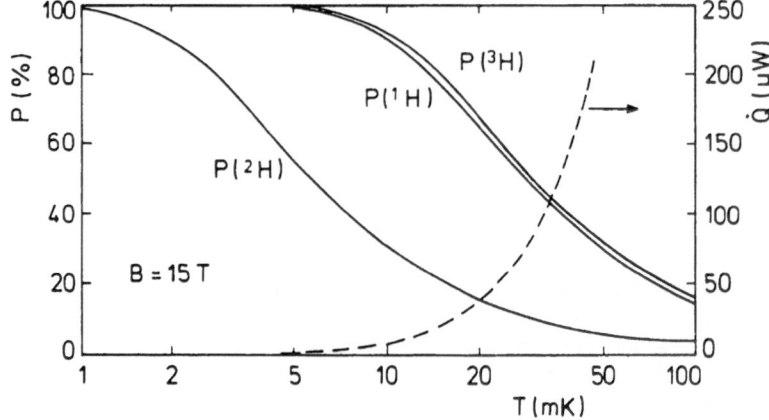

Fig.1: Brute-force polarizations of the hydrogen isotopes in a 15 T magnetic field. The dashed line represents the cooling power of the most powerful dilution refrigerators to date.

Table 1. Spins and magnetic moments of the
hydrogen isotopes

Isotope	I	μ (nucl.magn.)
^{1}H=H	1/2	+ 2.793
^{2}H=D	1	+ 0.857
^{3}H=T	1/2	+ 2.979

The term brute force polarization is used when the nuclei are in thermal equilibrium with the lattice and when the magnetic field at the nucleus is due only to an external magnetic field. Alternative to the thermal equilibrium method a non-equilibrium method has been developed with relaxed requirements with respect to magnetic field and temperature. It is known as "dynamic polarization" and will be discussed in the next section.

An important quantity for polarizing nuclei in solids is the nuclear spin-lattice relaxation time. It is a measure for the strength, with which the nuclear spin is coupled to the lattice. In the case of brute-force polarization one needs a not too slow spin-lattice relaxation in order to reach the equilibrium polarization in a reasonable time. For non-equilibrium methods, on the other hand, one needs a slow spin-lattice relaxation in order to retain the non-equilibrium nuclear polarization once established. The relaxation time has the tendency to increase towards lower temperatures, but it varies enormously from subtance to substance. At 0.01 K, e.g., some metals have a spin-lattice relaxation time of the order of a second, whereas pure dielectrics can have relaxation times up to years.

Dynamic Polarization

This method started to be developed in the early sixties to produce polarized proton targets for nuclear physics scattering experiments. The basic principle will be outlined here. The first step is to introduce some paramagnetic electrons into the sample. Because of their high magnetic moment μ_B compared to nuclear magnetic moments μ_n such electrons can be polarized quite easily (Fig.2a). For example a thermal equilibrium polarization of 93% is attained at B = 2.5 T and T = 1.0 K. The next step is to irradiate the sample with microwaves with frequency ν obeying $h\nu = 2 (\mu_B B - \mu_n B)$, see Fig.2b. A quantum of this radiation can be absorbed by an electron-nucleus pair, causing a spin exchange between the electron and the nucleus (Fig.2c). The electron flips back quickly (Fig.2d) due to its large magnetic moment and its strong coupling to the lattice. The nucleus depolarizes only slowly due to its weak coupling to the lattice. The nuclei can transfer their polarization due to mutual spin flips. In this way a small number of paramagnetic electrons can pump the nuclear spins of the whole sample.

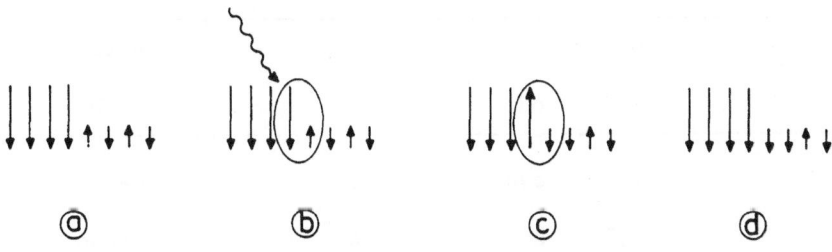

Fig.2. Principle of dynamic polarization, see
text for details.

In practice two slightly different mechanisms exist. The first is called "solid effect" and proceeds as described above. The second mechanism is known as "DNP" (dynamic nuclear polarization). Here the electron and nucleus do not flip simultaneously, but they are coupled by an additional energy reservoir. For more details the reader is referred to the literature.[6]

PROPERTIES OF SOLID HYDROGEN

Ortho-para Conversion

The properties of solid hydrogen are crucial in studying the feasibility of the production of polarized fuel. A comprehensive review on solid hydrogen has been given by Silvera.[7] In the solid state hydrogen appears in the form of H_2-molecules, as in the gas. Because the protons in an H_2-molecule are identical fermions the total wavefunction of the molecule has to be anti-symmetric. This brings about a correlation between the rotational states and the wavefunction of the proton spins. For symmetric rotational states (rotational quantum number J is even) the nuclear spin wavefunction has to be anti-symmetric. This means the total molecular nuclear spin $I_{tot}=0$ and the protons are anti-parallel. For anti-symmetric rotational states (J is odd) the nuclear spin state is symmetric with $I_{tot} = 1$. The situation is the same for tritium. Deuterium is different because deuterons are bosons with spin 1, therefore the total molecular wavefunction has to be symmetric. The rotational states involving the largest nuclear spin-degeneracy are designated ortho, the others para. The allowed combinations for the hydrogen molecules are listed in Table 2.

In hydrogen the energy spacing between the J = 0 and J = 1 rotational states corresponds to a temperature of 170 K, in deuterium it is half this value. At thermal equilibrium at T = 1 K or below the molecules are almost purely in the J = 0 state, therefore. Such low temperatures are needed for nuclear polarization. Inspection of Table 2 shows, however, that it is not possible to have polarized protons in J = 0 H_2-molecules, because the protons are anti-parallel. The same holds for tritium, but polarization of deuterons in D_2-molecules is possible. The mixed molecules such as HD do not have the ortho-para distinction, because the two nuclei are not identical. Thus no such limitation exists for polarizing the nuclei in mixed hydrogen molecules.

Table 2. Allowed combinations of nuclear spin states and rotational molecular states for the mono-isotopic hydrogen molecules; I_{mol} is the total nuclear spin of the molecule, J is the rotational quantum number. The species with the largest spin degeneracy are designated ortho.

Molecule	I_{mol}	J	$\Sigma(2I_{mol}+1)$	Designation
H_2, T_2	0	even	1	para
	1	odd	3	ortho
D_2	1	odd	3	para
	0,2	even	6	ortho

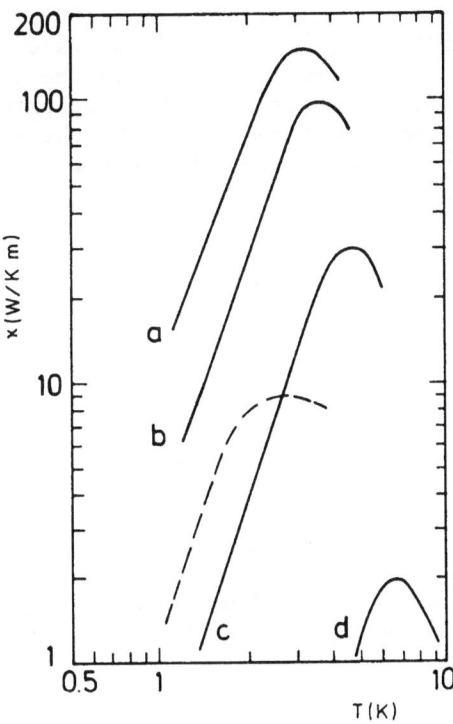

Fig.3. The solid lines are thermal conductivities of solid
hydrogen for various ortho-H_2 concentrations: a (0.3%),
b (1%), c (5%), d (30%), data from Refs.8 and 9. The
dashed line is the conductivity of pure solid HD.[10]

The transition from the J = 1 to the J = 0 rotational state is called
conversion. The conversion rate is quadratic in the concentration C_o of the
odd species:

$$\frac{dC_o}{dt} = -KC_o^2. \tag{2}$$

For hydrogen in the solid state the constant has the value K = 1.9%/h. The
conversion equation of solid deuterium additionally contains a linear term
in C_o. At C_o = 1 its conversion rate amounts to 0.06%/h. The ground-state
even rotational state is spherical, whereas the odd states are anisotropic
in charge distribution. By varying the concentration of the latter, one can
vary the amount of anisotropic interaction present in the sample. This has
large impact on heat conductivity and spin-lattice relaxation. Some heat
conductivity curves are shown in Fig.3 for various ortho-concentrations.
For a sample, which is left by itself, these features gradually change due
to the odd-even conversion.

Tritium Decay Heat

The difficulty that has to be overcome for polarizing a solid sample
containing tritium nuclei is the heat generated in the tritium decay. Tritium is a ß-emitter with a half-life of 12.3 y and an end-point energy of
18.6 keV. The average ß-energy is 5.7 keV. The range of these low-energy
ß's is very small, of the order of 0.01 mm in solid hydrogen. Thus for samples with larger dimensions all the ß-energy will be dissipated within the
sample itself.

The activity contained in 1 mol of DT is $A = 1.07 \times 10^{15}$ Bq (= 29 kCi). This corresponds to a heat load of $\dot{Q} = 1.07 \times 10^{15} \times 5.7 \times 10^3 \times 1.6 \times 10^{-19} = 0.98$ W. This heat load gives rise to a thermal gradient in the samples. Two gradients have to be considered. First we have the temperature jump between the cooling liquid and the bead surface due to the thermal boundary resistance (Kapitza resistance). The difference between the temperature of the surface T_s and of the liquid T_ℓ is given by:

$$T_s^4 - T_\ell^4 = \frac{\dot{Q}}{4\pi R^2 \alpha} , \qquad (3)$$

where $\dot{Q}$ is the heat load generated in the bead and R is the bead radius. The quantity α is the surface conductivity coefficient.

Apart from this temperature jump there is the radial gradient over the bead itself due to the heat flow to the surface. For a homogeneously distributed heat load $\dot{Q}$ it can be found from the following expression:

$$T_c^{\beta+1} = T_s^{\beta+1} + \frac{(\beta+1)\dot{Q}}{8\pi R\lambda} . \qquad (4)$$

Here T_c and T_s are the temperatures at the centre and at the surface, respectively, for a bead of radius R. The heat conductivity $\kappa(T)$ is supposed to have a temperature dependence of the form:

$$\kappa(T) = \lambda T^\beta, \qquad (5)$$

which appears to hold for many substances below 4 K. Quantitative examples are given below.

STATE OF THE ART OF POLARIZED HYDROGEN TARGETS

Hydrogen Compounds

Because of the ortho-para conversion in pure hydrogen physicists have looked for suitable hydrogen compounds as polarized targets. The only compound in which brute-force polarization of protons has been demonstrated up to now is TiH_2.[11] Polarizations up to 70% were achieved in a magnetic field of 9 T and at temperatures down to 10 mK.

Dynamically polarized hydrogen targets have been produced over two decades. The first target material was LMN (lanthanum-magnesium-nitrate), in which the protons of the crystal water were polarized. Since then one has searched for materials with a better ratio of polarizable free protons to unpolarizable protons bound in nuclei. The most favourable materials today are alcoholes and ammonia.[12] Proton polarizations over 90% are achieved at B = 2.5 T and T ~ 0.4 K in these materials. At these conditions the deuteron polarization is limited to 40 - 45%. For pushing it up one has to lower the temperature and/or to increase the magnetic field. A deuteron polarization of 49% in ND_3 has been reached in Bonn[12] at 3.5 T; polarizations of 64% and 71% in ^6LiD have been achieved in Saclay[13] at 4.8 T and 6.5 T, respectively.

A major issue is how to have stable paramagnetic electrons in the sample, because usually they are very reactive and pair off to spin 0 electron pairs. In the early days of polarized proton targets the 4f-electrons of the lanthanides were employed for this purpose, e.g. in LMN.[6] Later one has found special stable complexes containing a paramagnetic electron, which were employed to polarize protons in alcoholes.[14] The most recent method, applied e.g. in ammonia[15] and lithiumhydride,[13] is to create them by irradiation with electrons, protons or gammas. This is done at low temperature to prevent the recombination of the paramagnetic electrons. In ammonia the paramagnetic electrons appear to be situated at the radical $\dot{N}H_2$.

Pure Hydrogen

Brute-force polarization. In thermal equilibrium at low temperatures solid hydrogen occurs almost purely in the form of para-H_2 molecules in which the two proton spins are oriented anti-parallel to each other. In this state the protons can not be polarized. However, the ortho-para conversion is rather slow. This means that after quickly solidifying hydrogen gas from room temperature we will have a solid sample with a high temperature ortho-para distribution, i.e. 75% ortho and 25% para, which could be polarized up to 75%. The problematic factor here is the ortho-para conversion heat. It appears that the thermal gradient due to this heat in the sample is too high for brute-force polarization to be successful. In T_2 the situation is even worse because of the additional ß-decay heat.

Deuterium, on the other hand, can be polarized because its lowest molecular state is ortho, see Table 2. After solidifying and further cooling of the D_2 in a magnetic field one will have a slowly rising polarization due to slow conversion. The very high magnetic field and the very low temperature, that are required (see Fig.1) have prevented such samples from being produced up to now.

The para-ortho distinction does not apply for HD-molecules, because proton and deuteron are distinguishable particles. A brute-force proton polarization of 40% has been achieved by Bozler et al.[16] at T = 23 mK and B = 10 T. A small amount of ortho-H_2 had been added to the HD in order to decrease the proton spin-lattice relaxation time, which otherwise would have been on the order of many days.

By the subsequent conversion of the ortho-H_2 the spin-lattice relaxation time increases steadily in such a polarized sample, because the proton-lattice coupling in para-H_2 is very weak. This decoupling of the polarized protons from the lattice opens up the possibility to operate the sample at moderate fields and higher temperatures, e.g. 4.2 K, where relaxation times exceeding a day are expected if one has waited long enough.[17]

Dynamic Polarization. The process of dynamic polarization requires the presence of paramagnetic electrons. If these are introduced in ortho-H_2 (the polarizable, metastable specimen) they speed up the conversion rate enormously.[18] Because of this reason dynamic polarization of protons in solid H_2 can not be realized.

In D_2 the situation is just opposite to H_2, see Table 2. Here the lowest molecular state is the one in which the nuclei can be polarized. Hence the introduction of para-magnetic electrons has a favourable side-effect in accelerating the conversion to the desired state. Over 20 years ago about 0.35% polarization has been achieved[19] at B = 0.85 T and T = 4.2 K. Since then this has not been developed any further.

In the HD-molecule the nuclei are different, hence no symmetry requirement limits the polarization. Also the problems of conversion heat and conversion time are absent. This makes HD to an attractive substance. The dynamic polarization of the nuclei in HD has been studied by Solem.[18] Paramagnetic electrons were created by exposing solid HD-samples to bremsstrahlung. Their EPR (electron para-magnetic resonance)-spectrum has two lines, some 500 G apart. In pure HD the electrons relax too slowly for effective dynamic polarization. This problem can be solved by mixing in a faster-laxing paramagnetic impurity such as O_2. A concentration of some parts in 10^4 appears to be sufficient. The total radiation dose was about 1.5×10^7 R. After irradiation the proton relaxation time dropped to about 20 s and then increased to about 70 s (at 4.2 K) at a rate characterized by ortho-para conversion, because some of the HD dissociated during irradiation had recombined to H_2 and D_2.

Maximum polarizations of 3.75% for protons and 0.35% for deuterons were obtained at B = 1.24 Tesla and T = 1.2 K using one of the two EPR-lines. In an earlier experiment 1% deuteron polarization was achieved.[20] Improvement by a factor of two may be expected by driving both EPR-lines, going to a higher magnetic field of 15 Tesla may give a factor of 12 and decreasing the temperature to 0.5 K may provide for another factor of 2.4. These simple extrapolations lead to an increase by a factor 58. This number may be too optimistic, but extra improvements may be attained by optimizing the microwave irradiation and the density of paramagnetic electrons. Thus high nuclear polarizations seem to be realisable for HD and therefore also for DT if it can be cooled to 1 K or below. The latter problem is addressed in the next section.

DYNAMIC POLARIZATION OF DT

Brute-force polarization of DT would require a sample temperature below 5 mK, see Fig.1. It will be shown below that it is impossible to reach such low temperatures inside a DT sample due to the tritium decay heat. For example, the temperature jump between the liquid helium and the sample surface is expected to be of the order of 100 mK or more. Therefore we consider here only the dynamic polarization method.

Radical Creation

For dynamic polarization the ß-radiation of tritium also has a positive aspect: it causes the dissociation of a fraction of the DT-molecules, creating in this way the paramagnetic electrons that are necessary for the polarization process. The question now is, whether the rate of dissociation is adequate, i.e. whether the desired amount of free atoms is attained and whether this occurs in a reasonable time after solidification. Preliminary conclusions can be drawn from two experiments.

In the first experiment[21] a D_2 sample had been doped with 1% T_2. The EPR-spectra of the D- and T-atoms were studied. It was found that the concentration of free atoms of both species reached saturation after about 5 days. For the free atoms concentration a lower limit was deduced of 3×10^{-5} per D_2-molecule. In first approximation the saturation time would scale down to 2.4 hours for a DT-sample. A higher saturation concentration of free atoms may be expected for DT, because the recombination rate will be similar at the above concentration of free atoms, whereas the dissociation rate is 50 times higher. No polarization measurements were reported in this work.

The second experiment[18] has been discussed already above. An HD-sample was irradiated by 7×10^5 R/h bremsstrahlung. The exposition time was 21 h, resulting in a total dose of 1.5×10^7 R. The free H-atom density was estimated to be $\sim 10^{-4}$ per HD-molecule. An equivalent dose will be reached in DT in 13 min. after solidification.

The results of both experiments indicate that sufficient free atoms will be generated within a few hours after solidification. The polarization build-up time is shorter for higher free atom concentrations. Therefore, a high concentration is favourable. On the other hand, the increasing amounts of free D- and T-atoms, together with the ortho-T_2 and para-D_2 that are formed from it, will give rise to a significant reduction of the nuclear relaxation time. This may, of course, not become shorter than the time necessary to transport the sample from the polarizing cryostat to the reactor vessel. The free atom concentration will have to be such that the polarization build-up time is not too long and the nuclear relaxation time is not too short. If necessary, the polarization build-up time can be decreased by increasing the microwave power.

Temperature Gradient in a Solid Bead

Let us consider a DT-bead with a radius of 1 mm. It contains 0.21×10^{-3} mol of DT, assuming a molar volume of 20 cm^3. The heat released in such a bead is 0.21×10^{-3} x 0.98 W = 0.21 mW. This heat can be carried away easily by dilution refrigerators above 0.1 K. For example, the largest existing dilution refrigerator[22] has a cooling power of 0.2 W at 0.2 K, allowing to cool 1000 of such beads simultaneously. Assume the beads are cooled by liquid helium of 0.2 K. The surface temperature T_s of the bead can be found by Eq.3. We adopt the value $\alpha = 2.5 \times 10^{-3}$ W/cm^2 K^4, which is adequate for light dielectric solids. The result is $T_s = 0.90$ K. This value is mostly due to the surface conductivity coefficient α. Increasing the liquid temperature T_ℓ from 0.2 to 0.4 K, e.g., yields $T_s = 0.91$ K. Increasing the surface conductivity with a factor 2 gives $T_s = 0.76$ K, decreasing it with a factor 2 yields $T_s = 1.08$ K.

The thermal gradient over the bead appears to contribute only little. If we assume $T_s = 1.0$ K, and employ the heat conductivity curve shown for HD in Fig.3, we find from Eqs.4 and 5 for the temperature of the centre of the bead $T_c = 1.008$ K. Hence, bead temperatures around or below 1 K seem to be attainable. Of course, these results have to be verified by determining the thermal properties of DT itself. If these would appear to prevent the cooling of a 1 mm radius DT bead to below 1 K, one might think of producing solid DT in shapes with a larger surface/volume ratio, e.g. porous material or snowflake structures.

Temperature Gradient in a Multilayered Pellet

In the concept of Inertial Confinement Fusion the nuclear fuel is in the centre of a small pellet, surrounded by one or more layers of different material. Early laser fusion pellets, e.g., consisted of a thin glas balloon filled with high pressure deuterium - tritium gas. A recent design for a heavy ion fusion pellet, the HIBALL-2 pellet[23] is shown in Fig.4. The innermost shell is a thin layer of frozen DT. The interior of the pellet is empty. This construction allows for a very high density after compression. It is also favourable from a thermodynamic point of view, because the temperature jump, due to the Kapitza resistance to liquid helium, is reduced by the significant increase in surface area compared to a fully solid bead. This advantage, however, is reduced by the occurrence of extra thermal resistancies in the pellet. Quantitative estimates are given below.

The outer shells of the HIBALL pellet are metallic. This is intolerable in view of the dynamic polarization mechanism. The required microwave radiation of about 400 GHz can not penetrate metallic layers do to the skin effect. At this frequency the skin depth of lead at low temperature is of the order of 0.1 μm, three orders of magnitude below the thickness of the lead layer. Hence non-metallic tamper and pusher layers have to be employed. In view of the fact that pellet design is still an open question and many types are discussed, this limitation should not be regarded as serious at this stage.

We will consider now a pellet with the dimensions of the HIBALL-pellet, but having non-conducting tamper and pusher shells. It contains 4 mg DT, corresponding to 0.8×10^{-3} mol. This generates a ß-decay heat load of 0.8×10^{-3} x0.98 W = 0.78 mW. The pellet is cooled with liquid helium of

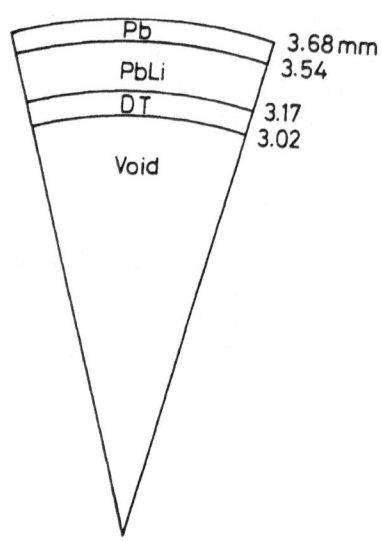

Fig.4. Design of the HIBALL-2 ICF pellet

T_ℓ = 200 mK. In Fig.5 temperature profiles in the pellet are shown, calculated with Eqs. (3) - (5). The upper curve is obtained by assuming 10 times
lower values for the heat conductivities than for the lower curve. The surface conductivity coefficient α was varied from 1.5 x 10^{-3} (upper curve)
to 1.5×10^{-2} W/cm^2 K^4 (lower curve). At the interior boundaries the surface

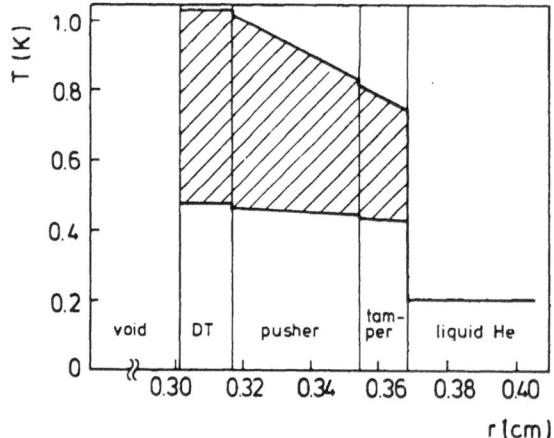

Fig.5. Calculated thermal gradients over the Hiball-2 pellet
cooled by liquid helium of 0.2 K. The upper curve is
obtained with 10 times worse thermal conductivities
than the lower curve

conductivities were assumed to be 10 times higher. The bulk thermal conductivity κ of tamper and pusher was varied between 0.01 and 0.1 W/Km. These values range in the lower and middle region of thermal conductivities.[24] The largest temperature jump is found at the outer surface of the pellet due to the surface boundary (Kapitza) resistance to liquid helium. The resulting average temperature of the DT-layer appears to vary between 0.5 and 1.0 K.

Prospects for Polarizing DT

From the results obtained in HD, that were discussed a few sections back, we have concluded that high polarizations can be expected in solid DT if it can be cooled to 1 K or below. The above sections have shown that such temperatures seem to be realistic, based on reasonable assumptions for the Kapitza resistance and heat conductivity of DT and of the tamper and pusher materials in the case of a multilayered pellet. Hence the chances for polarizing solid DT are good. Experiments on DT itself finally will have to show what polarizations can be achieved.

An important aspect, which has not been discussed yet, is the nuclear spin-lattice relaxation time. It denotes the life-time of the polarization after the polarizing procedure has been stopped. In hydrogen the dominant relaxation mechanisms are interactions with electron paramagnetic impurities and with the $J = 1$ impurities ortho-H_2 (or T_2) and para-D_2.[25] In DT such impurities are created continuously by the radiation damage of the β's from tritium.

In HD relaxation times have been measured after irradiations with γ-rays,[18] electrons and protons.[17] The results vary strongly as a function of radiation dose. A lowest value of 20 s was measured for the highest dose: 1.5×10^7 R bremsstrahlung, accumulated in 21 h at a sample temperature somewhat above 4.2 K in zero magnetic field. In DT the relaxation time decreases with time because of the continuous accumulation of radiation damage. The above dose of 1.5×10^7 R will be reached in 13 min after solidification. At that instance a relaxation time can be probably expected of the order of the above 20 s, though deviations will occur due to different ionization density, irradiation time, sample temperature and magnetic field.

Because of the continuously decreasing relaxation time both the polarization procedure and the transfer to the reactor of the polarized pellet should be made as fast as possible. The fastest transfer would be achieved by a system in which the pellet is solidified an polarized inside the injector right in front of the gun. For unpolarized pellets such systems already exist. Injection velocities, reached with a burst of pressurized gas, are close to 2000 m/s.[26] This means flight times of the order of 10-20 ms. As soon as the polarization procedure has stopped the nuclear polarization starts to decay. Hence to the flight time one has to add the time elapsing between stopping the polarization and firing the gun. The apparatus has to be designed to minimize this step. It is clear that the exact value of the relaxation time immediately after finishing the polarization is of crucial importance. If it is around 1 s or less, it will be very difficult to inject fast enough to prevent a considerable polarization decay of the sample.

CONCLUSIONS

The employment of polarized fuel is an interesting topic for fusion reactors. Completely polarized DT-fuel would provide a 50% increase in fusion cross section, meaning relaxed confinement conditions and earlier break-even. The accompanying anisotropic emission of the neutrons may also be advantageous. The price to be paid for polarization is an extra amount of technical equipment: mainly cryogenics if it is considered to polarize solid fuel as was done in this paper. A favourable circumstance is that cryogenic techniques are considered anyway for fuel injection, apart from polarization arguments. Density and confinement time of magnetically confined plasmas can be improved with solid bead injection. The most recent designs for ICF pellets conceive a multilayered hollow structure with an inner layer of frozen DT. Hence the concept of polarizing solid fuel fits well within the present ideas of fuel injection.

Polarized proton and deuteron targets are well established in the nuclear and particle physics community. As target material, however, always hydrogen compounds have been taken up to now. The reason being that $^1H_2$ (2 protons) can not be polarized, because at low temperature the molecules are in the para-state with proton spins aligned anti-parallel to each other. The same is the case for tritium molecules. Polarization is possible for D_2 and the bi-isotopic molecules. Hence, polarization of a DT-mixture can only be achieved if DT-molecules are employed. No polarization data on DT are known, but some experience has been made with HD, in which polarizations of a few % have been achieved over a decade ago. It may be assumed that the thermodynamic properties of HD and DT are similar. Extrapolation of the HD results to what is technically achievable nowadays leads to the conclusion that high polarizations in DT are possible in spite of the self heating by the ß-decay of tritium.

Polarization experiments on DT samples will have to reveal, whether high polarizations can be achieved and retained. Not much investment will be necessary, because the basic research on polarizing DT can be carried out with existing polarization equipment. Such a program seems to be well justified because it may bring the goal of a working fusion reactor closer to its realization using only well established techniques.

REFERENCES

1. R.M. Kulsrud, H.P. Furth, E.J. Valeo and M. Goldhaber, Fusion Reactor Plasmas with Polarized Nuclei, Phys. Rev. Lett. 49: 1248 (1982).
2. R.M. More, Nuclear Spin-polarized Fuel in Inertial Fusion, Phys. Rev. Lett. 51: 396 (1983).
3. B. Goel and W. Heeringa, Spin polarised ICF targets, in: "Emerging Nuclear Energy Systems", G. Velarde and G. Mínguez, ed., World Scientific, Singapore (1987).
4. M. Greenwald et al., Energy Confinement of High-density Pellet-fueled Plasmas in the Alcator C Tokamak, Phys. Rev. Lett. 53: 352 (1984).
5. L.L. Lengyel, Ignition and Fuel Scenario Calculations for Neutral-beam-heated Tokamak Reactors Based on Pellet Injection, Fusion Techn. 10: 354 (1986).
6. A. Abragam and M. Goldman, Principles of Dynamic Nuclear Polarisation, Rep. Prog. Phys. 41: 395 (1978).
7. I.F. Silvera, The Solid Molecular Hydrogens in the Condensed Phase: Fundamentals and Static Properties, Rev. Mod. Phys. 52: 393 (1980).
8. R.G. Bohn and C.F. Mate, Thermal Conductivity of Solid Hydrogen, Phys. Rev. B2: 2121 (1970).
9. J.E. Huebler and R.G. Bohn, Thermal Conductivity of Solid Hydrogen with Ortho-hydrogen Concentrations between 20 and 70 at. %, Phys. Rev. B17: 1991 (1978).
10. J.H. Constable and J.R.Gaines, Low-temperature Heat Transport in Solid HD, Phys. Rev. B8: 3966 (1973).
11. R. Aures, W. Heeringa, H.O.Klages, R. Maschuw, F.K. Schmidt and B. Zeitnitz, A Brute-force Polarised Proton Target as an Application of a Versatile Brute-force Polarisation Facility, Nucl. Instrum. and Methods 224: 347 (1984).

12. W. Meyer, Review of Operational Polarized Targets, Helv. Phys. Acta 59: 728 (1986).
13. W. Bouffard, Y. Roinel, P. Roubeau, A. Abragam, Dynamic Nuclear Polarization in ^{6}LiD, J. Phys. Paris 41: 1447 (1980).
14. M. Krumpolc, B.G. DeBoer and J. Rocek, A Stable Chromium (V) Compound, J. Am. Chem. Soc. 100: 145 (1978).
15. T.O. Niinikoski and J.M. Rieubland, Dynamic Nuclear Polarization in Irradiated Ammonia Below 0.5 K, Phys. Lett. 72A: 141 (1979).
16. H.M. Bozler, J.A. Brown and E.H. Graf, Proton Resonance in Highly Polarized Hydrogen Deuteride, Bull. Am. Phys. Soc. 18: 545 (1973).
17. H. Mano and A. Honig, Resistance of Solid HD Polarized Proton Targets to Damage from High-energy Proton and Electron Beams, Nucl. Instrum. and Methods 124: 1 (1975).
18. J.C. Solem, Dynamic Polarization of Protons and Deuterons in Solid Deuterium Hydride, Nucl. Instrum. and Methods 117: 477 (1974).
19. G.A. Rebka and M. Waine, Dynamic Orientation of the Nuclei of Solid Deuterium, Bull. Am. Phys. Soc. 7: 538 (1962).
20. G.A. Rebka and J.C. Solem, Dynamic Polarization of Nuclei in Solid Deuterium Hydride, Bull. Am. Phys. Soc. 12: 1064 (1967).
21. M. Sharnoff and R.V. Pound, Magnetic Resonance Studies of Unpaired Atoms in Solid D_2, Phys. Rev. 132: 1003 (1963).
22. T.O. Niinikoski, Dilution Refrigerator for a Two-litre Polarized Target, Nucl. Instrum. and Methods 192: 151 (1982).
23. R. Fröhlich, B. Goel, D.L. Henderson, W. Höbel, K.A. Long and N.A. Tahir, Heavy Ion Beam Driven Inertial Confinement Fusion Target Studies and Reactor Chamber Neutronic Analysis, Nucl. Eng. and Design 73: 201 (1982).
24. O.V. Lounasmaa, "Experimental Principles and Methods Below 1 K", Academic Press, London (1974).
25. M. Bloom, Nuclear Spin Relaxation in Hydrogen, Physica (Utrecht) 23: 767 (1957).
26. M. Kaufmann, Review on Pellet Fuelling, Plasma Phys. 28: 1341 (1986).

III. MAGNETIC CONTROLLED THERMONUCLEAR FUSION

CONTROLLED THERMONUCLEAR FUSION BY MAGNETIC CONFINEMENT - STATE OF THE ART AND STRATEGY

G. Grieger

Max -Planck-Institut für Plasmaphysik
Euratom-IPP Association
8046 **Garching**, Federal Republic of Germany

This paper was prepared on short notice (overnight) upon request of the organizers and without any background material available. The idea of the organizers was to use the development path of magnetic fusion as an example and as a basis for an assessment of the status of muon catalized fusion. With this idea in mind, this paper is not addressing the expert of magnetic fusion but rather our colleagues of muon catalized fusion. It is trying to start "ab ovo", to extract the essentials, to give arguments by order of magnitude and to avoid unnecessary details. Where appropriate, comparison with muon catalized fusion is made.

INTRODUCTION

Not only muon catalized but also magnetic fusion aims at exploiting the reaction

$$D + T \rightarrow {}^4He + n + 17.6 \; MeV$$

because its reaction cross-section is by far the highest among its potential competitors. This conclusion is not drawn lightheartedly because the fuel component T is not available in nature but has to be bred, and among the reaction products there are high energy neutrons which are certainly needed for T-breeding but which also produce radiation damage and activation in the reactor structural material. There exist fusion processes which are free of such undesired side-effects but, unfortunately, their cross-section is too small to be considered as the first choice. For most (but not all) of them it is even a question whether they can be considered at all for economic energy production by nuclear fusion.

Nuclear fusion of the electrically charged fuel nuclei can only occur if the particles have enough energy to overcome the mutually repelling Coulomb potential. Under the conditions of interest is happens that

$$\sigma \; (Coulomb) \approx 100 \; \sigma \; (fusion)$$

which means that all proposals for achieving nuclear fusion by colliding beams of sufficient energy have no chance to work because the particles will be scattered before a sufficient number of fusion processes can occur. Therefore, the conditions have to be arranged such that, on the average, the kinetic energy invested into the fuel particles is preserved during all occurring Coulomb collisions. This leads to the concept of a high temperature reacting gas or, better, plasma. The task is then to provide the required fusion conditions in a volume of limited size, i.e. one needs a pressure containment with good thermal insulation.

Here is a fundamental difference between muon catalized and magnetic fusion. In magnetic fusion one aims at fuel ignition followed by controlled burn. This is achieved by first heating the fuel by externally applied power, until the temperature is high enough for sufficient fusion reactions to occur. From then on the burn process is maintained by exploiting the energy of the fusion α-particles which couple their energy to the plasma by Coulomb collisions. This way the process can be kept going by internal compensation of the power losses. The fusion neutrons do not interact with the plasma but are moderated in the blanket surrounding the reaction chamber and are also used for T-breeding. In contrast, muon catalized fusion can only work as a power amplifier. High quality (electric) energy has to be continuously invested to produce muons. Their excess energy occurs as heat which is a low grade energy and can only be re-circulated to the accelerator with a conversion efficiency of about 1/3. For a fusion reactor to become practical the recirculating power should probably not exceed 20 % or so of the systems output power.

Confinement of charged particles can best be achieved by application of a magnetic field. Unfortunately, a homogeneous magnetic field only acts in the two dimensions perpendicular to B. Allowing free streaming of the plasma parallel to B would lead to a length of the device of more than 10^3 km, which is not practical. Mirror machines increase the magnetic field strength at both ends of the device and act on the particle's magnetic moment, $\mu \nabla B$. Comparison with the above numbers shows that very large reflection coefficients are needed to arrive at tolerable system lengths. Closing the magnetic field toroidally avoids the end-loss problem in principle but since

$$\oint B dl = \mu_0 I$$

the magnetic field is not constant everywhere but decays $\sim R^{-1}$ and thus leads to a drift motion of the particle perpendicular to both B and ∇B,

$$v_D = kT/RB$$

where R is the radius of curvature of the magnetic field. This velocity is by orders of magnitude smaller than the particle's thermal velocity but is still unacceptably large. In addition, it has opposite sign for opposite charges, thus leads to charge separation of electrons and ions, and to excessive ExB losses in turn. The solution to this remaining problem is to twist the magnetic lines by superimposing a poloidal magnetic field upon the toroidal one, as indicated in fig. 1. The result are nested magnetic surfaces each one generated in an ergodic fashion by the repeated toroidal and poloidal revolution of one single field line. This method allows fast communication within each magnetic surface and guarantees the highest possible thermal insulation from surface to surface. Also in this case one has to pay because the currents balancing the charge separation flow along a small but finite plasma resistance and dissipate energy. The plasma has to generate this energy by a somewhat faster expansion velocity than necessary otherwise. But this factor is easy to tolerate. In the European Community this toroidal magnetic confinement concept is attributed the highest probability for final success and, already long time ago, the European Community has concentrated all its efforts on this scheme. In order to be complete it should be mentioned that the twist of the field can be introduced by two means either by a toroidal plasma current (Tokamak) or by currents in an external coil system (Stellarator).

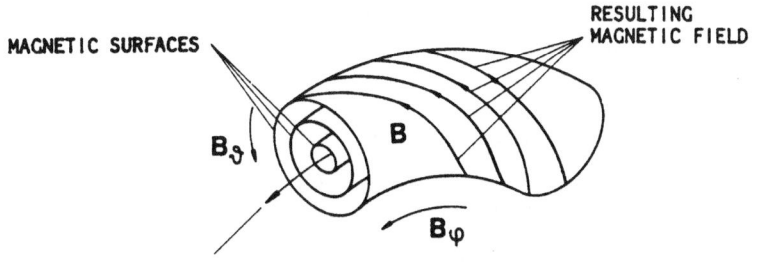

Fig. 1. Nested magnetic surfaces formed by superposition of toroidal and poloidal magnetic field components

PROBABLE REACTOR PARAMETERS

After having developed the basic principles for the plasma core of a fusion reactor, one has to ask which are the desired parameters of the reactor plasma? These are easy to imagine already from simple-minded considerations:

As already said, DT will be considered as fuel. Its reaction cross-section is strongly increasing with temperature up to 50 keV or so pointing towards high temperature operation. The practical problem, however, is the pressure containing capacity of the confinement system which is a certain fraction β of the magnetic field energy density, $B^2/2\mu_0$. The fusion power should therefore be maximized with the side-condition for the plasma pressure, nkT, to be constant, i.e. when raising the temperature, the density has to be reduced for compensation. From the temperature dependence of $\langle\sigma\cdot v\rangle$ and

$$P_{fus} = n_D \cdot n_T \cdot \langle\sigma\cdot v\rangle \, E_{fus}$$

an optimum operating temperature of

$$T_{opt} \approx 12.5 \text{ keV}$$

can then be derived. In a rather wide range around this optimum the fusion power varies according to

$$P_{fus} \sim n^2 T^2$$

The absolute value of the operating pressure is determined by the technically achievable strength of the magnetic field which for the present superconductors and structural materials is around 5 - 6 T at the magnetic axis, and by the fraction β of its energy useful for confinement, $\beta \approx 5$ %. This immediately leads to a practical value for the plasma density of

$$n \approx 2 \cdot 10^{20} \text{ m}^{-3}.$$

The final quantity to be determined is the thermal insulation of the plasma required for achieving controlled burn. This is usually expressed by the energy confinement time which is the time constant of the plasma energy decay if the heating power maintaining the burn (α-power) were suddenly switched off. Losses by heat conduction depend on the confinement properties and can thus in principle be managed to be small but an unavoidable loss is Bremsstrahlung, its density being proportional to $n_e^2 \cdot \sqrt{T}$. This power density is independent of size because the fusion plasma is optically thin. Its consideration leads to the so-called Lawson criterion

$$n \cdot \tau \approx 2 \cdot 10^{20} \ m^{-3} \ s.$$

A typical value of τ is one second.

PROBLEMS

Plasma physics is governed by Coulomb collisions between the charged particles. These are caused by long-range forces leading to many-body problems. Unfortunately, the related physics cannot be written in dimensionless quantities without severe simplifications. Thus there is no similarity law in fusion plasma physics, and as long as physics is not fully understood the uncertainty of extrapolation increases with the extrapolation factor. It is for this reason that the fusion programme proceeds in steps of carefully chosen width. The present step is based on a considerable variety of medium-size experimental devices, each one concentrated on a particular subset of problems.

About a decade ago, time was considered ripe to integrate the results so obtained into one single machine of next step size. This conclusion was drawn by each of the four major fusion programmes separately and four large-size Tokamaks JET (EC), TFTR (USA), JT-60 (Japan), and T-15 (USSR) were built. Except for T-15, these devices have been in operation for some time and are producing results by using H or D plasmas which are not too far from reactor parameters otherwise. These results include that fusion temperatures have been reached though at lower densities. Fusion densities have been achieved at temperatures lower than the reactor values and/or at magnetic fields higher than the reactor ones. Even $n \cdot \tau$ has been reached in high field machines. In total, β, τ, and $n \cdot T$ are lacking factors of only 3 or so and $n \cdot T \cdot \tau$ is lacking a factor 20. The chances look good for JET to achieve significant α-production when it will later be operated with DT.

It is now time to talk about the probable size of fusion reactors. Although the thermal insulation by the magnetic field is higher by orders of magnitude than that of the best material insulators in daily life, the large temperature difference requires plasma radii of 1 - 2 m. Economic considerations request high fusion power densities so that a typical output power of a fusion plant would be 3 GW thermal or 1 GW electric. This is about the right unit size for non-US countries. Only the particular conditions of the US electric power market favour smaller unit sizes. The power needed to heat the plasma to ignition is quite a fraction of the fusion burn alpha power, e.g. about 70 MW for INTOR.

Here a remark should be made: If for magnetic fusion the goal of a burning fusion plasma would be missed by a not too large factor, one always could compensate by, for instance, increasing the size of the device. Certainly, this would also increase the unit-size power output and there are limitations. But it would not provide a principle problem like a too large μ-sticking to helium for the case of muon catalized fusion.

254

Until recently, the fusion programmes were concentrated nearly entirely on solving physics problems and also the JET generation is still designed and built for solving physics questions. This strategy was chosen because it was in pyhsics were the largest hurdles were expected. Now the definition of the Post-JET generation (INTOR, NET, ETR ...) has started. Apart from demonstrating the viability of the plasma physics, their task will be to serve as a test bed for the development of fusion technology. This transition has also affected the orientation of the physics work: It is no longer sufficient to produce "solutions in principle" but the solutions offered have to be compatible with other criteria. Examples are

(i) Nested magnetic surfaces were chosen as confinement principle with one magnetic line densely forming a magnetic surface. Realistically, however, the twist of a magnetic line is a function of the plasma radius, e.g. as indicated in fig. 2. If this is so, then for certain radii the twist number becomes rational, m/n, which means that the magnetic line is closing upon itself after n revolutions rather than forming the whole magnetic surface. Surfaces with such rational twist numbers are very sensitive to error fields wich lead to the formation of magnetic islands as indicated in fig. 3. These islands are susceptible to MHD activity, loss of stability, and, perhaps, finally to current disruption.

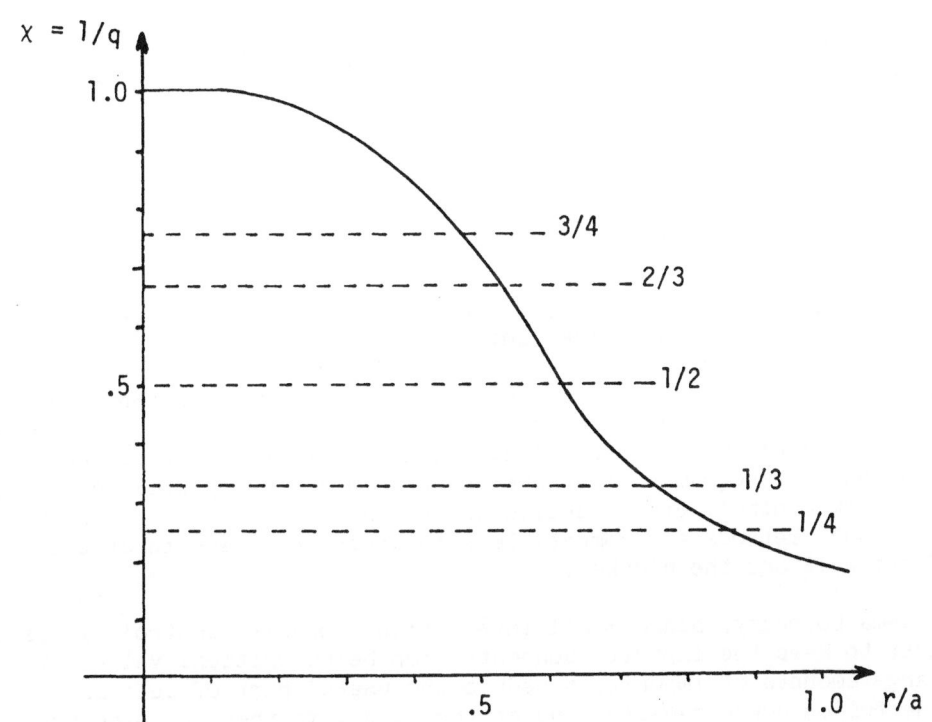

Fig. 2. Typical profile of the twist of magnetic lines, $X = 1/q$, vs. the normalized plasma radius, r/a. The low-order resonances, m/n, are indicated. At these points of the profile the magnetic line closes upon itself after m revolutions the long way and n revolutions the short way rather than filling this surface ergodically.

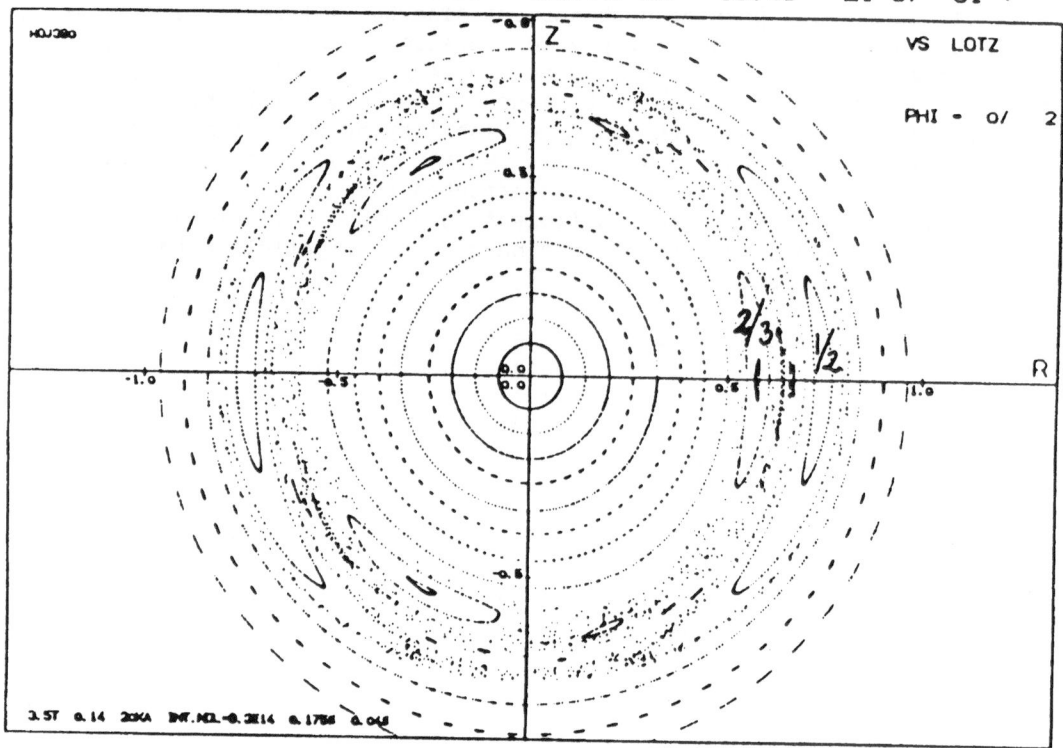

Fig. 3. Cross-section through magnetic surfaces as an example. Points
are transitions of a magnetic line through a plane perpendicular
to the magnetic axis. Circular figures are ergodically formed
by one and the same magnetic line each. In between islands are
clearly visible at t = 1/2 and 2/3.

Disruptions have to be avoided under reactor conditions because the
plasma energy will be dumped into the chamber wall on a fast timescale and will there produce surface melting. On a slightly slower
timescale also the plasma current will be induced into the wall and
lead to large electromagnetic forces which easily can become excessive. Reliable means to avoid disruptions have to be developed in
the running programme. They are using methods like profile control,
feedback control, etc.. Particular care has to be taken to ensure
that such methods are compatible with other requirements of the
first wall and the blanket.

(ii) Plasma boundary, plasma-wall interaction, impurity control. It is a
must to keep the impurity concentration below critical values. Too
large amounts of impurities reduce the useful part of beta and thus
the fusion power density, and at the same time they increase the
plasma power losses via radiation. A rule of thumb says that oxygen-
like impurities have to stay below 1 %, iron-like ones below 0.1 %,
and tungsten-like impurities below 0.01 %. The main generation of
impurities is via sputtering from the wall. Under INTOR conditions,
the α-power arriving at the wall by heat conduction, energetic particles and radiation amounts to about 300 kW/m² if it were uniform-
ly distributed. In reality, however, one will have rather large

peaking factors. For the present experiments the issue is the limitation of the radiation losses and not yet to avoid dilution of the
fuel. This can be achieved by carbonization of the wall. For a
reactor it might be necessary to use a divertor (see fig. 4) for
keeping the plasma clean. This device introduces a scrape-off layer
around the plasma. All particles entering the scrape-off layer have
a high chance to move towards divertor chambers by flowing parallel
to B. There, their energy can be extracted and the particles be
pumped away. To make this process efficient the temperature of the
boundary layer should be kept below the sputtering threshold.

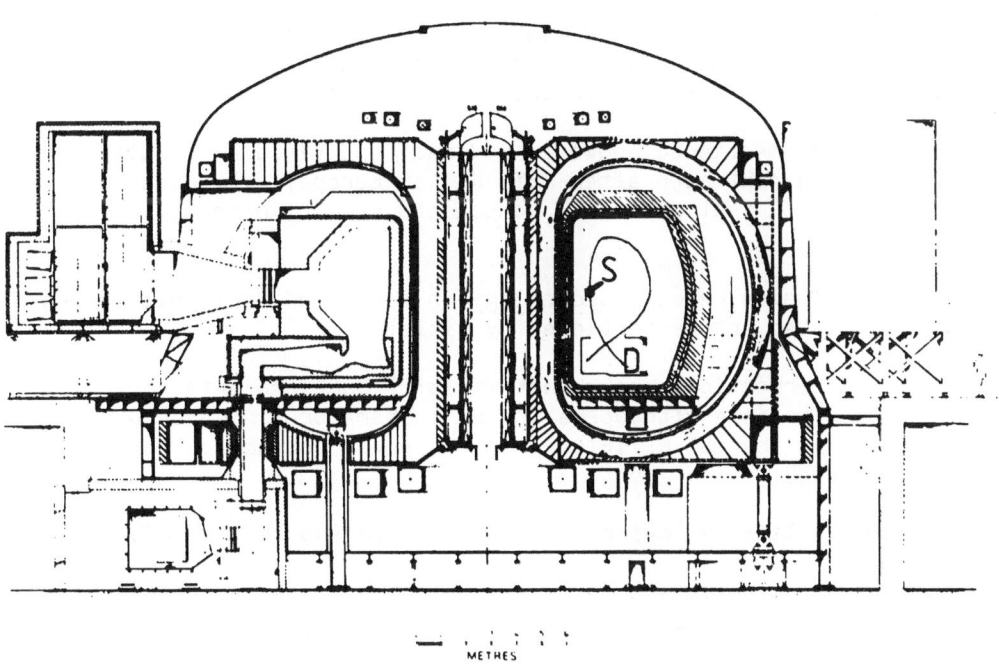

METRES

Fig. 4. Elevation view of the INTOR Phase I design configuration. One
recognizes the divertor (D) into which the magnetic lines
outside the shown separatrix (S) are led so that the particles
diffusing out of the plasma volume can there be collected.

Another challenging task is to cope with the continuous helium production
by the fusion processes. Although it dilutes the fuel, burn-up fractions
of 5 % or so have to be accepted to arrive at tolerable pumping powers
and tolerable throughputs through the mass separation, fuel clean-up
system. In this connection, it is useful to remember that the reactor
plasma does not contain more than 1 g of tritium and that any fusion
reactor of 3 GW thermal power produces about 2 kg He per day. For muon
catalyzed fusion the tritium content of the reaction chamber seems to be
much higher and the tolerable helium concentration much lower.

Further critical issues in magnetic fusion are the

- development of efficient heating methods for achieving ignition. These methods have to be compatible with the plasma edge conditions.

- Optimization of the configuration for confining high plasma pressure.

- Refuelling by DT pellets.

In this connection, the following remark might be useful: If the burning state is reached in magnetic confinement, the system burns upon refuelling and exhaust alone. Some means for controlling the burn might be needed, however. If the twist of the magnetic lines is generated entirely by currents flowing in external coils, the system has the full potential for steadystate operation. If, on the other hand, the twist is generated by a plasma current, one would be led to a pulsed operation by transformer action or would need means for non-inductive current drive, e.g. by travelling waves. The optimization of the confinement concept is done in parallel with the development of fusion reactor technology.

All the work mentioned above is underway and yielding results. The transition to the next step after JET, NET in Europe, will occur once there is conviction that all basic components of a DT operated device will work.

As far as fusion reactor technology is concerned, critical issues will only be listed and be mentioned by title:

- Tritium permeation through hot walls: T-barriers have to be developed compatible with operation at temperatures high enough to allow high thermal efficiency of the power conversion into electric energy.

- Divertor plates allowing high power loads (up to 5 MW/m^2).

- Structural materials to balance the magnetic forces. These materials should have high yield strength, low sensitivity to activation and radiation damage by the fusion neutrons and be fatigue resistant.

- Blankets are needed for tritium breeding and power conversion. These blankets should possess high breeding efficiency, allow easy extraction of heat and of tritium, lead to a small tritium inventory, and have a small electric conductivity.

- Superconductors for the magnet should allow high current density at high magnetic fields, should be easy to stabilize (high operating temperature) and show low sensitivity to radiation damage. Equally important is the development of non-brittle insulators with high resistance against radiation damage.

All these points are under investigation and solutions are being developed. There are two inherent timescales for these tasks. Some of the technologies are needed already for the construction of NET. Those for which NET will serve as a test bed may use a longer time for development.

FINAL REMARKS

The magnetic fusion programme fully concentrates on the utilization of the DT fusion process. In fact, there is no other choice at present since the DT fusion cross-section overwhelmingly exceeds all the competing ones. One therefore accepts the need for a tritium breeding and neutron power converting blanket. Exploitation of the DD reaction would drop the need for fuel breeding but still involve tritium and neutrons via the reaction products. A nearly ideal reaction would be

$$D + {}^3He \rightarrow {}^4He + p$$

which, in principle, would not involve any neutron. In addition, the fusion energy would occur via the kinetic energy of stable, charged particles which might even make it possible to convert the kinetic energy of the charged particles directly into electric power. The simultaneous occurrence of the DD reaction would be suppressed sufficiently by operating the reactor with an excess concentration of ${}^3He$ (e.g. $D:{}^3He = 1:9$) and by proper spin polarization of the fuel atoms. The plasma pressure, however, would have to be increased by an order of magnitude or more in order to get the same fusion power density as with DT. Perhaps, new superconductor developments will allow increasing the magnetic field intensity and thus the plasma operating pressure.

${}^3He$ is not available on earth but there are sufficient resources on the moon and on other planets. To get a feeling, one space ship with ${}^3He$ per year would allow generation of electric power equivalent to the electric power consumption of the United States. To get this ${}^3He$ from the moon is probably not more difficult than building a $D-{}^3He$ fusion reactor. This task might be tackled after DT-reactors are working successfully and constitute a long-term goal of the fusion programme. Its solution would give the ultimate benefit of fusion power.

ROUND TABLE

ROUND TABLE

Chairman: *G. Grieger*

Members: *S. Eliezer, S. Jones, J.D. Jukes, R. Kulsrud, K. Nagamine, D. Palumbo and C. Petitjean*

The conclusive Round Table was held on the last day of the workshop and followed the course suggested by the Chairman.

In order to avoid a *scattered* discussion, the Round Table started with a synthetic presentation of what had been the factual outcome of the workshop. The issues on *muon-catalyzed fusion* were summarized by Jukes and Eliezer, those on *Fusion with polarized nuclei* were summarized by Kulsrud.

After each presentation observations from the *panelists* were collected followed by a discussion open to the floor.

In the following the two summaries are reported taking into account the observations of the panelists and of the audience.

MUON CATALYZED FUSION

J.D. Jukes and S. Eliezer

1. INTRODUCTION

Nuclear fusion reactions can be catalysed in a suitable fusion fuel by muons (heavy electrons) which can form temporarily very tightly bound meso-molecules. The muons can be produced by the decay of negative pions, which in turn have been produced by an accelerated beam of light ions (p, d, t..., at 1 GeV/nucleon) impinging on a target. Muon catalyzed fusion (MCF) is appropriately called cold fusion because the meso-molecules cannot exist in the hot plasma state. For practical fusion energy generation, however, it seems necessary to have a fuel mixture of DT ($\sim$ 50/50) at liquid or greater density and temperature $\sim$ 1000 K. It is important to note that despite the relatively large quantities of thermonuclear fuel involved there is no risk of thermonuclear explosions because muons cannot form an exponentiating chain reaction.

Conceptually, a MCF reactor is seen to be an energy amplifier, increasing by fusion reactions the energy invested in the nucleon-pion-muon beams. There is an analogy here with certain hot plasma fusion systems, e.g., the beam heated mirrors. In fact there is a very close synergism between certain MCF reactor concepts already proposed (Petrov, Tajima, Eliezer et al.) and magnetic mirror machines. It is appropriate to use certain ideas to describe the progress of MCF which are already very familiar in magnetic fusion, e.g., terms like scientific break-even, technical feasibility, economic demonstations, etc.

The present status of MCF is almost entirely limited to demonstrating scientific break-even by showing that is possible in principle with demonstrable physics to just sustain an energy balance between muon production (input) and catalysed fusion (output). The important physical quantity determining this balance is X_μ, the number of fusion reactions each muon can catalyse before it is lost.

One can express X_μ as

$$X_\mu = \left[\frac{1}{\lambda_c \tau_\mu} + w \right]^{-1}$$

where:

λ_c = cycling rate for producing a meso-molecule and nuclear fusion;

τ_μ = lifetime of muon (2.2 10⁻⁶s);

w = probability of loss per fusion event.

It turns out that w is determined largely by the probability of a muon sticking to the alpha particle, although there are other smaller and partly quantifiable losses. Thus a crucial parameter is

$$\omega_s = \text{sticking probability per fusion.}$$

The issues most under discussion at the workshop were therefore the experimental and theoretical determination of λ_c, ω_s and thus X_μ.

Feasibility for useful power production is equivalent to showing that X_μ can exceed a sufficiently large number, which is estimated to be $\sim 10^4$ if standard technology only is used, or may be as small as 10^3 if more advanced physics and technology can be developed (Eliezer). Since a muon can be produced with present technology for an expenditure of ~ 5000 MeV and ~ 20 MeV are produced per fusion event, it follows that even $X_\mu \sim 250$ would be a significant demonstration of scientific break-even.

2. THEORY

The developments in theory were summarised by Fiorentini and Lane. For λ_c Lane has derived either

$$\lambda_c \tau_\mu = 176 \, \phi$$

or

$$\lambda_c \tau_\mu = 880 \, \phi^2 \qquad (\phi < 1.5)$$

according as $E_r \stackrel{<}{>} 0$, where E_r is the energy needed for resonance formation and ϕ is the density relative to liquid hydrogen density.

Recent, not yet published (1986) work by Cohen (LANL) gives ω_s^{eff}:

ϕ	1.2	0.10
$10^2\ \omega_s^{eff}$	0.53	0.59

3. EXPERIMENTS PAST AND PRESENT

Experiments have so far been based on one or other of two different methods:

a) Neutron measurements of the absolute yield and time development of the neutron spectrum performed at LAMPF, SIN, KEK as described by Jones, Breunlich, Nagamine, respectively. By this method ω_s^{eff} can only be deduced indirectly and rather inaccurately at low density, which is just where the data from the different groups diverge.

b) X-ray measurements of the photons (8 keV) from the de-excitation of $\alpha\mu$ atoms, performed at SIN and KEK, described by Petitjean and Nagamine, respectively. The merit of this experimental method is the possible more direct and accurate determinations of $\omega_s^o = \omega_s^{eff}/(1\text{-}R)$, where R = reactivation (*ionization*) parameter = 0.25. In particular KEK has a pulsed beam and works with larger tritium fraction ($C_T = 0.3$).

A third method (direct measurement of $(\alpha\mu)^+$ is not yet conclusive. All experiments are concerned to measure or deduce the variation of λ_c and ω_s^{eff} with temperature T and densities (absolute and ratios).

The various results cover a range of ϕ, C_T/C_D and T. We show in the table only *typical* (i.e., best) values which are observed. The last column shows an *extrapolated* value suggested by the experimenter himself as perhaps attainable under optimum conditions.

4. FUTURE EXPERIMENTS

The following are planned or proposed experiments aimed at determining the key parameter ω_s^{eff}, which is probably the only controversial one, especially its dependencies on density and temperature, by as direct a method as possible.

a)   LAMPF plans to proceed to very high density $\phi \rightarrow 2.3$, although at low temperature 40 K

	$\lambda_c \tau_\mu$(max)	100 $\omega_s$eff(min)	X_μ(max)	X_μ(extrap)
LAMPF	310 ± 30 *	0.35 ± 0.07	150 ± 24	350 **
SIN	320 ± 25 +	0.45 ± 0.05	113 ± 10	220
KEK		<0.36 ± 0.20 ‡		<277

Footnotes:
LAMPF: Collaboration at LANL with Brigham Young Univ. and Idaho NEL
SIN: Vienna, SIN (Switzerland), UCB, LANL- Munich collaboration
KEK: Univ. Tokyo and Inst. of Phys. and Chem. Res. Wakoh, Japan, and LANL *et al*.
RAL: Rutherford - Appleton Lab., Chilton, U.K.

Conditions:

* ϕ = 1.2   T < 125 K

\+ ϕ = 1.0   T ~ 30 K

‡   from X-ray method which assumes K (prob of $\alpha\mu$ excited states) = 0.25

**   assumes ϕ is 2.3 and assuming $\omega_s$eff decreases with density according to trend established at lower density

b)   LANL/RAL/ Birmingham Univ. etc.. A Proposal is to use RAL's pulsed beam to determine ω_s on essentially tha LAMPF experiment, at low density

c)   SIN/LLNL.   A proposal is to use DT at extreme conditions of T → 2000 K, P → 2000 bar

d)   KEK. Before end of 1987 KEK will do improved high statistics measurements of X-rays at high ϕ, C_T. Using 70/30 DT at 2 K the aim is to determine the lowest limit of ω_s and make spectral width determination of Doppler broadening.

5. ACCELERATOR TECHNOLOGY

Weiss has outlined a scheme for producing an initial ion beam suitable for producing pions. It is based on coupling different and existing types of accelerators in sequence of ascending energy: RF quadrupole, Alvarez drift tubes and a superconducting cavity unit. The CW duty requirements are of course very different from present accelerators which operate on a duty cycle ~ 10^{-4}.

For MCF typically one needs 1 GeV/nucleon (say 2 GeV for a deuteron, 100 mA beam current, 200 MW beam power) and 400 MW mains power. Interestingly, already Weiss has began an economic assessment of MCF, since for the accelerator part he estimates the beam cost at 2 M$/MW.

266

6. ADVANCED PHYSICS AND TECHNOLOGY FOR MCF

Since it appears unlikely that realistic *break-even* is achievable with standard technology even using present best estimated of X_μ (see X_μ^{extrap} in table) it is necessary to consider more exotic solutions. For example, Eliezer proposed laser-induced stripping of the $\alpha\mu$. The Texas group has also shown that MCF is perhaps feasible already with known physics if one uses a hybrid or fertile blanket. However, this negates one of the principal claims of fusion, to be free of fission products and fissile materials. We observe there is nothing technically wrong with the hybrid blanket, indeed it is intrinsically safe since control rods are not required.

7. INTERNATIONAL COLLABORATION

The growth of international collaboration to make use of special facilities has already been noted (see 3, 4) in relation to present proposed or planned experiments. On the other hand, the theoretical effort (outside the USSR, at least) is presently dispersed and weaker, in spite of an obvious need for improving theoretical understanding. Almost no consistent technological study of reactor problems has yet started to the knowledge of this workshop. Little use so far has been made of super computers, except for Monte-Carlo calculations.

8. CONCLUSIONS

The workshop brought together a notable gathering of experimental and theoretical physicists to discuss progress towards demonstrating scientific break-even. However, the practical demonstration of MCF as an energy source (if it be possible) lies in further improved physical understanding and optimized engineering design, involving possibly novel technologies.

FUSION WITH SPIN-POLARIZED NUCLEI

R. Kulsrud

2.1 Introduction

In nuclear collision events the cross section for reactions, and the direction of emission of the reaction products subsequent to the reaction depend on the direction of the nuclear spins just prior to the event. This makes it possible to control the reactivity of a plasma to some extent, as well as to influence the fate of the reaction products, by orienting the spins of the nuclei in various directions relative to the confining magnetic field B. This may be done in such a way so as to improve the plasma performance. A plasma is a *spin polarized plasma* if its nuclei are so oriented.

There are three aspects to be considered:

a)   The spin nuclear physics and the gains to be achieved by modifying it

b)   Depolarization mechanisms

c)   Sources of the polarized nuclei and mechanisms for injection into plasma.

2.2 Nuclear Physics

In a D-T plasma the nuclear spins can be polarized to increase the nuclear cross section by fifty per cent by polarizing the spins of both D and T along B. Alternatively the other types of polarization leave the cross section unmodified or reduce it by a factor of two.

In the first case the angular distribution of the velocity v of the reaction products n and α is proportional to $\sin^2\theta$ where θ is the angle between v and B. This enhancement mode can be used to increase the nuclear reactivity or to

269

decrease B for the same reactivity, which should reduce the reactor cost. On the other hand, in a toroidal device the alpha particle loss is increased slightly as is the wall damage by neutrons. This offsets the gain somewhat. For a mirror machine reactor the directional effects improve the alpha loss and decrease the n damage.

For the other two modes the angular distribution is $1 + 3 \cos^2\theta$, the alpha loss and n damage are improved for tori and made worse for mirrors.

The gain due to polarization for reactors is relatively small. On the other hand the gain for devices near ignition such as the Compact Ignition Torus (CIT) could be decisive.

There exist neutron free reactions such as $D + {}^3He \rightarrow p + \alpha$ on which a reactor could be based. In such a reactor there is still the D-D reaction that produces n and T. Even this reaction might be suppressed if the D spins are polarized along B. (However, there is no consensus concerning this suppression). A reasonable argument shows that the cross section is reduced by about four. Combining this with a composition ratio $[D]/[{}^3He] = 1/q$, it is seen that the neutron reaction can be reduced to a rate 10^{-3} times that of the D-^{3}He charged particle reaction. Such a reactor would be essentially neutron free. This employment of spin polarization is potentially a dramatic one.

In addition the D-^{3}He reaction is enhanced by a factor 1.5, similar to the DT reaction enhancement.

2.3 Depolarization

The most dangerous depolarization mechanism is wall interaction during plasma recycling. To suppress this mechanism one must find a nonmagnetic nonmetallic wall material that does not rapidly depolarize the recirculating nuclei. A possible choice is amorphous graphite. Experiments are necessary to decide if this is a correct choice.

A second mechanism for depolarization is the excitation of waves by the anisotropically produced alpha particles. It turns out these waves can be suppressed in a torus by choosing an aspect ratio less than four.

The toughest hurdle for spin polarized fusion to overcome is the production of a sufficient number of polarized nuclei per second. Present sources for accelerators produce from one hundred microamps to a few milliamps. This is a factor of 10^6 to 10^5 too small to fuel a gigawatt reactor. One might hope to improve these sources by such a factor, but it is more likely that alternative ways of polarizing nuclei could be developed that could produce such required quantities of polarized nuclei.

These ways include optical pumping, which polarizes nuclei in atomic form at room temperature. Alternatively, two cryogenic methods are either dynamic polarization or to go to ultrahigh fields and ultralow temperatures and to achieve nuclear polarization in the Boltzmann equilibrium.

Optical pumping would be useful if the reactor is to be fueled by gas puffing or possibly by neutral beam injection. Cryogenic methods are more suitable for pellet injection (or pellet fusion) but cryogenic methods are difficult to apply to radioactive tritium.

At present an optical pumping experiment at Princeton is the only source that is being pursued.

2.5 Conclusion

Because of several important obstacles, it is entirely possible that the polarization of a reactor plasma will not be achieved. However, there are ways that people are considering to overcome these obstacles. At the present funding for research in nuclear polarization of plasmas is scanty. This is perhaps because no real need for polarization has been perceived. This is partly because its role in D-T fusion reactors is only quantitative. However, it could play an important role in helping to achieve ignition in CIT. Its role for D-^{3}He could be decisive.

It is to be expected that when convincing progress in the sources is made, interest in polarization for fusion will revive and funding will emerge.

Although there is no cooperative international effort in spin polarized fusion, the papers and discussions at this workshop have clarified its problems and possibilities.

PARTICIPANTS

PARTICIPANTS

Forty-one attendees, named in the following list, took part in the 8th Course of the International School of Fusion Technology held in ERICE, APRILE 3-9, 1987.

Participants on the balcony adjacent to the San Domenico Lecturing Hall

PARTICIPANTS

E. BITTONI
ENEA - CRE " E. Clementel "
Via Mazzini, 2
40138 BOLOGNA
Italy

S. BIVONA
Dip. Energetica ed
Applicazioni di Fisica
Parco d'Orleans
90100 PALERMO
Italy

V.R. BOM
Physics Department
Delft Univ. of Technology
P.O.Box 5046
2600 GA DELFT
The Netherlands

W.H. BREUNLICH
Austrian Acad. of Sciences (ÖAW)
Institute for Medium
Energy Physics
Boltzmanngasse 3
1090 WIEN
Austria

B. BRUNELLI
ENEA - CRE Frascati
CP 65
00044 FRASCATI, Roma
Italy

R. BURLON
Dip. Energetica ed
Applicazione di Fisica
Parco d'Orleans
90100 PALERMO
Italy

A. CALABRO'
ENEA - CRE Casaccia
S.P. Anguillarese km.301
CP 2400
00100 ROMA A.D.
Italy

A. CARDELLA
The NET team
Max - Planck - Institut
für Plasmaphysik
Boltzmannstraße 2
8046 GARCHING bei München
Federal Republic of Germany

A. DAINELLI
Via Flabanico, 10 / 29
35127 PADOVA
Italy

O. DONZELLI
Dip. Fisica
Università di Ferrara
Via Paradiso, 12
44100 FERRARA
Italy

J. DUCLOS
CEA - DPHN / HE, CEN Saclay
91191 GIF - SUR - YVETTE Cedex
France

S. ELIEZER
Plasma Physics Department
SOREQ N.R.C.
70600 YAVNE
Israel

R. FELDBACHER
Inst. for Theoretical Physics
Technical University Graz
Petergasse 16
8010 GRAZ
Austria

G. FIORENTINI
INFN Sezione di Pisa
Dip. Fisica
Università di Pisa
56100 PISA
Italy

M.L. GAMEZ
Instituto Fusion Nuclear
E.T.S. Ingenieros Industr.
Universidad Politecnica
Paseo de la Castellana 80
28046 MADRID
Spain

M. GOLDSWORTHY
Nuclear Engineering Group
University of New South Wales
P.O. Box 1, KENSINGTON
New South Wales 2033
Australia

G. GRIEGER
Max - Planck - Institut
für Plasmaphysik
Boltzmannstaße 2
8046 GARCHING bei München
Federal Republic of Germany

W. GRÜEBLER
Institute for Intermediate
Energy Physics
ETH - Hönggerberg
8093 ZURICH
Switzerland

M. HAEGI
ENEA - CRE Frascati
CP 65
0044 FRASCATI, Roma
Italy

W. HEERINGA
KfK, Postfach 3640
7500 KARLSRUHE 1
Federal Republic of Germany

S. JONES
Department of Physics and
Astronomy
296 ESC, Brigham and Young
University
PROVO, Utah, 84602
USA

J. JUKES
UKAEA Culham Laboratory
ABINGDON
Oxfordshire OX14 3DB
Great Britain

H. Th. KLIPPEL
ECN P.O. Box 1
1755 ZG PETTEN
The Netherlands

R.M. KULSRUD
PPPL, Princeton University
J. Forrestal Campus
P.O. Box 451
PRINCETON, New Jersey 08544
USA

A.M. LANE
UKAEA Harwell Laboratory
HARWELL - DIDCOT
Oxfordshire OX11 ORA
Great Britain

G.G. LEOTTA
Commission of the
European Communities
Rue de la Loi, 200
B - 1049 BRUSSELS
Belgium

S. MERCURIO
Dip. Fisica
P. le delle Scienze
90100 PALERMO
Italy

K. NAGAMINE
Meson Science Laboratory
University of Tokyo
Hongo 7 - 3 - 1
Bunkyo - ku, TOKYO 113
Japan

T. NIINIKOSKI
CERN
1211 GENEVE 23
Switzerland

S. NUZZO
Dip. Fisica
Via Archirafi, 36
90123 PALERMO
Italy

J. OXENIUS
Université Libre de Bruxelles
Service Chimie Physique II
Campus Plaine CP 231
B - 1050 BRUXELLES
Belgium

D. PALUMBO
Accademia Nazionale di
Scienze, Lettere e Arti
90100 PALERMO
Italy

E. PEDRETTI
ENEA - CRE Casaccia
S.P. Anguillarese km. 301
CP 2400
00100 ROMA A.D.
Italy

F. PEGORARO
Scuola Normale Superiore
P.zza dei Cavalieri
56100 PISA
Italy

C. PETITJEAN
SIN
5234 VILLIGEN
Switzerland

M. PICCININI
Dip. di Fisica
Università di Bologna
Via Irnerio
40100 BOLOGNA
Italy

C. PONTI
CCR Ispra
21020 ISPRA, Varese
Italy

P. ROCCO
CCR Ispra
21020 ISPRA, Varese
Italy

C.W.E. VAN EIJK
Physics Department
Delft Univ. of Technology
P.O.Box 5046
2600 GA DELFT
The Netherlands

M. WEISS
CERN Div. PS
1211 GENEVE 23
Switzerland

G.H. WOLF
IPP, KFA Jülich
Postfach 1913,
5170 JÜLICH
Federal Republic of Germany

V. ZAMPAGLIONE
ENEA - CRE Frascati
CP 65
00044 FRASCATI, Roma
Italy

M. ZARCONE
Dip. Fisica
Via Archirafi, 36
90123 PALERMO
Italy

INDEX

INDEX

A

accelerator(s)
— chain for μCF, 148—150;
  — parameters of the, 149; 150;
— cost of, 149; 150;
— effective shunt impedence per unit length of, 146;
— for μ^- production from π^- decay by beam impingement on target, 141—150;
  — power requirements, 146—148;
  — superconducting cavities, 146—148;
— polarized ion sources for use in, 213—234;
— power dissipated in, 146;
— technological aspect of, 4;

advanced fuel, nuclear reactivity of, 167; 174; 180;
— the d-d reaction, 161;
— the d-^{3}He reaction, 161;
— the d-t reaction, 161;
— the ^{6}Li (p ^{3}He)^{4}He reaction, 167;
— the p^{11}B reaction, 167;

alpha particle, 48;
— cyclotron and spin-precession frequencies of, 181;
— CTR confinement
  — increase of by 12% with plasma transverse polarization in, 195;
— directionality of emission relative to the confining magnetic field, 171; 183;
— fusion burn power for ignition,
  — calculated for INTOR, 254;
— heating in the *ETR* scheme, 48;
— of 3.5 MeV released during *lasing* emission, 133;
— requirements for μCF confinement, 50;
— resonances in D-T toroidal plasma (or D-^{3}He) with the, 188—191;

alpha particle sticking, 10; 20; 54;

Alvarez linac, 144;

B

beams
— colliding beam method developed at the
  — Wisconsin laboratory, 230;

beams (continued)
  — Brookhaven AGS laboratory, 230;
  — BNL, 230;
— using ECR, 230;
— ionization processes of atomic, 227; 228; 229;
  — ionization cross-sections for, 227; 228; 229;
— neutral particle beam
  — based on a supersonic cesium jet, 230; 231;
  — for CTR applications, 230;
  — design developed at Grenoble, 230; 231;
— production of polarized atomic deuterium beams, 227; 228;

β-decay of tritium, 239; 240; 242;
— freezing problems created by, 236; 239;
— positive aspects of, 242;
  — in dissociation of the DT molecule, 242;
  — paramagnetic electron creation by, 242;

β-value
— in CTR magnetic confinement systems, 253;

Bohr radius, 3;
— of μ-atom, 21; 22;

C

CIT (Compact Ignition Torus) design PPPL—Princeton, USA, 205; 270; 271;

clean fusion, 36;
— by d-d reactivity suppression through spin polarization of reacting particles, 161—167;
— by d-^{3}He neutron-free reaction, 161; 162; 259;

cold fusion reactors, 37—57; 263;

confinement of fusion plasma particles
— by application of magnetic fields, 252;
  — β-value, 253;
  — disruption phenomena, 255; 256;
  — magnetic island formation, 255;
  — MHD stability problems, 255;
  — nested magnetic surfaces, 252; 253; 255;
— impurity control in, 256;
— mirror approach to, 252;
— plasma boundary phenomena in, 256;
— plasma-wall interactions in, 193;

confinement of fusion plasma particles (continued)
- with amorphous graphite wall material, 174;
- recycling, 173; 174; 193;
- triton spin-polarization destruction in, 193;
- stellarator approach to, 252;
- thermal insulation requirements, 252; 253; 254;
- tokamak approach to, 252;

cost, estimated
- for creation of muons, 37; 38;
- ICF apparatus, reduction in cost of, by spin-polarization of reacting nucleons, 174;
- magnetic confinement system, reduction in cost of, by spin-polarization of reacting nucleons, 174; 175; 176;
- of a spin polarization system, 176;
- of one-watt laser, 176;

CTR
abbreviation used for *Controlled Thermonuclear Research*, VI;

D

d-d reaction, 180;
- spin dependence in, 161—166;
- cyclotron and spin-precession frequencies, of D, 181;
- relative cross-section calculations for, 164; 165; 166;

d-^{3}He reaction, 161; 162; 180; 182; 259;
- ^{3}He availability
- resources available on the moon, 259;
- low values of magnetic fluctuations in the, 183;
- spin dependence in, 161;
- cyclotron and spin-precession frequencies of ^{3}He, 181; 183;
- d-d reactivity suppression in, 161; 162;
- neutronless (clean) reactor, 161; 162;

d-t reaction, 180; 251; 271;
- angular distribution of alpha emission relative to confining magnetic field, 171; 182;
- angular distribution of neutron emission in the, 171;
- cyclotron and spin-precession frequencies of T, 181; 183;
- resonant interactions with magnetosonic waves, 182;
- leading to enhanced magnetic fluctuations, 182;
- spin dependence in the, 161; 169; 170;

depolarization of nucleon spins, 171; 172; 173; 174;
- averaged depolarization rate, spatially, 191; 192;
- collisional rate of depolarization
- by electrons, 172;
- leading to randomization of spins, 172;
- by interaction with the wall materials, 173; 174;
- in the case of amorphous graphite wall, 174;
- during recycling, 173; 174; 176;
- by magnetic fluctuations due to plasma collective modes, 179; 181; 182;

depolarization of nucleon spins (continued)
- by magnetic interaction of spins with the plasma confining field, 172;
- spin precession frequency resonating with inhomogeneities of the confining magnetic field, 173; 181;
- physical processes leading to, 172; 179—194;
- by resonant waves, 179; 182;
- frequency of modes depolarizing the spin of trapped and circulating ^{3}He nuclei, 190;
- role of energy convection in the, 186;

deuterium cyclotron frequency, 181;

deuterium spin precession frequency, 181;

disappearance rates, MCF,1
- of fusion events, 104;

DNP (dynamic nuclear polarization), 237; 238; 241;
- to produce polarized nuclei in solidified gases, 237; 238;
- by non-simultaneous spin exchange between a paramagnetic electron and the nucleus of the sample, 237; 238;

D-T pellet
- velocities of, 245;
- polarized
- cooling problems arising from tritium β-decay, 236; 243; 271;
- requirements for the dynamic polarization mechanism, 237; 238; 243; 244;
- temperature gradient
- in a multilayered bead, 243;
- in a solid bead, 243;
- unpolarized
- HIBALL-2 pellet, 243; 244;
- injection in the ICF devices, 236;
- injection in the magnetically confined fusion plasmas, 236;
- multilayered, 236;

E

electron, paramagnetic
- necessary for the polarization process of a D_2 and T_2 sample, 242;
- resulting from tritium β-decay of the DT molecule, 242;

energy breakeven, 84; 85;

energy cost
- for the creation of muons, 37; 38;

ETR μ**CF reactor (Eliezer-Tajima-Rosenbluth hybrid reactor),** 43;
- alpha particle generation in, 48;
- composed of, 37;
- an *injector* for a deuterium (or tritium) beam, 37; 43;
- a *magnetic mirror configuration* containing the d-t fuel (target) and where pions and muons are created, trapped and catalyzed *in situ*, 43; 46; 47;

μCF reactor (continued)
— technical feasibility of, 263;

μCF studies,
— main steps of development at:
— JINR - Dubna (USSR), 15; 102;
— KEK (Japan), 14; 78; 117—137;
— LAMPF (USA), 14; 74—87; 92; 114;
— LNPI - Gatchina (USSR), 14; 94;
— Rutherford Laboratory (UK), 15; 117;
— SIN (Switzerland), 14; 78; 80; 89—97; 99—115;
— TRIUMF (Canada), 15;

N

nested magnetic surfaces, 252; 253; 255; 256;
— magnetic island formation, 255; 256;
— disruptions in plasma, 256;

NET (Next European Torus) design, Max-Planck Institut für Plasmaphysik — Garching, Federal Republic of Germany, 4; 5; 255; 258;

neutron
— anisotropy of emission of in the D-T and D-^{3}He reactions, 183; 195;
— damage to chamber wall from neutron fluences, 183; 195—206;
— directionality of emission relative to confining magnetic field, 171;
— disappearance rate, λ_n, 93; 103; 104; 105;
— neutron pulses, 133;
— neutron suppression, 161;
— by polarizing the spin of nuclear fusion fuels , 161—167;
— neutron yield, X_μ, 33; 74; 91; 263;

nuclear reactions
— in a fusing plasma, 180; 251;
— with spin-polarized nuclei, 161; 169; 179; 195; 207; 213; 235; 269;

nuclear spin-lattice relaxation, 245;
— in HD molecule, 245;

P

paramagnetic electron
(see: electron, paramagnetic)

Pbμ (lead muon), 22;
— binding energy of 1s level of, 22;
— fine structure splitting of, 22;
— radius of 1s level of, 22;

p ^{11}B reaction
— spin dependence in the, 167;

Petrov's μCF reactor, 42; 43; 85;
— composed of, 38; 39;
— an *accelerator* for a deuterium (or tritium) beam, 38; 39; 42;

Petrov's μCF reactor (continued)
— a *target* (of tritium or beryllium) where pions are created, 39; 42;
— a *converter* in strong magnetic field confinement of pions, 39; 42;
— a *synthesizer* provided with the d-t fuel, 39; 42;
— a *blanket* where driven breeder materials are produced by fusion created neutrons, 39; 42; 49;
— considered as a neutron factory, 43; 50;
— description of, 37; 42;
— energy cost of muon creation in, 37; 38;
— muon losses in the catalysis cycle of, 38
— μ^- production in, 39; 40; 41; 42;
— resonant formation of dtμ molecule in, 37; 38;

pion, 45; 46;
— behaviour of in the magnetic mirror field, 47;
— cost of production, 151—158;
— with different impinging particle-target configurations, 151; 155;
— evaluation of with a Montecarlo code, 151; 156;
— dynamics of, 45; 46;

plasma, fusion
— definition of, 252;
— density required in CTR, 253;
— polarization of nuclear spins in, 161; 169; 179;
— pressure containment in the CTR approach, 252; 253;
— thermal insulation requirements in the CTR approach, 252; 253; 254
— by application of magnetic fields, 252;
— nested magnetic surfaces, 252; 253; 255;

polarization of nucleon spins, 161; 169; 174; 179; 180;
— cost of a polarization system, 176;
— fusion plasma reactivity by, increase of, 161; 171;
— by amount of 50%, 171;
— in inertially confined fusion (ICF) plasmas, 174;
— in magnetically confined fusion plasmas, 161; 169;
— advantages of, 171; 174; 175; 176; 271;
— mode frequency and stability of spin polarized toroidal
— D-^{3}He plasmas, 188—191;
— D-T plasmas, 188—191;
— polarization values attainable during 10 *sec* of plasma confinement, 172;
— requiring ignition values for:
— plasma confinement time, 193;
— plasma temperature, 193
— in (cryogenic) solid gases, 236; 271;
— brute-force polarization, 241;
— of D_2, 238; 241;
— of H_2, 238; 241;
— of the HD molecule, 241;
— of T_2, 238; 241;
— comparative evaluation of polarization HD and DT molecule, 245; 246;
— dynamic nuclear polarization (DNP), 237; 238; 241;